CAMBRIDGE LIBRARY COLLECTION

Books of enduring scholarly value

Earth Sciences

In the nineteenth century, geology emerged as a distinct academic discipline. It pointed the way towards the theory of evolution, as scientists including Gideon Mantell, Adam Sedgwick, Charles Lyell and Roderick Murchison began to use the evidence of minerals, rock formations and fossils to demonstrate that the earth was older by millions of years than the conventional, Bible-based wisdom had supposed. They argued convincingly that the climate, flora and fauna of the distant past could be deduced from geological evidence. Volcanic activity, the formation of mountains, and the action of glaciers and rivers, tides and ocean currents also became better understood. This series includes landmark publications by pioneers of the modern earth sciences, who advanced the scientific understanding of our planet and the processes by which it is constantly re-shaped.

A System of Mineralogy

Robert Jameson (1774–1854) was a renowned geologist who held the chair of natural history at Edinburgh from 1804 until his death. A pupil of Gottlob Werner at Freiberg, he was in turn one of Charles Darwin's teachers. Originally a follower of Werner's influential theory of Neptunism to explain the formation of the earth's crust, and an opponent of Hutton and Playfair, he was later won over by the idea that the earth was formed by natural processes over geological time. He was a controversial writer, accused of bias towards those who shared his Wernerian sympathies such as Cuvier, while attacking Playfair, Hutton and Lyell. He built up an enormous collection of geological specimens, which provided the evidence for his *System of Mineralogy*, first published in 1808 and here reprinted from the second edition of 1816. Volume 3 deals with metal ores such as gold, iron and lead.

Cambridge University Press has long been a pioneer in the reissuing of out-of-print titles from its own backlist, producing digital reprints of books that are still sought after by scholars and students but could not be reprinted economically using traditional technology. The Cambridge Library Collection extends this activity to a wider range of books which are still of importance to researchers and professionals, either for the source material they contain, or as landmarks in the history of their academic discipline.

Drawing from the world-renowned collections in the Cambridge University Library, and guided by the advice of experts in each subject area, Cambridge University Press is using state-of-the-art scanning machines in its own Printing House to capture the content of each book selected for inclusion. The files are processed to give a consistently clear, crisp image, and the books finished to the high quality standard for which the Press is recognised around the world. The latest print-on-demand technology ensures that the books will remain available indefinitely, and that orders for single or multiple copies can quickly be supplied.

The Cambridge Library Collection will bring back to life books of enduring scholarly value (including out-of-copyright works originally issued by other publishers) across a wide range of disciplines in the humanities and social sciences and in science and technology.

A System of Mineralogy

VOLUME 3

ROBERT JAMESON

CAMBRIDGE UNIVERSITY PRESS

Cambridge, New York, Melbourne, Madrid, Cape Town,
Singapore, São Paolo, Delhi, Tokyo, Mexico City

Published in the United States of America by Cambridge University Press, New York

www.cambridge.org
Information on this title: www.cambridge.org/9781108029759

This edition first published 1816
This digitally printed version 2011

ISBN 978-1-108-02975-9 Paperback

SYSTEM

OF

MINERALOGY.

A

SYSTEM

OF

MINERALOGY.

BY

ROBERT JAMESON,

REGIUS PROFESSOR OF NATURAL HISTORY, LECTURER ON MINERALOGY, AND KEEPER OF THE MUSEUM IN THE UNIVERSITY OF EDINBURGH;
Fellow of the Royal and Antiquarian Societies of Edinburgh; President of the Wernerian Natural History Society; Honorary Member of the Royal Irish Academy, and of the Honourable Dublin Society; Fellow of the Linnean and Geological Societies of London, and of the Royal Geological Society of Cornwall; Member of the Physical and Mineralogical Societies of Jena; of the Society of Natural History of Wetterau,
&c. &c. &c.

SECOND EDITION.

VOL. III.

EDINBURGH:

Printed by Neill & Company,

FOR ARCHIBALD CONSTABLE AND COMPANY, EDINBURGH; AND LONGMAN, HURST, REES, ORME & BROWN, LONDON.

1816.

TABLE OF CONTENTS

OF

VOLUME THIRD.

SYSTEM OF ORYCTOGNOSY.

CLASS IV.—METALLIC MINERALS.

I. Order, Platina.

II. Order, Gold.

III. Order, Mercury.

IV. Order, Silver.

V. Order, Copper.

VI. Order, Iron.

6. Magnetic

VII. Order, Manganese.

Foliated

VIII. Order, Titanium.

IX. Order, Lead.

X. Order, Zinc.

XI. Order, Tin.

XII. Order, Bismuth.

XIII. Order, Tellurium.

XIV. Order,

XIV. Order, Antimony.

XV. Order, Molybdena.

XVI. Order, Cobalt.

XVII. Order, Nickel.

XVIII. Order, Arsenic.

XIX. Order,

XIX. Order, Tungsten or Scheelium.

XX. Order, Uranium.

XXI. Order, Tantalum.

XXII. Order, Cerium.

MINERAL

CLASS IV.

METALLIC MINERALS.

The Minerals belonging to this Class exhibit many bright colours: they are frequently crystallised; their lustre is often metallic; they are in general either sectile or malleable; their specific gravity exceeds that of the minerals of the other classes; and they feel cold.

I. ORDER.—PLATINA.

This Order contains three species, viz. Platina, Palladium, and Iridium.

1. Native Platina.

Gediegen Platin, *Werner.*

Platina aurum album, *Wall.* t. ii. p. 365.—Platine, *Romé de Lisle,* t. iii. p. 487. *Id. De Born,* t. ii. p. 479.—Platin, *Werner,* Pabst. b. i. s. 31. *Id. Wid.* s. 661. *Id. Kirw.* vol. ii. p. 109. *Id. Emm.* b. ii. s. 106. *Id. Lam.* t. i. p. 96.—Le Platine natif, *Broch.* t. ii. p. 86.—Platin natif ferrifere, *Hauy,* t. iii. p. 368.—Platin, *Reuss,* b. iii. s. 234. *Id. Lud.* b. i. s. 210. *Id. Suck.* 2ter th. s. 97. *Id. Mohs,* b. iii. s. 3. *Id. Hab.* s. 98.—Platin natif, *Brong.* t. ii. p. 275.—Platin natif ferriferé, *Brard,* p. 234. *Id. Lucas,* p. 101.—Gediegen Platin, *Leonhard,* Tabel. s. 51. *Id. Karsten,* Tabel. s. 60.—Platina, *Kid,* vol. ii. p. 73.—Platin natif ferrifere, *Hauy,* Tabl. p. 72. Platin, *Oken,* b. i. s. 462.—Native Platina, *Aikin,* p. 18.

External Characters.

Its colour is very light steel-grey, which approaches to silver-white.

It occurs in flat, small, and very small grains, having pretty smooth surfaces; seldom in small angular or roundish grains, with impressions of other minerals.

Externally it is shining, glistening, or glimmering, and the lustre is metallic.

The fracture, on account of the smallness of the particles, cannot be determined. It is probably hackly.

The streak is more shining than the true lustre.

It is intermediate between semi-hard and soft: it is nearly as hard as iron.

It is malleable *.

It

* It is so ductile, that Dr Wollaston has succeeded in drawing it into a wire $\frac{1}{18750}$ part of an inch in diameter.

It is uncommonly heavy.

Specific gravity, 15.601. 18.947, *Tralles.* Purified, 23.0, *Thomson.*

Chemical Characters.

It is soluble in the nitro-muriatic acid. It is infusible, without addition, excepting in the focus of a burning-glass, or when exposed to the action of flame urged by oxygen gas. It is the least fusible of the metals. It does not amalgamate with mercury.

Constituent Parts.

The variety in grains, with a granulated surface, consists of Platina, with a very minute portion of Gold, and of Palladium: the variety in flat and angular grains, consists of Platina, alloyed with small proportions of Iron, Copper, Lead, Palladium, Iridium, Rhodium, and Osmium.

Geognostic and Geographic Situations.

Europe.—Platina has not hitherto been discovered, either pure, or forming a principal constituent part of any ore in Europe. The only ore in which it occurs, is the grey silver-ore of Guadalcanal in Spain.

America.—It has never yet been discovered to the north of the isthmus of Panama, on the continent of North America. Platina in grains is only found in two places in the Spanish South American dominions; in Choco, one of the provinces of the kingdom of New Granada; and near the shores of the South Sea, in the province of Barbacoas, between the 2° and 6° of north latitude. It is peculiar to an alluvial tract of 600 square leagues, where it is associated with grains of native gold, zircon, spinel, quartz, and magnetic ironstone. It is not true that this metal occurs near Carthagena, or Santa Fé, or on the islands of Porto Rico and Barbadoes, or in Peru, although these different localities are mentioned by authors *. The platina in granulated grains is found in alluvial soil, along with grains of gold, in gold-workings in Brazil †.

Uses.

Its property of remaining unaltered in the air, or when exposed to high heats, of resisting the action of many salts, and of receiving a fine polish, have rendered this metal useful for various chemical and physical instruments, as pyrometers, crucibles, pendulums, reflecting telescopic mirrors, and for wheels in the construction of watches. Reflecting mirrors made of glass, although they preserve their lustre and polish well, are inconvenient, because they form a double image: mirrors made with metallic alloys, which were substituted in their place, give but a single image, but tarnish on exposure to the air: mirrors of platina possess the advantage of not tarnishing, and they give but one image, and, owing to their great density, augment the reflecting power. Of all metals it expands the least by heat, and follows the most regular course in its expansion: hence it is admirably fitted for measures. The geometers Delambre and Mechain, in measuring the arc of the meridian contained between Dunkirk and Barcelona, used, in their operations, rods made of this metal. Klaproth has shewn, that it may be used with great advantage in painting and ornamenting

---

* Humboldt's New Spain, vol. iii. p. 150. Black's translation.

† Wollaston, Phil. Trans. for 1809.

namenting porcelain ; and although when burnt in and burnished, it has nearly the same colour as silver, yet it is not, like it, liable to be tarnished by sulphureous effluvia, or to be affected by alterations of the atmosphere. The platina used for these purposes is repeatedly melted with arsenic; without its aid, we could only have obtained it in very small masses, owing to the intense heat required for its fusion, and the small quantity fused.

Observations.

1. The name Platina is originally Spanish, and is the diminutive of *plata*, silver, probably on account of its resemblance to it in colour, owing to its occurring in small grains.

2. It is distinguished from *Silver*, by its colour, external shape, greater hardness, and specific gravity.

3. It is chemically distinguished from *Silver*, by its infusibility without addition, and its insolubility in nitric acid.

4. In the cabinet of the Academy of Bergaria in Biscay, there is said to be a mass of platina, the size of a pigeon's egg *. Humboldt has lately presented the King of Prussia with a mass which is still larger, and which weighs 1088.8 grains, and has a specific gravity of 18.947, according to Professor Tralles.

 2. Palladium.

* Mr Alaman, a Mexican gentleman, informs me, that there is no specimen of platina of the size mentioned in the text, in the Academy of Bergaria.

2. Palladium.

Palladium, *Wollaston.*

Philosophical Transactions for 1809, p. 192.

External Characters.

Its colour is pale steel-grey, passing into silver-white.
It occurs in small grains.
The lustre is metallic.
The fracture is diverging fibrous.
It is opaque.
Specific gravity, 11.8, 12.148 *, *Wollaston.*

Chemical Characters.

It is infusible; but on the addition of sulphur, it melts with ease; by continuance of the heat, the sulphur is dissipated, and a globule of malleable palladium remains. It forms a deep red solution with nitric acid.

Constituent Parts.

It consists of Palladium, alloyed with a minute portion of Platina, and of Iridium.

Geognostic and Geographic Situations.

It is found in grains, along with grains of native platina, in the alluvial gold districts in Brazil.

Observations.

This mineral was first discovered, described, and analysed by Dr Wollaston.

3. Iridium.

* The second pecific gravity was communicated to me by Mr Lowry.

3. Iridium.

Iridium, *Wollaston.*

Philosophical Transactions for 1805.

External Characters.

Its colour is very pale steel-grey.
It occurs in very small irregular flat grains.
The lustre is shining and metallic.
The fracture is foliated.
It is brittle.
It is harder than platina.
Specific gravity, 19.5.

Chemical Characters.

By fusion with nitre, it acquires a dull black colour, but recovers its original colour and lustre by heating with charcoal.

Constituent Parts.

It is composed of Osmium and Iridium, alloyed together.

Geognostic and Geographic Situations.

It occurs in alluvial soil in South America, along with platina.

Observations.

It was first examined, and introduced to the notice of naturalists, by Dr Wollaston.

II. ORDER.—GOLD.

This Order contains only two species, viz. Native Gold, and Electrum.

1. Native Gold.

Gediegen Gold, *Werner*.

This species is divided into three subspecies, viz. Gold-yellow Native Gold, Brass-yellow Native Gold, and Greyish-yellow Native Gold.

First Subspecies.

Gold-yellow Native Gold.

Gold-gelbes Gediegen Gold, *Werner*.

Id. Wer. Pabst. b. i. s. 3. *Id. Emm.* b. ii. s. 111. *Id. Estner*, b. iii. s. 215.—L'Or natif, jaune d'Or, *Broch.* t. ii. p. 89.—Gold-gelbes gediegen Gold, *Reuss*, b. iii. s. 246. *Id. Mohs*, b. iii. s. 11. *Id. Leonhard*, Tabel. s. 51. *Id. Karsten*, Tabel. s. 60.

External Characters.

Its colour is perfect gold-yellow, which varies in intensity; in some varieties it inclines to brass-yellow.

It occurs seldom massive, often disseminated, in membranes, in roundish and flattish pieces, in grains which are large, coarse, small and fine, in leaves, filiform, reticulated,

culated, dentiform? amorphous, ramose, and crystallised.

The surface of the grains is rough, and of the leaves delicate, drusy or smooth.

The external lustre is glistening.

Internally it is glimmering, passing into glistening, and the lustre is metallic.

The fracture is fine hackly.

The fragments are indeterminate angular, and very blunt-edged.

The streak is shining.

It is soft.

It is completely malleable.

It is common flexible.

It is uncommonly heavy.

Specific gravity, from 17.000 to 19.000.

Chemical Characters.

It is fusible into a globule, which is not altered by continuance of the heat.

Constituent Parts.

It contains only a very minute portion of Silver and Copper.

Geognostic Situation.

It occurs disseminated, in veins, and mineral beds, in granite, gneiss, mica-slate, clay-slate, clay-porphyry, and sandstone; also in grains and masses in alluvial deposites, in the beds of rivers, or in the alluvial soil in the flat country through which rivers occasionally flow. It is generally associated with quartz and iron-pyrites. Its other accompanying minerals are felspar, hornstone, cal-

careous-

careous-spar, heavy-spar, red silver-ore, brittle silver-glance, copper-pyrites, copper-green, variegated copper-ore, malachite, brown ironstone, galena or leadglance, red lead-ore, blende, grey antimony-ore, white cobalt, manganese, copper-nickel, arsenical-pyrites, and orpiment.

Geographic Situation.

Europe.—It is found in alluvial soil in the mining field of Leadhills. In the time of Queen Elizabeth, extensive washings were carried on in that district, for the purpose of collecting this precious metal; and it is reported that three hundred men were employed in searching for it, and that in the course of a few summers a quantity was collected equal in value to £100,000 Sterling. It also occurs in Glen Turret in Perthshire *; in stream-works in Cornwall; and in a ferruginous sand near Arklow, in the county of Wicklow, where a mass weighing twenty-two ounces, the largest piece hitherto met with in Europe, was found †. It occurs in granite at Gastein in Salzburg; at Gardette in France; in gneiss in Upper Hungary; in mica-slate in Salzburg and the Tyrol; in clay-porphyry in Transylvania; in hornblende rock, along with auriferous iron-pyrites, in veins of quartz, at Edelfors in Sweden. Rich mines of gold were formerly worked in Spain, and the most important of these were situated in Gallicia, where the gold occurred in regular veins. These mines, according to Diodorus Siculus, were worked by the Phœnicians, and afterwards by the

* I am informed that gold has been found at Cumberhead in Lanarkshire.

† The sand of any river is worth washing for the gold it contains, provided it will yield twenty-four grains in a hundred weight; but the sand of the African rivers often yield sixty-three grains in not more than five pounds weight; which is in the proportion of fifty times as much.—*Kid*, vol. ii. p. 76.

the Romans, who derived great wealth from them. The island of Thasos in the Mediterranean was celebrated for its mines of gold; and Thrace and Macedonia afforded much gold to the ancients. The sand of the Danube, Rhine, Rhone, Tagus, and many other European rivers, afford gold, and have been at different periods washed for this metal.

Asia.—There are few considerable mines of gold at present worked in this quarter of the globe. In Siberia, native gold occurs at Schlangenberg, in veins that traverse hornblende rock: auriferous pyrites is met with in quartz at Beresof, in the same country. In the southern parts of Asia, the sands of many rivers afford gold. The Pactolus, a small river in Lydia, formerly afforded so much gold, that it is alleged to have been one of the chief sources of the riches of Crœsus. The numerous islands in the Indian ocean, as Java, Japan, Formosa, Borneo, and the Philippines, afford considerable quantities of gold.

In the island of Sumatra, 15,400 ounces of gold are collected annually. It is obtained, either from veins, where it is associated with quartz, or from alluvial soil, where it occurs in the form of dust, or in masses that sometimes weigh upwards of nine ounces *.

There are considerable mines of gold in Cochinchina, of which the most important are those in the provinces of Cham and Naulang, where the gold occurs in dust or grains, and in pieces that sometimes weigh fully two ounces. Gold-mines are also worked in the kingdom of of Siam.

Africa.—This continent affords a considerable quantity of gold, which is always obtained in the form of dust or rolled masses, which is found in the sand of rivers, or the

* Marsden's Sumatra, p. 165,—172. 3d edition.

the alluvial soil of valleys or plains. The northern parts of Africa afford but little gold, but in the middle and southern regions, there are several tracts remarkable for the quantity of gold they afford. The first is Kardofan, situated between Darfur and Abyssinia. The gold collected there, is brought to market by the Negroes in quills of the ostrich and vulture. This territory, it would appear, was known to the ancients, who regarded Æthiopia as a country rich in gold.

The second principal tract lies to the south of the great Desart of Zara, and in the western part of Africa. The gold is collected in that extensive flat which stretches from the foot of those mountains in which are situated the sources of the rivers Gambia, Senegal, and Niger. Gold is also found in the sands of all these rivers. Bambouck, which is situated to the north-west of these mountains, furnishes the greatest part of the gold which is sold on the western coast of Africa, as well as that which is brought to Morocco, Fez, Algiers, and to Cairo and Alexandria in Egypt.

The third principal tract where gold is abundant, lies on the south-east coast, between 15° and 22° of south latitude, and nearly opposite Madagascar. The gold of that country, it is said, is found not only in the state of dust, but also in veins; and it is supposed, that Ophir, from which Solomon obtained gold, was a country on the same coast. Nearer to the equator, the Gold Coast supplied the Portugueze, and afterwards the Dutch, with great quantities of gold dust *.

America.—In modern times, this continent is considered the richest country of the world in gold. There the gold is chiefly collected in alluvial soil, and in the beds

* Brongniart's Mineralogie, t. ii. p. 271, 272, 273.

beds of rivers, and sometimes also from veins. In Mexico, the gold is for the most part extracted from alluvial soil by means of washing; and the particles vary in size, from that of dust to the weight of from five to six pounds. Another part of the Mexican gold is extracted from veins which traverse primitive mountains. The veins of native gold are most frequent in the province of Oaxaca, either in gneiss or mica-slate. This last rock is particularly rich in gold, in the celebrated mines of Rio San Antonio. These veins are about a foot and half wide, and contain besides the gold common quartz. The same metal occurs, either pure, or mixed with silver-ore, in the greatest number of veins that have been wrought in Mexico; and there is scarcely a single silver-mine which does not also contain gold.

On the coast of California, there is a plain of fourteen leagues in extent, covered with an alluvial deposite, in which lumps of gold are dispersed.

In the kingdom of New Granada in South America, gold is found in considerable quantity. It is obtained by the washing of the alluvial deposites in which it is contained. Gold veins have been found in the mountains of Guamoco and Antioquia, but their working is almost entirely neglected. The greatest riches in gold obtained by washing, are deposited to the west of the central Cordillera, in the provinces of Antioquia and Choco, in the valley of the Rio Cauca, and on the coast of the South Sea, in the *Partido de Barbacoas.* The alluvial grounds which contain the greatest quantity of gold in dust and grains, disseminated among fragments of greenstone and porphyry-slate, extend from the western Cordilleras almost to the shores of the South Sea.

The province of Antioquia, into which we can only enter on foot, or on the shoulders of men, contains veins of of gold in mica-slate, also wash-gold, or gold dust, as it is sometimes called, in alluvial deposites.

The largest piece of gold ever found in Choco, weighed 25 lbs. It is said that a piece of gold was found in Peru, near La Paz, in the year 1730, of the weight of 45 lbs.

Humboldt, to whom we are indebted for the preceding particulars, in regard to the gold of Spanish America, informs us, that the total annual produce of the gold-mines of the Spanish American colonies, amounts to 25,026 lbs. Troy.

A very considerable proportion of the gold of commerce comes from the Portugueze possessions in the Brazils. In that country, it is collected by washing from the sand of rivers, and the other alluvial deposites. Gold is found almost every where throughout that vast country, along the foot of the immense chain of mountains which lies nearly parallel with the coast, and extends from 5° to 30° of south latitude. From this country nearly 30,000 marcs of gold are annually exported to Europe; so that the total produce of gold from the Spanish and Portugueze colonies in the Americas, may be stated at 45,580 lbs. Troy *.

A

* If we understand correctly the accounts given by early writers, the quantity of gold amassed by the ancients must have been prodigious. Thus, in the 1st Book of Kings, chap. x. verse 14. we are told, that Solomon received 666 talents of gold (more than 27 tons weight) in one year; and in the 21st verse of the same chapter, it is said, "And all King Solomon's drinking vessels were of gold, and all the vessels of the forest of Lebanon were of pure gold; none were of silver; it was nothing accounted of in the days of Solomon." Diodorus says, that the tomb of King Simandius was environed with a circle of gold three hundred and fifty cubits about, and a foot and a half thick. Semiramis erected in Babylon three statues of gold, one of which was forty feet high, and weighed a thousand Babylonian talents. For these statues there was a table or altar of gold forty feet long, and twelve feet broad, weighing fifty talents.

A considerable quantity of gold has been of late years collected in North Carolina. It is there found in alluvial land.

It would appear from the preceding statement, that most of the gold of commerce comes from America and Africa, and that by far the greatest proportion of this is collected from an alluvial land, which is frequently ferruginous. The only considerable gold-mines in Europe, are those of Hungary, but the gold is principally the brass-yellow subspecies.

Second Subspecies.

Brass-yellow Native Gold.

Messing-gelbes gediegen Gold, *Werner.*

Id. Werner's Pabst. b. i. s. 5. *Id. Emm.* b. ii. s. 113.—L'Or natif d'un jaune de Laiton, *Broch.* t. ii. p. 91.—Messing-gelbes gediegen Gold, *Reuss,* b. iii. s. 258. *Id. Mohs,* b. iii. s. 16. *Id. Leonhard,* Tabel. s. 51. *Id. Karsten,* Tabel. s. 60.

External Characters.

Its colour is brass-yellow, which is more or less light or pale, and sometimes inclines to silver-white.

It occurs disseminated, rarely massive, capillary, moss-like, reticulated, and in leaves; also crystallised in the following figures:

1. Cube.
2. Octahedron: sometimes cuneiform.
3. Rhomboidal or garnet dodecahedron.
4. Leucite form.
5. Double six-sided pyramid.
6. Very acute six-sided table, in which the terminal planes are set on alternately straight and oblique.

Specific gravity, 12.713, *Karsten.*

Constituent

Constituent Parts.

Eula in Bohemia.	
Gold, - - -	96.9
Silver *, - -	2.0
Iron, - - -	1.1
	100

Lampadius, Handbuch zur Chem. Annal. d. Min. 251.—253.

Geognostic and Geographic Situations.

This mineral is found principally in Hungary and Transylvania. It generally occurs in small veins in porphyry and grey-wacke. The minerals with which it is most frequently associated in these veins, are quartz, iron-pyrites, and grey antimony-ore. Besides these, the following also occur, viz. various ores of silver; also ores of copper, and of these copper-pyrites, grey copper-ore, copper-glance, variegated copper-ore, and copper-green, are the most frequent; almost the only iron-ores are brown ironstone; the ores of zinc are yellow and brown blende; of lead, leadglance or galena, and green lead-ore; traces of copper-nickel, and white cobalt-ore; native arsenic, arsenical-pyrites, and red orpiment. Besides quartz, the following earthy minerals are met with in these veins, viz. brown-spar, calcareous-spar, heavy-spar, selenite, common garnet, and lithomarge.

It occurs in bituminous-wood in Transylvania,—is found in Bohemia,—and is mentioned as a native of Siberia.

Third

* It is probable that this subspecies contains more silver than appears from the analysis of Lampadius.

Third Subspecies.

Greyish-yellow Native Gold.

Graugelbes gediegen Gold, *Werner.*

Id. Emm. b. ii. s. 114.—L'Or natif d'un jaune grisatre, *Broch.* t. ii. p. 92.—Fahlgelbes gediegen Gold, *Reuss,* b. iii. s. 260. —Graugelbes gediegen Gold, *Leonhard,* Tabel. s. 51. *Id. Karsten,* Tabel. s. 60.

External Characters.

Its colour is brass-yellow, which verges on steel-grey.

It occurs in very small flattish grains, like those of platina.

Its surface is glistening.

It is never crystallised.

It is heavier than brass-yellow native gold, but lighter than gold-yellow native gold.

In other characters it does not differ from the preceding.

Geognostic and Geographic Situations.

It occurs, along with platina, in South America.

Uses.

The numerous and important uses of this metal, will be considered in another work. We may here only remark, that for whatever purpose gold is used, it is mixed with a quantity of copper, which is usually about $\frac{1}{24}$, and never exceeds $\frac{1}{4}$, which gives the gold a consistence and a hardness it does not possess when pure.

Observations.

Iron-pyrites is sometimes auriferous, but the richest varieties, at Facebay in Transylvania, do not afford more 0.02 to 0.03 of gold. Auriferous pyrites is also met with at Adelfors in Smoland in Sweden, in the Valais and Grisons in Switzerland, in Dauphiny, Siberia, and Mexico. This variety is distinguished from copper and iron pyrites, by colour, specific gravity, and malleability.

2. Electrum, or Argentiferous Native Gold.

Electrum, *Klaproth.*

Electrum, *Plin.* Hist. Nat. xxxiii. cap. iv. § 23.—Naturliches Electrum, v. *Veltheim's* Grundriss einer Mineralogie Braunscher, 1781, fol. 11.—Elektrum, *Klap.* b. iv. s. 1.—Argentiferous Native Gold, *Aikin,* 2d edit. p. 76.

External Characters.

Its colour is pale brass-yellow, passing into silver-white.

It occurs in small plates, dentiform, and in imperfect cubes.

The other characters are not stated by Klaproth, to whom we are indebted for what is known of this mineral.

Chemical Characters.

It is not soluble either in nitrous or nitro-muriatic acids.

Constituent

Constituent Parts.

Gold, - - -	64
Silver, - - -	36
	100

Klaproth, Beit. b iv. s. 3.

Geognostic and Geographic Situations.

It occurs, along with massive heavy-spar, or ash-grey splintery hornstone, at Schlangenberg in Siberia.

Observations.

1. The ancients applied the name *Electrum*, not only to amber, but also to a particular mixture of gold and silver, as appears from the following passage of Pliny: "*Omni auro inest argentum vario pondere. Ubicunque quinta argenti portio est,* electrum *vocatur* *." Hence Klaproth applies the name Electrum to this mineral.

2. As this ore is not acted on, either by nitrous or nitro-muriatic acid, it follows, that the gold and silver are more than mechanically mixed.

 III. ORDER.

* Plin. Hist. Nat. Lib. xxxiii. cap. iv. § 23.

III. ORDER.—MERCURY.

This Order contains five species, viz. Native Mercury, Native Amalgam, Mercurial Horn-Ore, Mercurial Hepatic Ore, and Cinnabar.

1. Native Mercury or Quicksilver.

Gediegen Quecksilber, *Werner.*

Argentum vivum, *Plin.* Hist. Nat. xxxiii.—Mercurius virgineus; Hydrargyrum nativum, *Wall.* t. ii. p. 148.—Mercure natif, *Romé de Lisle,* t. iii. p. 152.—Gediegen Quecksilber, *Wid.* s. 719. *Id. Wern.* Pabst. b. i. s. 6 —Native Mercury, *Kirw.* vol. ii. p. 223.—Gediegen Quecksilber, *Emm.* b. ii. s. 129.—Mercure natif, *Lam.* t. i. p 166. *Id. Broch.* t. ii. p. 96. *Id. Hauy,* t. iii. p. 423.—Gediegen Quecksilber, *Reuss,* b. iii. s. 269. *Id. Lud.* b. i. s. 205. *Id. Suck.* 2ter th. s. 109. *Id. Bert.* s. 432. *Id. Mohs,* b. iii. s. 93.-96. Mercure natif, *Lucas,* p. 109.—Gediegen Quecksilber, *Leonhard,* Tabel. s. 51.—Mercure natif, *Brong.* t. ii. p. 241. *Id. Brard,* p. 253.—Gediegen Quecksilber, *Karsten,* Tabel. s. 60. *Id. Haus.* s. 69.—Native Quicksilver, *Kid,* vol. ii. p. 93.—Mercure natif, *Hauy,* Tabl. p. 77.—Native Quicksilver, *Aikin,* p. 23.

External Characters.

Its colour is pure tin-white.

It occurs perfectly fluid; and in larger or smaller particles or globules in the cavities of other ores of mercury.

It is splendent, and the lustre is metallic.

It

It does not wet the finger.

It feels very cold.

It is uncommonly heavy.

Specific gravity, in its fluid state, 13.581, *Hauy.* When solid, 15.61, *Biddle.*

Chemical Characters.

It volatilises entirely before the blowpipe, at less than a red heat.

Constituent Parts.

According to Klaproth, it contains no intermixture of any other metal.

Geognostic Situation.

This mineral occurs principally in rocks of the coal formation, and either disseminated, or in veins traversing them. It is associated with ores of mercury, and often also with iron-pyrites, heavy-spar, calcareous-spar, and quartz. Small veins of it are rarely met with in primitive rocks, where it is accompanied with native silver, grey manganese-ore, and flexible asbestus.

Geographic Situation.

Europe.—It is found at Idria in the Friaul; Niderslana in Upper Hungary; Morsfeldt and Wolfstein in the Palatinate; Moschellandsberg and Stahlberg in Deux-Ponts; Leogang in Salzburg; Horzowitz in Bohemia; Almaden in Andalusia, and Albaracia in Arragon; in slate-clay at Paterno in Sicily; and at Oristani in Sardinia.

America.—Guancavelica in Peru.

Uses.

This metal is used in the construction of barometers and thermometers; also for collecting gases absorbable in water; and its property of amalgamating, enables the metallurgist to extract, at a small expence, minute portions of gold and silver from poor ores *. When amalgamated with tin, it is used for silvering mirrors: amalgams of gold and silver are employed for plating other metals †; and the amalgam of mercury and bismuth is used for the rubbers of electrical machines. In the oxidated and saline states, it acts as a powerful medicine.

Observations.

1. The greater part of the mercury of commerce is obtained by distilling native cinnabar, not from native mercury, which occurs but in small quantity.

2. When rendered solid by artificial freezing mixtures, it is found to be malleable, and to crystallise in octahedrons.

3. The fracture of congealed mercury is hackly.

2. Native.

---

* The amalgamation of gold and silver appears to have been known to the ancients.—Vid. Plin. Hist. Nat. xxxiii.; Vitruv. viii. 8.

† The process of gilding is mentioned by Pliny.—Vid. Plin. l. c. ed. Bip. p. 101.

2. Native Amalgam.

Natürliches Amalgam, *Werner*.

Amalgame natif d'Argent, *Romé de Lisle*, t. iii. p. 162.—Natürliches Amalgam, *Wern*. Pabst. b. i. s. 7. *Id. Wid.* s. 722. Natural Amalgama, *Kirw*. vol. ii. p. 225.—Naturliches Amalgam, *Emm.* b. ii. s. 134.—Amalgame natif d'Argent, *Lam.* t. i. p. 432. *Id. Broch.* t. ii. p. 99.—Mercure argenteal, *Hauy,* t. iii. p. 432.—Amalgam, *Reuss,* b. iii. s. 273. *Id. Lud.* b. i. s. 205. *Id. Suck.* 2ter th. s. 111. *Id. Bert.* s. 433. *Id. Mohs,* b. iii. s. 99. *Id. Leonhard,* Tabel. s. 51.—Mercure argental, *Brong.* t. ii. p. 242. *Id. Brard,* p. 254.—Amalgam, *Karsten,* Tabel. s. 60.; *Id. Haus.* s. 69.—Quicksilver alloyed with Silver, *Kid,* vol. ii. p. 94.—Mercure argental, *Hauy,* Tabl. p. 77.—Silver Amalgam, *Aikin,* p. 24.

This species is divided into two subspecies, viz. Fluid or Semi-fluid Amalgam, and Solid Amalgam.

First Subspecies.

Fluid or Semi-fluid Amalgam.

Flüssiges oder halbflüssiges Amalgam, *Werner*.

External Characters.

Its colour is intermediate between tin-white and silver-white, according as it contains more or less silver, but usually inclines more to the first.

It occurs in small massive pieces, and in balls; sometimes also in membranes, and disseminated; also crystallised in the following figures:

1. Octahedron,

1. Octahedron, more or less deeply truncated on the edges. When the truncating planes become so large that the original planes disappear, the
2. Garnet dodecahedron is formed. The edges of this figure are also frequently truncated, and often so deeply, that it passes into the
3. Leucite crystal, or the double eight-sided pyramid, flatly acuminated on both extremities by four planes, which are set on the alternate lateral edges.

The crystals are small and very small; usually imbedded, and sometimes two or three occur adhering together.

Externally it is shining and splendent, with a metallic lustre.

The fracture is imperfect conchoidal.

The fragments are rather blunt-edged.

It is soft.

When pressed between the fingers, or cut with a knife, it emits a creaking sound like artificial amalgam.

It is uncommonly heavy.

Second Subspecies.

Solid Amalgam.

Festes Amalgam, *Werner.*

External Characters.

Its colour is silver-white, which in some varieties falls into reddish-white.

It occurs massive and disseminated.

It is shining, approaching to glistening.

The

The fracture is imperfect flat conchoidal, sometimes passing into fine-grained uneven.

The fragments are indeterminate angular.

It is semi-hard.

It is brittle.

It creaks strongly when cut.

It is uncommonly heavy.

Chemical Characters.

Before the blowpipe, the mercury is volatilised, and a bead of pure silver remains. It whitens the surface of copper when rubbed warm on it.

Constituent Parts.

Mercury, - -	74	64
Silver, - - -	25	36
	99	100
	Heyer.	*Klaproth*, Beit. b. i. s. 183.

Geognostic Situation.

It is usually accompanied with native mercury and cinnabar; it also occurs along with native silver, and iron-pyrites; and the earthy minerals with which it is associated, are calcareous spar, quartz, heavy-spar, hornstone, &c.

Geographic Situation.

It is found at Rosenau and Niderslana in Hungary; Morsfeld in the Palatinate; Moschellandsberg and Stahlberg in Deux Ponts; in the Leogang in Salzburg; and Sahlberg in Sweden.

Observations.

Observations.

1. It is distinguished from *Native Silver*, by colour, fracture, tenacity, specific gravity, and sound.

2. Native silver, when rubbed on copper, does not whiten it as amalgam does.

3. Mercurial Horn-Ore, or Corneous Mercury.

Quecksilber Hornerz, *Werner*.

Woulfe, in Phil. Trans. lxvi. ii. 618.—Mercure corné, ou Mercure doux volatile, *Romé de L.* t. iii. p. 161.—Quecksilber-hornerz, *Wern.* Pabst. b. i. s. 7. *Id. Wid.* s. 724.—Mercury mineralized by the Vitriolic and Marine Acids, *Kirw.* vol. ii. p. 266. —Quecksilber-hornerz, *Estner*, b. iii. s. 275. *Id. Emm.* b. ii. s. 136.—Mercure corné, *De Born*, t. ii. p. 399. *Id. Lam.* t. i. p. 168.—La Mine de Mercure cornée, ou le Mercure muriaté, *Broch.* t. ii. p. 101.; Mercure muriaté, t. iii. p. 447. Quecksilber Hornerz, *Reuss*, b. i. s. 277. *Id. Lud.* b. i. s. 206. —Salziges Quecksilber, *Suck.* 2ter th. s. 112.—Quecksilber-hornerz, *Bert.* s. 434. *Id. Mohs*, b. iii. s. 91. *Id. Leonhard*, Tabel. s. 52.—Mercure muriaté, *Brong.* t. ii. p. 244. *Id. Brard*, p. 260.—Quecksilber-hornerz, *Karsten*, Tabel. s. 60. Horn-quecksilber, *Haus.* s. 134.—Muriate of Quicksilver, *Kid*, vol. ii. p. 97.—Mercure muriaté, *Hauy*, Tabl. p. 78.— Horn Quicksilver, *Aikin*, p. 25.

External Characters.

Its colour is ash-grey, of various degrees of intensity, which passes into yellowish-grey, and from this into greyish-white, and even sometimes inclines to greenish-grey.

It

It occurs very rarely massive, almost always in small vesicles, which are crystallised in the interior.

The crystals are the following:

1. Cube.
2. Cube truncated on the edges.
3. Rectangular four sided prism, acutely acuminated with four planes, set on the lateral edges.
4. Cube truncated on the angles.
5. Octahedron.

The crystals are always so minute, that it is with difficulty their forms can be determined Their external surface is sometimes smooth, sometimes drusy, and is in general splendent.

Internally it is splendent, with an adamantine lustre.

The fracture is straight foliated, and also fine and small grained uneven.

It sometimes occurs in fine granular distinct concretions.

It is faintly translucent, or only translucent on the edges.

It does not alter its colour in the streak.

It is soft, approaching to very soft.

It is sectile.

It is easily frangible.

Chemical Characters.

It is totally volatilised before the blowpipe, and emits a garlic smell. It is said to be soluble in water, and the solution mixed with lime-water gives an orange-coloured precipitate.

Constituent Parts.

Oxide of Mercury,	-	76.00
Muriatic Acid,	-	16.40
Sulphuric Acid,	- -	7.60
		100.00 *Klaproth.*

Geognostic Situation.

In the quicksilver mines of the Palatinate, and the Dutchy of Deux Ponts, it is accompanied with native mercury, cinnabar, ochry brown ironstone, seldomer with fibrous malachite, massive and crystallised azure copper-ore, massive grey copper-ore, in cavities of an iron-shot clayey sandstone, sometimes in clay-ironstone, and red ironstone. That at Idria, occurs in cavities of an indurated clay, accompanied with crystals of cinnabar; sometimes in slate-clay, which is traversed with small veins of cinnabar.

Geographic Situation.

It occurs at Horzowitz in Bohemia; Moschellandsberg in Deux Ponts; Morsfeld in the Palatinate; Rutha in Upper Hessia, and at Almaden in Spain.

Observations.

It was discovered about thirty years ago in the mines of the Palatinate by Mr Woulfe.

4. Mercurial

4. Mercurial Hepatic-Ore or Mercurial Liver-Ore.

Quecksilber Lebererz, *Werner.*

This species contains two subspecies, viz. Compact Mercurial Hepatic Ore, and Slaty Hepatic Mercurial Ore.

First Subspecies.

Compact Mercurial Hepatic-Ore.

Dichtes Quecksilber Lebererz, *Werner.*

Id. Werner, Pabst. b. i. s. 8.—Compact Hepatic Mercurial-Ore, *Kirw.* vol. ii. p. 224.—Dichtes Quecksilber Lebererz, *Estner,* b. iii. s. 281. *Id. Emm.* b. ii. s. 140.—Mine de Mercure hepatique, *Broch.* t. ii. p. 104.—Dichtes Lebererz, *Reuss,* b. iii. s. 282. *Id. Lud.* b. i. s. 207. *Id. Mohs,* b. iii. s. 88.—Mercure sulphuré hepatique, *Brong.* t. ii. p. 243.—Dichtes Lebererz, *Karsten,* Tabel. s. 60. *Id. Haus.* s. 76.—Mercure sulphuré bituminifere compacte, *Hauy,* Tabl. p. 78.

External Characters.

Its colour is intermediate between dark cochineal-red and dark lead-grey.

It occurs massive.

Internally it is glistening or glimmering, and the lustre is semi-metallic.

The fracture is even; it rarely approaches to fine-grained uneven, but sometimes passes into imperfect flat conchoidal.

The fragments are indeterminate angular, and rather sharp-edged.

The

The streak is shining, and of a cochineal-red colour.

It is opaque.

It is soft.

It is sectile.

It is easily frangible.

It is uncommonly heavy.

Specific gravity, 7.937, *Gellert*. 7.186 to 7.352, *Kirwan.*

Constituent Parts.

Mercury, - - -	81 80
Sulphur, - - -	13.75
Carbon, - - -	2.30
Silica, - - - -	0.65
Alumina, - - -	0.55
Oxide of Iron, - - -	0.20
Copper, - - -	0.02
Water, - - -	0.73
	100

Klaproth, Beit. b. iv. s. 24.

Second Subspecies.

Slaty Mercurial Hepatic-Ore.

Schriefriges Quecksilber Lebererz, *Werner.*

Schiefriges Lebererz, *Reuss,* b. iii. s. 284. *Id. Lud.* b. i. s. 207. *Id. Mohs,* b. iii s. 89 *Id. Karsten,* Tabel. s. 60. *Id Haus.* s. 76.—Mercure sulphuré bitumine testacé, *Hauy,* Tabl. p. 78.

External Characters.

Its colour is nearly the same with the preceding subspecies, only rather more inclining to red.

It

It occurs massive.

The lustre of the principal fracture is shining; that of the cross fracture is glimmering, and both have a semi-metallic lustre.

The principal fracture is curved and thick slaty; the cross fracture is even.

It occurs in globular, and concentric lamellar concretions.

The fragments are slaty.

It is uncommonly easily frangible.

In other characters, it agrees with the preceding sub-species.

Geognostic Situation.

This mineral occurs in considerable masses in slate-clay and bituminous-shale. It is sometimes intermixed with cinnabar and iron-pyrites; and veins of native mercury and of cinnabar occasionally traverse it. Both species occur together.

Geographic Situation.

It occurs most abundantly in Idria: it is also met with at Almaden in Spain, Nertschinsk in Siberia, and in Deux Ponts.

Observations.

1. The variety in globular and concentric lamellar concretions is named *Corallenerz*.

2. When exposed for some time to the air, it acquires a liver-brown tint of colour: hence the name *Hepatic* or *Liver-ore* given to it.

5. Cinnabar.

5. Cinnabar.

Zinnober.

Of this species, there are two subspecies, Dark Red Cinnabar, and Bright Red Cinnabar.

First Subspecies.

Dark Red Cinnabar.

Dunkel-rother Zinnober, *Werner.*

Minium, *Plin.* Hist. Nat. xxxiii. 7. s. 38. (ed. Bip. 5. v. 204.) Dunkel-rother Zinnober, *Werner,* Pabst. b. i. s 8. *Id. Wid.* s. 728.—Dark Red Cinnabar, *Kirw.* vol. iii. p. 228.—Dunkel-rother Zinnober, *Estner,* b iii. s. 290. *Id. Emm.* b. ii. s. 144.—Le Cinnabre d'un rouge foncé, ou le Cinnabre commun, *Broch.* t. ii. p. 107.—Mercure sulphuré, *Hauy,* t. iii. p. 437.—Dunkelrother Zinnober, *Reuss,* b. iii. s. 287. *Id. Lud.* b. i. s. 207. *Id. Suck.* 2ter th s. 118. *Id Bert.* s. 436. *Id. Mohs,* b. iii. s. 76. *Id. Leonhard,* Tabel. s. 52.—Mercure sulphuré compacte, *Brong.* t. ii. p. 243.—Gemeiner Zinnober, *Karsten,* Tabel. s. 60.—Blättriger Zinnober, & Dichter Zinnober, *Haus.* s. 76.—Sulphuret of Quicksilver, *Kid,* vol. ii. p. 94.—Mercure sulphuré couleur rouge foncé, *Hauy,* Tabl. p. 78.—Cinnabar, *Aikin,* p. 24.

External Characters.

Its principal colour is perfect cochineal-red, which in some varieties inclines very much to lead-grey ; in others passes into carmine-red.

Besides massive, disseminated, in blunt-cornered pieces, in membranes, amorphous, stalactitic, dendritic, and fruticose,

ticose, it occurs also crystallised. Its crystallisations are the following:

1. Regular six-sided prism, sometimes flatly acuminated with three planes, which are set on the alternate lateral planes.
2. Rhomb, rather oblique, and slightly elongated.
3. The preceding figure, truncated on the diagonally opposite acute angles. When these truncating planes become so large as to assume the size of lateral planes, there is formed an
4. Octahedron. It sometimes ends in a line.
5. Rhomb truncated on the diagonally opposite obtuse angles. When these truncating planes increase much in size, there is formed a
6. Six-sided table.
7. Flat lenticular rhomb.

The crystals are small, and very small; occur in druses, on one another, side by side, and promiscuous.

Externally the crystals are splendent.

Internally it is shining, which, according to the differences in the fracture, passes into glistening, and sometimes into glimmering, with an adamantine, verging on semi-metallic lustre.

The fracture is sometimes fine-grained uneven, sometimes even and conchoidal. It occurs also more or less perfect foliated, with a threefold oblique angular cleavage *.

The fragments are indeterminate angular, and blunt-edged.

It occurs in granular and lamellar distinct concretions.

The massive varieties are opaque, or translucent on the edges,

* The foliated varieties have the strongest lustre.

edges; the crystals are translucent, sometimes semi-transparent, and even verging on transparent.

It yields a scarlet-red shining streak.

It is very soft, passing into soft.

It is sectile.

It is easily frangible.

It is uncommonly heavy.

Specific gravity.—The specific gravities, as given by authors, vary very much. Thus, according to *Brisson*, the massive from Almaden, is only 6.9022; whereas the crystallised from the same place is 10.2185: According to *Klaproth*, that of Japan is 7.710, and of Neumarktel 8.160; and the Cinnabar of Almaden, according to *Kirwan*, is 7.786.

Chemical Characters.

Before the blowpipe, it melts, and is volatilised with a blue flame, and sulphureous odour.

Constituent Parts.

	Japan.	Neumarktel in Carniola.
Mercury,	84.50	85.00
Sulphur,	14.75	14.25
	99.25	99.25

Klaproth, Beit. b. iv. s. 17. & 19.

For Geognostic and Geographic Situations, see the following subspecies.

Second

Second Subspecies.

Bright Red Cinnabar*.

Hochrother Zinnober.

Id. Wern. Pabst. b. i. s. 11. *Id. Wid.* s. 727. *Id. Estner,* b. iii. s. 297. *Id. Emm.* b. ii. s. 146.—Le Cinnabre d'un rouge vif, ou le Cinnabre fibreux, *Broch.* t. ii. p. 111.—Mercure sulphuré rouge vif, *Hauy,* t. iii. p. 440.—Lichtrother Zinnober, *Reuss,* b. iii. s. 293.—Hochrother Zinnober, *Lud.* b. i. s. 208. *Id. Mohs,* b. iii. s. 86. *Id. Leonhard,* Tabel. s. 52.—Mercure sulphuré fibreux, *Brong.* t. ii. p. 243.—Zerreiblicher Zinnober, *Karsten,* Tabel. s. 60.—Erdiger Zinnober, *Haus.* s. 76.—Mercure sulphuré couleur rouge vif, *Hauy,* Tabl. p. 78.

External Characters.

Its colour is bright scarlet-red.

It occurs massive and disseminated.

Internally it is glimmering; the cross fracture is dull.

The fracture is intermediate between earthy and fibrous; the cross fracture is earthy.

The fragments are indeterminate angular, and blunt-edged.

It is opaque.

The streak is shining.

It soils.

It is very soft, passing into friable.

It is sectile.

 It

* It is also named *Native Vermilion.*

It is very easily frangible.

It is heavy, inclining to uncommonly heavy.

Geognostic Situation.

This mineral occurs in small quantities in beds and veins, in rocks of clay-slate, talc-slate, and chlorite-slate, and is associated with quartz, calcareous-spar, sparry ironstone, and with minuter portions of iron-pyrites, copper-pyrites, iron-glance or specular iron-ore, and iron-mica or micaceous iron-ore; also in veins that traverse trap-porphyry, pitchstone and hornstone porphyries, and alpine limestone, but most abundantly along with sandstone, slate-clay, &c. in the coal formation.

Geographic Situation.

Europe.—It occurs in veins at Horzowitz in Bohemia, where it is associated with red ironstone, sparry ironstone, galena or lead-glance, yellow blende, and straight lamellar heavy-spar; at Idria in the Friaul, in a coal formation; at Rosenau in Upper Hungary, in clay-slate, chlorite-slate, and talc-slate; in veins, along with ironstone, and ores of mercury, at Schemnitz and Kremnitz in Lower Hungary; Transylvania; Carinthia; Carniola; Salzburg; Tuscany; Sicily; in a coal formation at Wolfstein and Morsfeld in the Palatinate; at Moschellandsberg and Stahlberg in Deux Ponts; Allemont and Pellançon in France; at Almaden in Spain, in the coal formation; and in the neighbourhood of Conna in Portugal.

Asia.—Nertschinsk and Terentui in Siberia; also in the peninsula of Taygonos, near the mouth of the river Topolefka in Kamtchatka; and in Japan.

America.

America.—Mines of cinnabar occur in different parts of New Spain. At Durasno, between Terra Neuva and San Luis de la Paz, cinnabar, mixed with globules of mercury, forms a horizontal bed, which rests on porphyry. This bed is covered with strata of slate-clay, impregnated with nitrate of potash, which include a bed of slate-coal, and contain fragments of petrified vegetables. The cinnabar vein of San Juan de la Chica, is six, nine, and even sometimes twenty feet in width. It occurs in pitchstone-porphyry, which is disposed in globular and concentric lamellar concretions, of which the centre is occupied with hyalite. The cinnabar, and a little native mercury, are sometimes observed in the middle of the porphyritic rock, at a very considerable distance from the vein. The cinnabar extracted from the veins of the mountain del Fraile, near the Villa de San Felipe, is found in porphyry, with a hornstone base, which is traversed by veins of tinstone.

In the kingdom of New Granada, cinnabar occurs in three different places, namely, in the province of Antioquia, in the Valle de Santa Rosa, east from the Rio Cauca; in the mountain of Quindiu, in the pass of the central Cordillera, between Ibaque and Carthago, at the extremity of the ravine of Vermellon; and lastly, in the province of Quito, between the village of Azogue and Cuenca. The cinnabar is not only found in round fragments, mixed with small grains of gold, in the alluvial soil with which the ravine de Vermellon, at the foot of the table land of Ibague Viejo, is filled; but they know the vein also from which the torrent appears to have detached these fragments, and which traverses the small ravine of Santa Anna. Near the village of Azogue, to the N.W. of Cuenca, the mercury is found, as in the department Mont Tonnerre, in a formation of quartzy sandstone, with a clay base or cement. This sandstone is nearly 4592 feet in thickness, and contains bituminous wood and mineral pitch.

In Peru, cinnabar is found near Valdivui, in the province of Pataz, between the eastern bank of the Maranon and the missions of Guailillas; at the foot of the great Nevado de Pelagato, in the province of Conchucos, to the east of Santa; near Huancavelica, in the intendancy of that name; near Guaraz, in the province of Guailas; and at the Baths of Jesus, in the province of Guamalies, to the south-east of Guacarachuco. The famous mine of Huancavelica, as to the state of which so many false ideas have been disseminated, is in the mountain of Santa Barbara, to the south of the town of Huancavelica, at a horizontal distance of 7606 feet. The height of the town above the level of the sea is 12,308 feet. If we add to this the height of the mountain Santa Barbara above the level of Huancavelica, we shall find the absolute height of his mountain 14,506 feet. The cinnabar is found in the vicinity of this town, in two very different repositories, in beds, and in veins. In the great mine of Santa Barbara, the cinnabar is contained in a bed of sandstone, of upwards of 1200 feet in thickness. This sandstone is analogous to that of the environs of Paris; and the mountains of Aroma and Cascas, in Peru, resemble pure quartz. The quartz rock which contains the cinnabar, forms a bed in a limestone conglomerate, from which it is only separated by thin layers of slate-clay. This conglomerate is covered with a flœtz limestone, and the fragments of compact limestone in the conglomerate seem to indicate, that the whole mass of the mountain of Santa Barbara itself reposes on what is called alpine limestone.

The

The cinnabar does not fill the whole quartz bed of the great mine of Santa Barbara: it forms particular layers, and sometimes it is found in small veins, that occasionally unite into *stock-werke*. Hence, the metalliferous mass is only in general from 196 to 229 feet in breadth. Native mercury is very rare; but the cinnabar is accompanied with red ironstone, magnetic iron-ore, galena, and iron-pyrites, and also with calcareous-spar, sulphate of lime, and fibrous alum. The metalliferous bed, at great depths, contains a good deal of orpiment. Cinnabar is also found near to Sillacasa, in small veins which traverse the alpine limestone; but these veins, which are frequently full of calcedony, do not follow regular directions: they cross each other, and form nests, often of considerable magnitude. It is these veins that at present furnish all the mercury of Peru, the metalliferous bed of the great mine of Santa Barbara having been completely abandoned, owing to the works having fallen in *.

The most important mercury mines at present in a state of activity are those near Almaden in Spain, which have been worked for upwards of 2000 years; at Idria in the Friaul; in the ci-devant Palatinate; Deux Ponts; and in Spanish America.

Uses.

It is the ore from which the greatest quantity of the mercury of commerce is obtained. It is also used by the painter as a pigment; but artificial cinnabar, on account of

* Vide Black's Translation of Humboldt's New Spain, from which the above particulars in regard to the Spanish American mercury mines have been obtained.

of the purity and brightness of its colour, is preferred. It is also used for tinting wax of a red colour.

Observations.

1. It is distinguished from *Red Silver Ore* by its scarlet-red streak, and the red trace it affords on paper; and also in being entirely volatilised when heated: From *Red Orpiment*, by the colour of its streak, that of red orpiment being orange-yellow; and from *Red Lead-Ore*, also by the streak, that of the lead-ore being lemon-yellow.

2. It appears from Vitruvius, that the term *Minium* was derived from the name of a river in Spain; and there are several passages in Pliny, which shew that the term minium was applied to a substance corresponding with our cinnabar. He says, that almost all the minium in use at Rome, came from Spain, and that the ore was sent over from Spain sealed. He also says, that those who were employed in reducing minium to powder, wore loose bladders over the face, lest they should inhale the dust; the effects of which were very pernicious. This custom is also observed at the present day, by those who are employed for a length of time in triturating preparations of mercury *.

3. The term *Cinnabar*, was originally applied to the drug commonly called *Dragon's Blood*, which is of a dull red colour: it was afterwards transferred to the ore of mercury now under consideration †

4. Sage, in the Journal de Physique for 1784, and Estner in his Mineralogie, B. iii. 2. s. 314. describe a native

* Kid's Mineralogy, vol. ii. p. 95.

† Ibid.

native red oxide of mercury found at Idria; but it has not been met with by succeeding naturalists.

5. Baron Born, in his *Catalogue rais. d. l. Collect. d. Mlle. de Raab* t. ii. p. 394 describes a mineral under the name *Cinnabre alcalin*, which is mentioned by Wideman, Estner, Reuss, and Hausmann, under the name *Stink Zinnober*. The following description is given of it:

Its colour is intermediate between crimson-red and blood-red. It occurs massive, disseminated, in vesicles, and indistinctly crystallised. Internally the lustre is shining and adamantine. The fracture is imperfect foliated, inclining to radiated. It is translucent. It is soft. When triturated, it emits a hepatic smell. It is said to be sulphuret of mercury, combined with sulphureted hydrogen. It is found at Idria, along with calcareous-spar, and iron-pyrites.

IV. ORDER.

IV. ORDER.—SILVER.

This Order contains eleven species, viz. Native Silver, Auriferous Silver, Antimonial Silver, Arsenical Silver, Bismuthic Silver, Corneous Silver-ore or Horn-ore, Silver-glance or Sulphureted Silver, Brittle Silver-glance or Brittle Sulphureted Silver, Red Silver-ore, White Silver-ore, and Grey Silver-ore.

1. Native Silver.

Gediegen Silber, *Werner.*

Id. Wern. Pabst. b. i. s. 12. *Id. Estner,* b. iii. s. 319. *Id. Emm.* b. ii. s. 156.—L'Argent natif ordinaire, *Broch.* t. ii. p. 116.—Argent natif, *Hauy,* t. iii. p. 384.—Gediegen Silber, *Reuss,* b. iii. s. 310.—Gemeiner gediegen Silber, *Lud.* b. i. s. 210. *Id. Suck.* 2ter th. s. 129. *Id. Bert.* s. 360. *Id. Mohs,* b. iii. s. 102. *Id. Hab.* s. 102.—Argent natif, *Lucas,* p. 103.—Gemeiner gediegen Silber, *Leonhard,* Tabel. s. 53.—Argent natif, *Brong.* t. ii. p. 248. *Id. Brard,* p. 240.—Gediegen Silber, *Karsten,* Tabel. s. 60. *Id. Haus.* s. 69.—Native Silver, *Kid,* vol. ii. p. 83.—Argent natif, *Hauy,* Tabl. p. 73.—Native Silver, *Aikin,* p. 18.

External Characters.

Its colour is pure silver-white; but the surface, by exposure to the air, becomes yellowish-brown, or greyish-black.

It occurs massive, disseminated, in blunt-cornered pieces, in plates, and in membranes: it is said also to occur in Spanish

Spanish America in rolled pieces *. Besides these, it presents the following particular and regular external shapes: dentiform, filiform, reticulated, in leaves, capil lary, which latter, when it is very much entangled passes into compact. The crystallizations are the following:

1. Cube †.
2. Cube, truncated on the angles ‡.
3. Octahedron, either common or cuneiform ||.
4. Three-sided pyramid, bevelled on the edges, and also truncated on the angles and the summit; the pyramid either single or double.
5. Double six-sided pyramid, either perfect, or truncated on the angles and summits.
6. Leucite form.
7. Three-sided table, either perfect, or truncated on the angles.
8. Six-sided table, in which the terminal planes are set on alternately straight and oblique, and are bevelled.
9. Rectangular four-sided prism, either perfect, or truncated on the angles; sometimes acuminated with four planes, set on the lateral edges, and the acumination sometimes truncated, or in place of the truncation, the summit of the acumination bevelled, and the edges of the bevelment truncated.
10. Six-sided prism.

The

* In the Imperial Cabinet of minerals at Vienna, there is a rolled piece of native silver from Spanish America, which weighs upwards of 36 pounds.

† Argent natif cubique, Hauy.

‡ Argent natif cubo-octaedre, Hauy.

|| Argent natif octaedre, Hauy.

The octahedrons are frequently obliquely aggregated in rows: the cubes often rectangularly aggregated, so as to form the reticular external shape; and the prisms are occasionally obliquely aggregated, and form the dendritic external shape.

The crystals are small and very small, and microscopic.

The surface of the crystals is smooth; that of the particular shapes longitudinally streaked; that of the external shapes in leaves is sometimes drusy, sometimes streaked.

The surface varies from splendent to glimmering, according to the kind of surface; that of the crystals being splendent and shining; of the particular and common external shapes glistening and glimmering, with a metallic lustre.

The fracture is fine hackly.

The fragments are indeterminate angular, and blunt-edged.

The streak is splendent, with the metallic lustre.

It is harder than gold, tin, or lead; but softer than iron, platina, and copper.

It is perfectly malleable.

It is common flexible.

It is uncommonly heavy.

Specific gravity, 10.4743, *Hauy*. 10.000, *Gellert*. 10.338, *Selb*.

Chemical Characters.

It is soluble in nitric acid at the common temperature of the atmosphere; but the sulphuric acid does not act on it until heated. It is precipitated from its solution in nitric acid by muriatic acid; and the precipitate, which is is *luna cornea*, is insoluble in water: if a plate of copper be immersed in a solution of nitrate of silver, the silver is deposited, in its metallic state, on the surface of the copper. It is fusible into a globule, which is not altered by continuance of the heat.

Constituent Parts.

Native Silver from Johangeorgenstadt.

Metallic Silver, - -	99
Metallic Antimony, -	1
With a trace of Copper and Arsenic.	
	100

John, Chem. Untersuchungen, b. i. s. 283.

Geognostic Situation.

It occurs principally in veins in primitive mountains. In Suabia, and in some places in the Saxon Erzgebirge, it occurs in granite; in gneiss, and mica-slate, in Saxony, Bohemia, and Norway; in clay-slate in Ireland, Saxony, and Bohemia; in syenite and porphyry in Saxony and Hungary; and in primitive trap in Norway. In veins in transition rocks, as in grey-wacke in the Hartz. In flœtz rocks, as in clay-porphyry at Alva, in the Ochil Hills; and in other districts in limestone, sandstone, clay-stone, and slate-clay. The native silver in these rocks, is accompanied with various metalliferous and earthy minerals. The following are the principal metalliferous minerals, viz. corneous silver-ore, silver-glance or sulphureted silver, brittle silver-glance or brittle sulphureted silver, red silver-ore; also antimonial and arsenical silver, native arsenic, white cobalt-ore, red cobalt-ochre, copper-nickel,

nickel, and native bismuth; further, galena or lead-glance, black and brown blende, copper-pyrites, iron-pyrites, brown ironstone, native mercury, &c. The following are some of the earthy minerals, viz. heavy-spar, brown-spar, calcareous-spar, fluor-spar, quartz, hornstone, flint; and less frequently asbestus, steatite, apatite, &c.

Geographic Situation.

Europe.—Many years ago, a vein of silver-ore was, for a short time, wrought with considerable advantage in the parish of Alva, in the county of Stirling. The metalliferous minerals were, native silver, and silver-glance, with ores of copper and cobalt; and the vein-stones were calcareous-spar and heavy-spar. It is said, that from £40,000 to £50,000 worth of silver was extracted from the ores, before the repositories were exhausted. We are told, that a mass of capillary native silver was found in the veins traversing the blue-coloured limestone of the island of Isla. Native silver has also been met with at St Mewan, St Stephen's, Huel-Mexico, and Herland, in Cornwall *. The most northern silver-mines in Europe, are those of Kongsberg in Norway. The predominating rocks of the district, which are mica-slate and hornblende-slate, are traversed by numerous veins containing native silver, silver-glance or sulphureted silver, also native gold, auriferous silver, red silver-ore, corneous silver-ore, galena, native arsenic, brown blende, copper-pyrites, iron-pyrites. The most abundant and frequent vein-stones, are

* In the second volume of the Transactions of the Geological Society, it is mentioned, that the native silver in Cornwall is associated with alena or lead-glance, iron-pyrites, bismuth, cobalt, and wolfram, in veins traversing clay-slate.

are calcareous-spar and heavy-spar. In former times, these mines afforded uncommonly beautiful and large specimens of native silver. In the year 1628, a mass of pure silver, weighing 68 lbs. was met with in the mine Siegen Gottes, and in the year 1630, in the same mine, one of 204½ lb. In another mine, named Nye-Forhaabning, was found in the year 1666 a mass of silver weighing 560 lb., and which is still preserved in the Royal Collection at Copenhagen. In the year 1695, the mine Neue Juels afforded a mass weighing upwards of 118 lb.; and in the year 1769, in the mine Gottes Hülfe in der Noth, a mass estimated at 500 lb. was extracted from one of the veins *. Native silver is also found at Sala, in Westmannland in Sweden; and in the mines in the Hartz, in small quantity, along with galena and calcareous-spar. In the kingdom of Saxony, as in the district of Freyberg, it occurs in veins, along with various ores of silver, arsenic, iron, lead, and nickel, along with calcareous-spar, heavy-spar, fluor-spar, and quartz, in veins that traverse gneiss. The masses are sometimes of great magnitude: thus, we are told, that in 1750, a mass of native silver, weighing upwards of 1¼ cwt. was dug out of the great vein named Himmelsfurst, situated within a few miles of Freyberg. It is also mentioned by Albini, in his "Meissnische Berg-Chronicke," p. 30. that at Schneeberg, in the year 1478, a rich silver vein was discovered, and so large a block of native silver and ore cut out, that Duke Albert of Saxony descended into the mine, and used this huge block, which smelted 400 centners of silver, (a centner is 110 lbs.), as a table to dine on.

* Hausmann's Reise durch Scandinavien in den Jahren 1806 & 1807, b. ii. s. 18.

on. Native silver also occurs in Bohemia, in veins in clay-slate, along with galena or lead-glance, blende, silver-glance, cobalt, nickel, sparry iron-ore, iron-pyrites, quartz, and calcareous-spar: At Rudelstadt in Silesia, along with red-silver-ore, quartz, calcareous-spar, and lamellar heavy-spar: At Furstenberg in Suabia, in calcareous-spar, with quartz, and lamellar heavy-spar: At Wittichen, also in Suabia, in granite, along with black cobalt-ochre, white cobalt-ore, seldom with native bismuth, red silver-ore, and iron-glanee: At Reinerzau in Wirtemberg, along with silver-glance, and red silver-ore, and ores of cobalt, bismuth, copper, iron, and manganese; and the vein-stones are lamellar heavy-spar and fluor-spar: in the mine named Herzog Frederick, in the same country, it is associated with uran mica, and lamellar heavy-spar, and fluor-spar. At Allemont in France, it occurs in veins, along with native silver, silver-glance, red silver-ore, corneous silver-ore, ores of cobalt, native antimony, and nickel; and the vein-stones are calcareous-spar, mixed with asbestus and epidote: At Guadaleanal in Spain, along with red silver-ore, and calcareous-spar: At Felsobanya in Hungary along with native gold and iron-shot quartz: At Schemnitz, with white and brown lead-ore, native gold, brittle silver-glance, and quartz; and in other mines, also in Hungary, associated with silver-glance, brittle silver-glance, red manganese-ore, brown-spar, and calcedony: At Kapnik in Transylvania, with silver-glance, red silver-ore, blende, brown-spar, and quartz.

Asia.—Native silver is collected in several parts of Siberia: thus, at Kolywan, in the mine of St Andreas, it occurs disseminated in hornstone, along with brittle silver-glance; at Schlangenberg, in various forms, along with

with azure copper-ore; at Nertschink, with copper-green, and heavy-spar. It is said to be mined in China; and it is known to occur at Pondang in Java.

North America.—The silver-mines of Mexico and Peru have long been celebrated. The greatest portion of the Mexican silver is obtained from silver-glance or sulphureted silver, grey copper-ore, corneous silver-ore, red silver-ore, argentiferous galena or lead-glance, and argentiferous iron-pyrites. In some parts of Mexico, however, as we are informed by M. Humboldt, the operations of the miner are directed to a mixture of ochry brown iron-stone and minutely disseminated native silver. This ochreous mixture, which is named *pacos* in Peru, is the object of considerable operations at the mines of Angangueo, in the intendancy of Valladolid, as well as of Yxtepexi, in the province of Oaxaca *. Massive native silver, which is much less abundant in America than is generally supposed, has been found in considerable masses, sometimes more than 444 lb. avoirdupois, in the veins of Batopilas in New Biscay These mines, which are not very briskly worked at present, are amongst the most northern of Mexico. Nature exhibits the same minerals there, that are found in the silver-mines of Kongsberg in Norway. Native silver is constantly accompanied by silver-glance or sulphureted silver, in the veins of Mexico as well as

 in

* The pacos, according to Klaproth, contains the following ingredients:

Silver,	14.00
Brown Oxide of Iron,	71.00
Silica,	3.50
Sand, &c.	1.00
Water,	8.50
	98.00

Klaproth, Beit. b. iv. s. 9.

in those of the mines of Europe. These very minerals are frequently found united, in the rich mines of Sombrerete, Madrona, Ramos, Zacatecas, Hapujaha, and Sierra de Penos. From time to time, small branches or filaments of native silver are also discovered in the celebrated vein of Guanaxuato; but these masses have never been so considerable as those which were formerly drawn from the mine Del Encino, near Pachuca and Tasco, where native silver is sometimes contained in selenite. At Sierra de Pinos, near Zacatecas, native silver is accompanied with radiated azure copper-ore.

Dr Schumacher informs us, that a Mr Ginge, a missionary, brought from West Greenland a specimen of capillary native silver, associated with calcareous-spar, and which he says was picked up on the shores of that country *.

South America.—The mines of Huantajaya, surrounded with beds of rock-salt, are particularly celebrated, on account of the great masses of native silver which they contain in a decomposed vein; and they furnish annually between from 45,942 to 52,505 lb. Troy of silver. The native silver is accompanied with conchoidal corneous silver-ore, silver-glance or sulphureted silver, galena or lead-glance, with small grains of quartz, and calcareous-spar. In 1758 and 1789, two masses of native silver were discovered in the mines of Coronel and Loysa, the one weighing eight, the other two quintals. The mines of Gualgayoc and Micuipampa, commonly called Chota, also in South America, afford native silver. Immense wealth, M. Humboldt, remarks, has been found even at the surface, both in the mountain of Gualgayoc, which rises like a fortified castle

* Verzeichniss der Danisch-Nordischen Mineralien, p. 147.

castle in the midst of the plain, and at Fuentestiana, at Caromolachc, and at La Pampa de Navar. In this last plain, for an extent of more than half a square league, wherever the turf has been removed, silver-glance has been extracted, and filaments of native silver adhere to the roots of the gramina. Frequently the silver is found in masses, as if melted portions of this metal had been poured upon a very soft clay. The mines of Gualgayoc have furnished to the treasury of Truxillo, between the month of April 1774 and the month of October 1802, the sum of 1,189,456 lb. Troy of silver; or at an average 44,095 lb. annually.

The mines of Pasco afford native silver, along with ores of this metal, and afford annually from 131,263 lb. Troy to 196,894 lb. Troy of silver.

Mr Helms is of opinion, that the Cordilleras of America, if properly investigated, will afford so great a quantity of silver, as to overturn our present commercial system,—by making silver as common as copper and iron.

Uses.

Its various uses, in coinage, and for other useful and ornamental purposes, will be considered in a separate work.

Observations.

Native silver is distinguished from *Antimonial Silver*, and *Native Antimony*, by fracture, and tenacity: it has a hackly fracture, and is completely malleable; but they are brittle, and have a foliated fracture.

2. Auriferous Native Silver.

Guldisches gediegen Silber, *Werner*.

Id. Werner, Pabst. b. i. s. 12. *Id. Estner*, b. iii. s. 315. *Id. Emm.* b. ii. s. 154.—L'Argent natif aurifere, *Broch.* t. ii. p. 114.—Guldisch Silber, *Reuss*, b. iii. s. 322. *Id. Lud.* b. i. s. 210. *Id. Suck.* 2ter th. s. 128. *Id. Bert.* s. 362. *Id. Mohs*, b. iii. s. 123. *Id. Leonhard*, Tabel. s. 53. *Id. Karsten*, Tabel. s. 60. *Id. Haus.* s. 69.

External Characters.

Its colour is intermediate between silver-white and brass-yellow.

It occurs disseminated, in membranes, which are pretty thick, passing into plates, capillary, in leaves, and sometimes crystallised in cubes.

Its specific gravity, on account of the quantity of gold which it contains, is greater than that of common native silver.

In other characters, it agrees with the preceding species.

Constituent Parts.

Silver, - -	72.00
Gold, - -	28.00
	100.00

Fordyce, Phil. Trans. 1779, p. 523.

Geognostic and Geographic Situations.

It occurs in veins in primitive rocks at Kongsberg in Norway; at Rauris in Salzburg; and at Schlaugenberg in Siberia.

3. Antimonial

3. Antimonial Silver-Ore.

Spiesglas Silber, *Werner*.

Mine d'Argent blanche antimoniale, *Romé de Lisle*, t. i. p. 460. —Antimonialisch gediegen Silber, *Wid.* s. 684.—Antimoniated Native Silver, *Kirw.* vol. ii. p. 110.—Spies-glanz Silber, *Estner*, b. iii. s. 337. *Id. Emm.* b. ii. s. 162.—Argent antimonial, *Broch.* t. ii. p. 119. *Id. Hauy*, t. iii. p. 391.—Spiesglanz Silber, *Reuss*, b. iii. s. 325. *Id. Lud.* b. i. s. 211. *Id. Suck.* 2ter th. s. 135. *Id. Bert.* s. 369. *Id. Mohs*, b. iii. s. 127.—Argent antimonial, *Lucas*, p. 104.—Spiesglanz Silber, *Leonhard*, Tabel. s. 53.—Argent antimonial, *Brong.* t. ii. p. 249. *Id. Brard*, p. 243.—Spiesglanz Silber, *Karsten*, Tabel. s. 60. *Id. Haus.* s. 70.—Silver alloyed with Antimony, *Kid*, vol. ii. p. 85.—Argent antimonial, *Hauy*, Tabl. p. 74. —Antimonial Silver, *Aikin*, p. 19.

External Characters.

Its colour is intermediate between silver-white and tin-white; sometimes inclining more to the one, sometimes more to the other, yet in general more to the first.

It occurs massive, disseminated, globular, tuberose, and crystallised. Its crystallisations are,

1. Rectangular four-sided prism.
2. Perfect six-sided prism; sometimes truncated on all the edges, so that it assumes a roundish aspect.
3. Six-sided table.
4. Cube, which is rarely perfect, usually truncated on one or more of its angles.
5. Double six-sided pyramid, generally truncated on the angles.

 The

The surface of the prisms is usually longitudinally streaked; of the massive, uneven; and of the globular and tuberose, uneven and rough.

Externally it is glistening, sometimes only glimmering.

Internally it is shining and splendent, with a metallic lustre.

The fracture is perfect foliated, but the number and direction of the cleavages hitherto unascertained; sometimes small and fine grained uneven, and scopiform radiated.

It occurs in coarse, small and fine granular distinct concretions.

It is sectile.

It is rather easily frangible.

It is soft.

It is uncommonly heavy.

Specific gravity, 9.4406, *Hauy*. 10.000, *Selb*. 9.820, *Klaproth*.

Chemical Characters.

Heated on charcoal before the blowpipe, the antimony is volatilised, with the odour which is peculiar to it, and there remains a mass of silver, surrounded with a brown slag, which colours borax green.

Constituent Parts.

				Fine Granular from Wolfach.	Coarse Granular from Wolfach.	From Andreasberg.
Silver,	89	78	75.25	84	76	75¼
Antimony,	11	22	24.75	16	24	24¾
	100	100	100.00	100	100	100
	Selb.	*Vauquelin.*	*Abich.*	*Klap.* Beit. b. ii. s. 301.	*Ibid.*	*Ibid.* b. iii. s. 176.

Geognostic

Geognostic Situation.

It occurs in veins, in granite and grey-wacke, and in both situations it is associated with arsenical silver, native arsenic, galena or lead-glance, brown blende, calcareous-spar; but in the granite, it is also accompanied with native silver, red silver-ore, iron-pyrites, straight lamellar heavy-spar, and fluor-spar; and in the grey-wacke, with native antimony, red silver-ore, silver-glance, and brown-spar.

Geographic Situation.

It occurs in veins that traverse granite, at Altwolfach in Suabia; in veins that traverse clay-slate, at Andreasberg in the Hartz; at Kasalla, near Guadalcanal in Spain; and, it is said, also at the Rathausberg in Gastein, and the Goldberg at Rauris in Salzburg; and at Allemont in France.

Observations.

It is distinguished from *Native Silver*, by its sectility and foliated fracture: from *White Cobalt-ore*, by its sectility, and inferior hardness: from *Arsenical Pyrites*, by its foliated fracture, and inferior hardness.

4. Arsenical Silver-Ore.

Arseniksilber, *Werner*.

Id. Wern. Pabst. b. i. s. 28.—Argent arsenical, *De Born,* t. ii. p. 417.—Arsenikalisch gediegen Silber, *Wid.* s. 687.—Arsenicated native Silver, *Kirw.* vol. ii. p. 111.—Arsenicsilber, *Estner,* b. iii. s. 342. *Id. Emm.* b. ii. s. 165.—L'Argent arsenical, *Broch.* t. ii. p. 122.—Argent antimonial, arsenifere, et ferrifere, *Hauy,* t. iii. p. 398.—Arsenik-silber, *Reuss,* b. iii. s. 499. *Id. Lud.* b. ii. s. 211. *Id. Suck.* 2ter th. s. 144. *Id. Bert.* s. 503. *Id. Mohs,* b. iii. s. 131. *Id. Leonhard,* Tabel. s. 53.—Argent arsenical, *Brong.* t. ii. p. 250.—Silber-arsenik, *Karsten,* Tabel. s. 74.—Silver alloyed with Arsenic and Iron, *Kid,* vol. ii. p. 86.—Argent antimonial ferro-arsenifere, *Hauy,* Tabl. p. 74.—Arsenical Antimonial Silver, *Aikin,* p. 19.

External Characters.

Its colour is tin-white, which passes into silver-white, and verges on light lead-grey.

It is always more or less tarnished with a blackish colour.

It occurs massive, disseminated, small reniform, globular, and crystallised in rectangular four-sided prisms.

Internally it is shining or glistening, which borders on glimmering; and the lustre is metallic.

The fracture is imperfect foliated, and is sometimes spherical, sometimes straight foliated, and in other directions even.

The fragments are indeterminate angular, and blunt-edged.

It occurs in small and fine granular distinct concretions; also in thin and curved lamellar distinct concretions,

tions, which are bent in the direction of the external surface.

It is harder than antimonial silver.

It is shining in the streak.

It is sectile, slightly inclining to brittle.

It is easily frangible.

It is uncommonly heavy.

Chemical Characters.

Before the blowpipe, the arsenic and antimony are volatilised, and emit a garlic smell; a globule of silver remains, which is more or less pure.

Constituent Parts.

Andreasberg.	
Arsenic, - -	35.00
Iron, - - -	44.25
Silver, - -	12.75
Antimony, - -	4.00
	96.00

Klaproth, Beit. b. i. s. 187.

Geognostic and Geographic Situations.

It generally occurs along with native arsenic, dark-red silver-ore, brittle silver-glance, galena or lead-glance, and brown blende, in massive white calcareous-spar. It is found in the Hartz; also at Altwolfach in Suabia; and at Guadalcanal and Kasalla in Spain.

Observations.

Observations.

1. It does not tarnish so quickly, and its colour is lighter than that of native arsenic, with which it has been confounded.

2. It passes into Native Arsenic and Native Silver.

5. Bismuthic Silver-Ore.

Wissmuth Silbererz, *Selb.*

Wismuthisches Silber, *Selb,* in Crell's Chem. Annal. 1793, 1. 10. *Id. Wid.* s. 716.—Wissmuthblei, *Reuss,* b. ii. 4. s. 191. *Id. Karsten,* Tabel. s. 68.—Wismuthsilbererz, *Selb,* in den Mineralogischen Studien, b. i. s. 79.—Bismuthic Silver, *Aikin,* p. 28.

External Characters.

Its colour is pale lead-grey, becoming deeper on exposure to the air.

It occurs disseminated; and rarely crystallised in acicular and capillary crystals.

Its lustre is glistening and metallic.

The fracture is fine-grained uneven.

It is soft.

It is sectile.

It is easily frangible.

Chemical Characters.

Before the blowpipe, metallic globules begin to oose out, and on the addition of borax, unite into one mass, the

the flux at the same time acquiring an amber colour: the metallic button is brittle, and of a tin-white colour.

Constituent Parts.

Bismuth, - -	27.00
Lead, - - -	33.00
Silver, - -	15.00
Iron, - - -	4.30
Copper, - -	0.90
Sulphur, - -	16.30
	96.50 *

Klaproth, Beit. b. ii. s. 297.

Geognostic and Geographic Situations.

It has hitherto been found only in the mine named Friedrich-Christian in the Schapbach, in the Black Forest, where it occurs in veins that traverse gneiss, along with copper-pyrites, and quartz, and a smaller quantity of iron-pyrites, and galena or lead-glance.

6. Corneous

* Selb, in his Mineralogical Studies, vol. i. p. 81. states, the quantity of silver in this ore at 20 *per cent.*

6. Corneous Silver-Ore or Horn-Ore.

Hornerz, *Werner.*

Minera Argenti cornea, *Wall.* t. ii. p. 331.—Argent corné, *Romé de L.* t. iii. p. 463. *Id. De Born,* t. ii. p. 420.—Hornerz, *Wern.* Pabst. b. i. s. 29. *Id. Wid.* s. 691.—Corneous Silver-ore, *Kirw.* vol. ii. p. 113.—Hornerz, *Estner,* b. iii. s. 348. *Id. Emm.* b. ii. s. 168.—Argent corné, *Lam.* t. i. p. 130.—La Mine cornée, ou L'Argent corné ou muriaté, *Broch.* t. ii. p. 127.—Argent muriaté, *Hauy,* t. iii. p. 418. 422.—Hornerz, *Reuss,* b. iii. s. 330. *Id. Lud.* b. i. s. 212. *Id. Suck.* 2ter th. s. 137. *Id. Bert.* s. 364. *Id. Mohs,* b. iii. s. 134.—Argent muriaté, *Lucas,* p. 108.—Silber Hornerz, *Leonhard,* Tabel. s. 54.—Argent muriaté, *Brong.* t. ii. p. 256. *Id. Brard,* p. 250.—Hornerz, *Karsten,* Tabel. s. 60.—Muriate of Silver, *Kid,* vol. ii. p. 91.—Argent muriaté, *Hauy,* Tabl. p. 76.—Horn-silver, *Aikin,* p. 20.

This species is divided into four subspecies, viz. Conchoidal Corneous Silver-ore, Radiated Corneous Silver-ore, Common Corneous Silver-ore, and Earthy Corneous Silver-ore.

First Subspecies.

Conchoidal Corneous Silver-Ore.

Muschlichtes Hornerz, *Karsten.*

Id. Karsten, in Magazin der Gesellschaft der Naturforschender Freünde zu Berlin, b. i. s. 156.

External Characters.

Its colours are greyish-white which passes into pale pearl-

pearl-grey, and greenish-white, which passes into pale olive-green.

It occurs massive.

The lustre is splendent and adamantine.

The fracture is sometimes flat and large conchoidal, sometimes small and imperfect conchoidal

The fragments are indeterminate angular, and rather sharp-edged.

It occurs in large and small granular distinct concretions.

It is generally semi-transparent, but sometimes passes into transparent.

The streak is resinous, and with a lower lustre than the fracture.

It is very soft.

It is malleable.

Specific gravity, 4.7488.

Constituent Parts.

Silver, - - -	76.0
Oxygen, - - -	7.6
Muriatic Acid, - -	16.4
	100.0

Klaproth, Beit. b iv. s. 12.

This is the purest and richest subspecies of corneous silver-ore.

Geognostic and Geographic Situations.

This mineral occurs in compact limestone, along with common corneous silver-ore, and has hitherto been found only at Guantahoyio (Huantajayo) in Peru.

Observations.

Observations.

1. This beautiful subspecies is characterised by its lustre, fracture, transparency, and shape of its distinct concretions.

2. It was first made known to mineralogists by Humboldt, who brought specimens of it with him from South America. The division into subspecies, and the descriptions, we owe to Karsten, and the analyses to Klaproth.

Second Subspecies.

Radiated Corneous Silver-Ore.

Strahliges Hornerz, *Karsten.*

Id. Karsten, in Magazin der Gesellschaft der Naturforschender Freünde zu Berlin, b. i. s. 157.

External Characters.

Its colour is dark pistachio-green, which passes into lemon-yellow.

It occurs massive, and in microscopic crystals.

Its lustre is splendent and resinous.

The fracture is straight and narrow radiated.

It occurs in thin prismatic concretions.

It is translucent.

It is shining in the streak.

It is very soft.

It is malleable.

Constituent Parts.

In the assay-furnace, 100 grains afforded 63½ of pure silver.—*Karsten.*

Geognostic

Geognostic and Geographic Situations.

It occurs in small veins, along with native silver, quartz, and heavy-spar, in gneiss or clay-slate. It has hitherto been found only in South America.

Third Subspecies.

Common Corneous Silver-Ore.

Gemeines Hornerz, *Karsten.*

Id. Karsten, in Magazin der Gesellschaft der Naturforschender Freünde zu Berlin, b. i. s. 158.

External Characters.

Its colours are pearl-grey, and violet-blue; rarely olive-green and leek-green. By the action of the weather, it acquires a brown, and more rarely a black colour.

It occurs massive, disseminated; and frequently crystallised in the following figures:

1. Cube.
2. Cube truncated on the angles.
3. Octahedron.
4. Single four-sided prism.

The crystals are generally small and very small, and are occasionally aggregated in rows, or in a scalar-like form. The very minute crystals form small flakes.

The external surface is smooth.

Externally the lustre is shining or splendent; internally glimmering. and often dull.

The fracture is fine earthy.

The fragments are indeterminata angular, and rather sharp-edged.

It

It is opaque, or translucent on the edges.

The lustre of the streak is resinous.

It is very soft.

It is malleable.

Specific gravity, 4.804.

Chemical Characters.

It is fusible before the flame of a candle: before the blowpipe, on charcoal, it is reducible to a metallic globule, giving out at the same time vapours of muriatic acid: when rubbed with a piece of moistened zinc, the surface becomes covered with a thin film of metallic silver.

Constituent Parts.

Silver, - - -	67.75
Oxygen, - - -	6.75
Muriatic Acid, -	14.75
Oxide of Iron, - -	6.00
Alumina, - - -	1.75
Sulphuric Acid, - -	0.25
	97.25
Loss, - - -	2.75
	100.00

Klaproth, Beit. b. iv. s. 13.

Geognostic Situation.

It occurs in silver veins, and generally in their upper part. These veins traverse gneiss, mica-slate, clay-slate, grey-wacke, porphyry, and limestone, and contain, besides the corneous silver-ore, the following metalliferous and

and earthy minerals, viz. silver-glance or sulphureted silver, and iron-ochre; more rarely native silver, earthy cobalt-ochre, red silver-ore, tile-ore, malachite, azure copper-ore, white lead-ore, iron-pyrites, galena, atacamite or muriate of copper, copper-green, grey copper-ore, and and hornstone; also calcareous-spar, heavy spar, and quartz.

Geographic Situation.

Europe.—At Huel-Mexico in Cornwall: in occurs in France and Saxony, in veins that traverse gneiss, mica-slate, and clay-slate, where it is associated with silver-glance or sulphureted silver, and iron-ochre, and more rarely with native silver; at Schemnitz in Hungary, in massive quartz, along with fibrous malachite, silver-glance or sulphureted silver-ore, and white lead-ore; in the mountain of Chalanches, in calcareous-spar, and accompanied with silver-glance, native silver, and more rarely with earthy cobalt-ochre, and red silver-ore.

Asia.—In Siberia, it occurs along with native gold, common and auriferous native silver, silver-glance, malachite, azure copper-ore, tile-ore, white lead-ore, hornstone, calcareous-spar, heavy-spar, quartz, and sometimes lithomarge. These veins appear to traverse limestone.

America.—This ore, which is so seldom found in Europe, is very abundant in the mines of Catorce, Fresnillo, and the Cerro San Pedro, near the town of San Luis Potosi That of Fresnillo is frequently of an olive-green colour, which passes into leek-green. In the veins of Catorce, the corneous silver-ore is accompanied with yellow lead-ore and green lead-ore.

Fourth Subspecies.

Earthy Corneous Silver-Ore.

Erdiges Hornerz, *Karsten*.

Id. Karsten, in Magazin der Naturforschender Freünde zu Berlin, b. i. s. 159. *Id. Karsten,* Tabel. s. 60.

External Characters.

Internally the colour is pale mountain-green, inclining to greyish-white ; externally it has a bluish-grey tarnish.

It occurs in thick crusts.

Internally it is dull.

The fracture is coarse and fine earthy.

The fragments are blunt-angular.

It is very soft, almost friable.

The streak is shining and resinous.

It is sectile.

It is heavy.

Constituent Parts.

Silver, - - -	24.64
Muriatic Acid, - -	8.28
Alumina, with a trace of Copper, - - -	67 08
	100.00

Klaproth, Beit. b. i. s. 137.

Geognostic

Geognostic and Geographic Situations.

It is found in veins that traverse transition rocks at Andreasberg in the Hartz.

Observations.

1. It sometimes occurs in a fluid form, in veins and drusy cavities, when it is said to resemble butter-milk: hence the German name (*Buttermilkerze*) given to it.

2. It appears to be an intimate mixture of corneous silver-ore and clay.

3. It was first discovered in the Hartz in the year 1576, and continued to be found until the year 1617; since that period, it has almost disappeared, and therefore is at present a very rare mineral.

7. Silver-Glance or Sulphureted Silver-Ore.

Glaserz, *Werner*.

This species is divided into two subspecies, viz. Compact Silver-Glance, and Earthy Silver-Glance.

First Subspecies.

Compact Silver-Glance or Compact Sulphureted Silver-Ore.

Dichtes Glanzerz, *Hausmann.*

Minera Argenti vitrea, *Wall.* t. ii. p. 329.—Mine d'Argent vitreuse, *Romé de Lisle,* t. iii. p. 440.—Argent vitreuse, *De Born,* t. ii. p. 424.—Glaserz, *Werner,* Pabst. b. i. s. 33. *Id. Wid.* s. 696.—Sulphurated Silver-ore, *Kirw.* vol. ii. p. 115.—Geschmeidiges Silberglanzerz, *Estner,* b. iii. s. 370.—Glaserz, *Emm.* b. ii. s. 175.—Argent vitreuse, *Lam.* t. i. p. 120. *Id. Broch.* t. ii. p. 134.—Argent sulphuré, *Hauy,* t. iii. p. 398, -402.—Glanzerz, *Reuss,* b. iii. s. 342. *Id. Lud.* b. i. s. 214. *Id. Suck.* 2ter th. s. 142. *Id. Bert.* s. 366. *Id. Mohs,* b. iii. s. 144.—Geschmeidiges Silber-glanzerz, *Hab.* s. 103.—Argent sulphuré, *Lucas,* p. 105.—Glanzerz, *Leonhard,* Tabel. s. 54.—Argent sulphuré, *Brong.* t. ii. p. 251. *Id. Brard,* p. 245.—Glanzerz, *Karsten,* Tabel. s. 60.—Dichtes Glanzerz, *Haus.* s. 71.—Sulphuret of Silver, *Kid,* vol. ii. p. 87.—Argent sulphuré, *Hauy,* Tabl. p. 74.—Sulphureted Silver, *Aikin,* p. 20.

External Characters.

The colour is dark blackish lead-grey. On exposure, its surface acquires a tempered-steel coloured tarnish.

It generally occurs massive, sometimes disseminated, and in membranes, but seldom in plates; also in several particular external shapes, as dentiform, filiform, capillary, reticulated, irregular dendritic, stalactitic, in leaves, with globular and pyramidal impressions, corroded and amorphous;

amorphous; also crystallised. Its crystallisations are the following:

1. Cube, which is either perfect or truncated on its edges or angles, or on both at the same time *.
2. Octahedron, which is the fundamental crystal †. It is either perfect or truncated on its angles or edges. When the edges of the common basis are very deeply truncated, it passes into the rectangular four-sided prism, acuminated on both extremities by four planes, which are set on the lateral planes.
3. Garnet dodecahedron, which is formed from the cube or the octahedron, by the truncation of their edges ‡. Its edges are sometimes truncated.
4. Double eight-sided pyramid, flatly acuminated on both extremities by four planes, which are set on the alternate lateral edges.
5. Three and six-sided tables, in which the terminal planes are set on alternately straight and oblique. These may be considered as varieties of the octahedron.

The two last-mentioned crystallisations are rare.

The crystals are seldom middle-sized; usually small and very small; superimposed, or aggregated in rows. The octahedron is usually aggregated in rows, the other crystallisations usually superimposed. The cubes are sometimes hollow.

The surface of the crystals is sometimes smooth, sometimes drusy.

* Argent sulphuré cubique; Argent sulphuré cubo-octaedre, Hauy.

† Argent sulphuré octaedre, Hauy.

‡ Argent sulphuré dodecaedre, Hauy.

Externally it is shining and glistening, and when drusy, faintly glimmering.

Internally it alternates from shining to glistening, and the lustre is metallic.

The fracture is commonly small-grained uneven; sometimes it inclines to imperfect small and flat conchoidal.

The fragments are indeterminate angular, blunt-edged.

Its lustre is increased in the streak.

It is soft.

It is completely malleable.

It is flexible, but not elastic.

It is difficultly frangible.

It is uncommonly heavy.

Specific gravity, 7.215, *Gellert.* 7.200, *Lametherie.* 6.9099, *Brisson.*

Chemical Characters.

Before the blowpipe it loses its sulphur, and a bead of pure silver remains. If heated gently in a furnace, the sulphur dissipates, and the silver appears in its metallic state, in dendritic and capillary forms, resembling some varieties of native silver.

Constituent Parts.

	From Himmelsfürst.	From Joachimsthal.		
Silver, -	85	84.81	84	75
Sulphur, -	15	14.19	16	25
	100	99.00	100	100
	Klaproth.	*Klaproth.*	*Sage.*	*Bergman.*

Geognostic Situation.

It is one of the most frequent of the ores of silver, and there are few formations of that metal which do not contain

tain it. It occurs principally in veins that traverse primitive and transition rocks, such as gneiss, mica-slate, clay-slate, and grey-wacke; less frequently in porphyry; and still seldomer in granite. In these veins, it is associated with various ores of silver, copper, and lead, and also of iron, zinc, cobalt, arsenic, and more rarely with native gold. Of the accompanying earthy minerals, the following may be enumerated, viz. quartz, calcareous-spar, brown-spar, heavy-spar, fluor-spar, and hornstone.

Geographic Situation.

Europe.—This ore was formerly met with in the workings for silver at Alva in Stirlingshire; and it has also been found at Herland in Cornwall. At Kongsberg, it occurs in veins, along with native silver, and various ores of that and other metals; in the Hartz, in veins that traverse grey-wacke; in Saxony, in veins in gneiss; in veins in the granite of Altwolfach in Suabia; at Annaberg in Lower Austria, in veins that traverse compact grey-coloured limestone; at Joachimsthal in Bohemia, in mica-slate and clay-slate; in porphyritic-syenite at Schemnitz in Hungary; and in Sardinia, along with corneous silver-ore, and native silver.

Asia.—At Schlangenberg in Siberia.

America.—This ore is very common in the mines of Guanaxuato and Zacatecas, as well as in the *Veta Biscaina* of Real del Monte in Mexico; but in Peru, where it also occurs, it is much less abundant.

Observations.

This mineral has received the name *Silver-glance* from its shining appearance: it is also named *Vitreous Silver-ore*, from the German name *Glaserz*, which, however, is but a corruption of glanz-erz.

Second Subspecies.

Earthy Silver-Glance or Earthy Sulphureted Silver-Ore.

Silberschwärze, *Werner.*

Silberschärze, *Wid.* s. 694.—Sooty Silver-Ore, *Kirw.* vol. ii. p. 117.—Silberschwärze, *Estner,* b. iii. s. 365. *Id. Emm.* b. ii. s. 173.—L'Argent noir, *Broch.* t. ii. p. 132.—Silberschwärze, *Reuss,* b. iii. s. 338. *Id. Lud.* b. i. s. 213. *Id. Suck.* 2ter th. s. 141. *Id. Bert.* s. 363. *Id Mohs,* b. iii. s. 141. *Id. Leonhard,* Tabel. s. 54. *Id. Karsten,* Tabel. s. 60.—Erdiges Glanzerz, *Haus.* s. 71.—Black sulphureted Silver, *Aikin,* p. 21.

External Characters.

Its colour is bluish-black, which slightly inclines to dark lead-grey.

It occurs massive, disseminated, as a coating or crust, corroded, perforated, amorphous, and filling up the cavities in the globular corneous silver-ore.

It varies from friable to solid.

Internally it is dull, passing into feeble metallic glimmering.

When friable, it occurs in dull dusty particles, but when solid, its fracture is fine earthy, inclining to uneven.

The fragments are blunt-edged.

It is weakly translucent.

It is very soft, sometimes passes into soft, sometimes into friable.

It

It affords a metallic shining streak.
It soils a little.
It is easily frangible.
It is sectile.
It is heavy.

Chemical Characters.

It is easily fusible: is converted into a slaggy mass, containing globules of impure silver.

Constituent Parts.

It appears to be a Sulphuret of Silver.

Geognostic Situation.

It occurs in veins in primitive mountains, in which it is generally associated with compact silver-glance, corneous silver ore, brittle silver glance, native silver, native gold, ochry brown ironstone, quartz, and straight lamellar heavy-spar.

Geographic Situation.

Europe.—It occurs principally in the Saxon Erzgebirge, in veins, along with other ores of silver; at Kremnitz, along with native gold, silver-glance, and amethyst; near Schemnitz, in ironshot quartz, with white lead-ore, and malachite; in Chalanches, near Allemont in Dauphiny, with native silver, earthy black cobalt-ochre, red cobalt-ochre, ochre of nickel, and calcareous-spar; and in the mines of Kongsberg.

Asia.—At Schlangenberg in Siberia, along with iron-pyrites, blende, auriferous native silver, and hornstone.

8. Brittle

8. Brittle Silver-Glance or Brittle Sulphureted Silver-Ore.

Sprödglaserz, *Werner.*

Argent fragile, *De Born,* t. ii. p. 429.—Sprödglaserz, *Wid.* s. 669. *Id. Werner,* Pabst. b. i. s. 41.—Sprödes Silber-glanzerz, *Estner,* b iii. s. 398.—Sprödglaserz, *Emm.* b. ii. s. 180.—L'Argent vitreux aigre, *Broch.* t. ii. p. 138.—Argent noire, *Hauy,* t. iii. p. 416.—Sprödglanzerz, *Reuss,* b. iii. s. 351. *Id. Lud.* b. i. s. 215. *Id. Suck* 2ter th. s. 148. *Id. Bert.* s. 370. *Id. Mohs,* b. iii. s. 160. *Id. Leonhard,* Tabel. s. 54.—Argent rouge aigre, *Brong.* t. ii. p. 254.—Sprödglanzerz, *Karsten,* Tabel. s. 60. *Id. Haus.* s. 71.—Argent antimonie sulphuré noire, *Hauy,* Tabl. p. 76.—Brittle Sulphureted Silver, *Aikin,* p. 21.

External Characters.

Its colour is intermediate between iron-black and dark lead-grey.

It occurs massive, disseminated, in membranes, and very often crystallised. Its crystallisations are the following:

1. Equiangular six-sided prism, with straight or convex terminal faces.
2. The preceding figure, rather acutely acuminated by six planes, which are set on the lateral planes, and the extremities of the acuminations sometimes very deeply truncated.
3. Equiangular six-sided table. In this figure the lateral edges are sometimes truncated.
4. Double six-sided pyramid.

5. When

5. When the convex terminal faces of the prism meet the lens, is formed
6. Rectangular four-sided table, truncated on the angles and edges.

The tabular crystals often intersect one another, and thus form the cellular external shape; sometimes they are superimposed. They are seldom middle-sized, usually small, and very small, and even microscopic.

The lateral planes, particularly of the prism, are longitudinally streaked; in the other figures the planes are sometimes smooth, sometimes drusy.

Externally it is highly splendent.

Internally it is shining, inclining to glistening, and the lustre is metallic.

The fracture alternates from small conchoidal to fine-grained uneven.

The fragments are indeterminate angular, and rather blunt-edged.

The lustre is not increased in the streak.

It is soft.

It is sectile.

It is easily frangible.

It is uncommonly heavy.

Specific gravity, 7.208, *Gellert.*

Chemical Characters.

Heated on charcoal before the blowpipe, it melts with difficulty, and the sulphur, arsenic and antimony, are in part volatilised. A globule of imperfectly malleable silver, accompanied with a brown scoria, remains behind.

Constituent

Constituent Parts.

According to Klaproth, the brittle silver-glance from the mine Hoffnung Gottes in Gross-Voightsberg near Freyberg, affords

Silver, - - -	66.50
Sulphur, - - -	12.00
Antimony, - - -	10.00
Iron, - - - -	5.00
Copper and Arsenic, -	0.50
Earthy substances, -	1.00
	95.00

Klaproth, Beit. b. i. s. 166.

Geognostic Situation.

It occurs in veins that traverse gneiss, clay-slate, and porphyry, and in these it is accompanied with other ores of silver; also ores of lead, zinc, copper, cobalt, iron, and more rarely gold, and also with quartz, calcareous-spar, and brown-spar.

Geographic Situation.

Europe.—This ore occurs in the district of Freyberg in Saxony, in veins, along with native silver, silver-glance or sulphureted silver, dark red silver-ore, white silver-ore, galena or lead-glance, black blende, &c. with brown-spar, calcareous-spar, and quartz, seldomer with heavy-spar and fluor-spar: in the Upper Erzgebirge, both on the Saxon and Bohemian sides, it is associated with light red silver-ore, white cobalt-ore, native arsenic, silver-glance; also native silver, iron-pyrites, copper-pyrites, brown-spar, and calcareous-spar, and these veins occur in gneiss and

and clay-slate. In Hungary, it occurs at Chemnitz and Kremnitz; at Chemnitz, it is associated with silver-glance, copper-glance, galena or lead-glance, iron-pyrites, brown and black blende, dark red silver-ore, and very rarely light red cinnabar, and native gold, calcareous-spar, brown-spar, and calcedony; in the mines at Kremnitz, it is usually accompanied with quartz, brown-spar, amethyst, iron and copper pyrites, dark red silver-ore, and seldomer with native gold: at Joachimsthal in Bohemia, the ores and earthy minerals with which it is accompanied are dark red silver-ore, hepatic pyrites, calcareous-spar, and brown-spar.

Asia.—In Siberia, it is accompanied with granular heavy-spar, copper-pyrites, azure copper-ore, and brown blende.

America.—In the mine of Ecateras in Mexico; and also in the silver mines of Peru.

Observations.

This mineral is distinguished from *Compact Silver-glance*, by its sectility, its lustre not being increased in the streak, and its splendent external lustre.

9. Red Silver-Ore.

This species is divided into two subspecies, viz. Dark Red Silver Ore, and Light Red Silver Ore.

First

First Subspecies.

Dark Red Silver-Ore.

Dunkles Rothgiltigerz, *Werner.*

Id. Wern. Pabst. b. i. s. 45. *Id. Wid.* s. 703.—Dark red Silver-Ore, *Kirw.* vol. ii. p. 123.—Dunkles Rothgiltigerz, *Estner*, b. iii. s. 410. *Id. Emm.* b. ii s. 185.—L'Argent rouge foncé, *Broch.* t. ii. p. 143.—Argent antimonié sulphuré, *Hauy*, t. iii. p. 402,–416.—Dunkles Rothgiltigerz, *Reuss*, b. iii. s. 358. *Id. Lud.* b. i. s. 215. *Id. Suck.* 2ter th. s. 153 *Id. Bert.* s. 372. *Id. Mohs*, b. iii. s. 168 *Id. Leonhard*, Tabel. s. 55.—Árgent rouge sombre, *Brong.* t. i. p. 254.—Dunkles Rothgiltigerz, *Karsten*, Tabel. s. 60. *Id. Haus.* s. 77.—Antimoniated Sulphuret of Silver, *Kid*, vol. ii. p. 89.—Argent antimonié sulphuré rouge obscur, *Hauy*, Tabl. p. 76.—Red or ruby Silver, *Aikin*, p. 22.

External Characters.

Its colour is intermediate between cochineal-red and lead-grey, which sometimes passes into iron-black.

It occurs massive, disseminated, dendritic, in membranes, in plates, corroded, cellular, small reniform, and small botryoidal; and crystallised in the following figures:

1. Equiangular six-sided prism, which is either perfect, or truncated on the terminal edges and angles, or on the alternate lateral edges.
2. Equiangular six-sided prism, flatly acuminated with three planes, which are set on the alternate lateral edges Sometimes the summit and the edges of the acumination are truncated.

3. Equiangular

3. Equiangular six-sided prism, flatly acuminated with six planes, which are set on the lateral planes. The summits of the acuminations are sometimes so deeply truncated, that the acuminating planes become truncations on the terminal edges.
4. The preceding figure, in which the six planed acumination is again acuminated with three planes, which are set on the alternate lateral edges of the first acumination.
5. Rectangular four-sided prism, flatly acuminated with four planes, which are set on the lateral planes.
6. Eight-sided prism, acuminated with eight planes, which are set on the lateral planes.

The crystals are sometimes acicular and capillary, and are middle-sized, small, and very small.

The surface of the crystals is generally smooth, the lateral planes sometimes longitudinally, but rarely transversely streaked.

Externally it is splendent, and the lustre is metallic.

Internally it alternates from shining to glimmering, and has sometimes a metallic, sometimes a semi-metallic lustre.

The fracture of the massive and other similar varieties is usually coarse and fine-grained uneven, that of the crystallised varieties imperfect and small conchoidal, and rarely imperfect foliated.

The fragments are indeterminate angular, and blunt-edged.

The massive varieties are opaque, those which are crystallised semi-transparent, passing into transparent, and sometimes only translucent.

The

The streak is cochineal-red, and its lustre unchanged.

It is soft.

It is sectile, but not in so high a degree as brittle silver-glance.

It is easily frangible; more easily frangible than brittle silver-glance.

It is heavy, approaching uncommonly heavy.

Specific gravity, 5.608, 5.684, *Gellert*. 5.5637, *Brisson*.

Chemical Characters.

Before the blowpipe, it first decrepitates, then melts with a slight effervescence, and the disengagement of sulphureous and antimonial yellow and white vapours, leaving behind a globule of silver.

Constituent Parts.

Silver, - - -	60
Antimony, - -	19
Sulphur, - -	17
Oxygen, - - -	1
	100

Klaproth, Beit. b. v. s. 200.

Geognostic Situation.

It occurs in veins in gneiss, mica-slate, porphyry, and grey-wacke, along with brittle silver-glance, white silver-ore, galena or lead-glance, iron-pyrites, sparry ironstone, blende, copper-pyrites, zeolite, cross-stone, calcareous-spar, and brown-spar.

Geographic

Geographic Situation.

Europe.—This ore occurs in the silver-mines of Kongsberg already so often mentioned; also in those of the Hartz, where it is sometimes associated with galena or lead-glance, iron-pyrites, quartz, and calcareous-spar; sometimes with galena or lead-glance, native arsenic, hepatic pyrites, quartz, and calcareous-spar: at Schemnitz, it is generally accompanied with brittle silver-glance: at Kremnitz, its accompanying minerals are brittle silver-glance, iron and copper-pyrites, brown-spar, quartz, and amethyst, and sometimes silver-glance and galena or lead-glance: at Boitza in Transylvania, it occurs along with iron-pyrites, yellow blende, galena or lead-glance, brittle silver-glance, brown-spar, common quartz, and amethyst: at Joachimsthal in Bohemia, it is found imbedded in quartz, calcareous-spar, or hepatic pyrites, and is accompanied with brittle silver-glance, and several other ores.

America.—This ore forms a principal part of the wealth of Sombrerete, Cosala, and Zolaga, near Villalta, in the province of Oaxaca in Mexico. From this ore more than 700,000 marcs of silver have been extracted, in the famous mine of *La veta Negra*, near Sombrerete. in the space of from five to six months. It is affirmed, that the mine which produced this enormous quantity of metal, the greatest which was ever yielded by any vein on the same point of its mass, was not ninety-eight feet in length *.

* Humboldt's New Spain, vol. iii. p. 155. Black's translation.

Second Subspecies.

Light Red Silver-Ore.

Lichtes Rothgiltigerz, *Werner.*

Id. Wern. Pabst. b. i. s. 52. *Id. Wid.* s. 706.—Light Red Silver-ore, *Kirw.* vol. ii. p. 122.—Lichtes Rothgiltigerz, *Estner,* b. iii. s. 426. *Id. Emm.* b. ii. s. 190.—L'Argent rouge clair, *Broch.* t. ii. p. 147.—Argent antimonié sulphuré rouge vif, *Hauy,* t. iii. p. 410.—Lichtes Rothgiltigerz, *Reuss,* b. iii. s. 365. *Id. Lud.* b. i. s. 217. *Id. Suck.* 2ter th. s. 102. *Id. Bert.* s. 375. *Id. Mohs,* b. iii. s. 193. *Id. Leonhard,* Tabel. s. 55. *Id. Karsten,* Tabel. s. 60. *Id. Haus.* s. 76.—Argent antimonié sulphuré rouge vif, *Hauy,* Tabl. p. 75.

External Characters.

Its colour is cochineal-red, which passes on the one side into blood-red, on the other into lead-grey. Its surface sometimes exhibits a tempered-steel tarnish.

It occurs massive, disseminated, frequently in membranes, dendritic, reniform, botryoidal; and crystallised in the following figures:

1. Equiangular six-sided prism.
2. Six-sided prism, flatly acuminated with three planes, which are set on the alternate lateral edges. Sometimes the summit of the acumination is truncated.
3. The preceding figure, in which all the edges are truncated.
4. Very acute double six-sided pyramid, with alternate obtuse and acute lateral edges, in which the lateral

lateral planes of the one are set obliquely on the lateral planes of the other, and very flatly acuminated on the extremities with three planes, which are set on the obtuse lateral edges.

4. Very acute six-sided pyramid, very flatly acuminated with six planes, which are set on the lateral planes.
5. Double six-sided pyramid, less acute than the preceding, in which the acute lateral edges are truncated.
6. Equiangular double six-sided pyramid, acuminated with six planes, which are set on the lateral planes, and the summit truncated.
7. Also in acicular crystals.

The crystals are seldom middle sized, usually small, and very small, and occur in druses.

The surface of the crystals is usually smooth, sometimes streaked; the streaks being longitudinal in the prisms, but oblique in the pyramids, and sometimes drusy.

Externally the lustre is splendent.

Internally it alternates from splendent to glistening, and has usually an adamantine lustre; the varieties that incline to the foregoing subspecies have a semi-metallic lustre.

The fracture is usually imperfect and small conchoidal, which sometimes passes into coarse and small grained uneven, sometimes into imperfect foliated, which latter has the greatest degree of lustre, and occurs only in the crystallised varieties.

The fragments are rather sharp-edged.

The massive varieties are generally translucent on the edges; the crystallised usually transparent.

It yields an aurora-red streak, and the lustre is somewhat increased.

It is soft, passing into very soft.

It is sectile.

It is uncommonly easily frangible.

It is heavy, approaching to uncommonly heavy.

Specific gravity, 5.443, *Gellert*. 5.5886, *Brisson*. 5.592, *Vauquelin*.

Chemical Characters.

On charcoal, before the blowpipe, it melts, blackens, and burns with a blue flame like sulphur, diffusing a white smoke, and a feeble garlic smell, and leaves a globule of nearly pure silver.—*Vauquelin.*

Constituent Parts.

From the mine called Catharina Neufang, at Andreasberg.		Mine called Churprinz Friedrich August, near Freyberg.	1st Analysis.	2d Analysis.
Silver,	60.0	62.00	56.6748	54.2713
Antimony,	20.3	18.50	16.13	16.13
Sulphur,	14.7	14.45	15.0666	17.75
Oxygen,	5.0	5.05	12.1286	11.8487
	100.0	100.0	100	100
	Klaproth.	*Klaproth.*	*Vauquelin.*	*Vauquelin.*

Oxide of Silver,	58.4
Oxide of Antimony,	23.5
Sulphur,	16.0
	97.9
	Thenard.

From Johanngeorgenstadt.	
Silver,	61.0
Antimony,	19.0
Sulphur,	11.1
Sulphuric Acid,	7.0
Arsenic,	2.9
	101
	Lampadius.

Geognostic

Geognostic Situation.

It occurs in veins in the same species of rocks as the dark red subspecies, but is remarkably distinguished from it by its accompanying minerals, which are, native arsenic, red orpiment, copper-nickel, white cobalt-ore, straight lamellar heavy-spar, calcareous-spar, and fluor-spar, and occasionally native silver, silver-glance or sulphureted silver-ore, copper-pyrites, and small quantities of galena or lead-glance, iron-pyrites, and sparry ironstone.

Geographic Situation.

Europe.—This ore occurs at Andreasberg in the Hartz, where it is accompanied with native arsenic, quartz, and calcareous-spar; in many of the mines in the kingdom of Saxony, as at Kurprinz Friedrich-August zu Grossscherma, along with native arsenic, and lamellar heavy-spar; at Himelsfürst, along with native arsenic, copper-pyrites, heavy-spar, brown-spar, and quartz; at Johanngeorgenstadt, with white cobalt-ore, nickel-ochre, silver-glance, and iron-pyrites; at Marienberg, with white cobalt-ore, native arsenic, galena or lead-glance, dark red silver-ore, iron-pyrites, heavy-spar, calcareous-spar, fluor-spar, sparry ironstone, and brown spar; at Schneeberg, with white cobalt-ore, dark red silver-ore, copper-nickel, iron-pyrites, sparry ironstone, calcareous-spar, and quartz; at Joachimsthal in Bohemia, it is accompanied with silver-glance, brittle silver-glance, white cobalt-ore, orpiment, copper-pyrites, sparry ironstone, brown-spar, calcareous-spar, heavy-spar, and hornstone; at Markirchen in Alsace, along with native arsenic, silver-glance, galena or lead-glance, copper-pyrites, brown-spar, calcareous-spar, and quartz; in the Sierra Morena in Spain, along with arsenical silver-ore, and calcareous spar; and at Schemnitz and Kremnitz in Hungary.

America.—It is found in the mines of Guanaxuato, in some veins, associated with native gold, silver-glance, brittle silver-glance, copper green, azure copper-ore, iron-pyrites, quartz, and calcareous-spar; in others with native gold, galena or lead-glance, blende, copper-pyrites, iron-pyrites, and sparry ironstone. In the mining-district of Porco, in Potosi in Peru, it is accompanied with dark red silver ore, native silver, blende, iron-pyrites, and calcareous spar.

Uses.

Both subspecies are smelted, on account of the silver they contain. The dark red is considerably more productive than the light red.

Observations.

1. Red Silver-ore, Cinnabar, and Red Copper-ore, have several characters in common: the following, however, sufficiently distinguish them from one another.

Cinnabar has a specific gravity of 7.0, and is almost always accompanied with native mercury and iron ochre; whereas the specific gravity of Red Silver-ore does not exceed 5 8; and its accompanying fossils, as already mentioned, are very different from those of cinnabar.

Red Copper-ore has a specific gravity of 3.9, and is usually accompanied with native copper, malachite, and brown iron ochre,—characters that distinguish it sufficiently from Red Silver-ore.

2. It has a slight resemblance to *Copper-glance* and *Red Orpiment* Copper-glance gives a blackish streak; red orpiment an orange-yellow streak, and its specific gravity is only 3 2,—characters that distinguish them at once from Red Silver-ore.

3. The

3. The colour of the streak distinguishes the two subspecies from one another: the dark red affords a cochineal or brick-red coloured streak; but the light red ore an aurora-coloured streak.

4. The Light Red Silver-ore, as already mentioned, occurs usually with native arsenic, and white cobalt-ore, also with orpiment and heavy-spar; but the dark, on the contrary, with galena or lead-glance, white silver-ore, brittle silver-ore, quartz, calcareous-spar, and iron-pyrites. They are thus, by these geognostic characters, well distinguised from one another.

5. In the Hartz and Hungary, it is principally the dark red silver-ore which occurs.

10. White Silver-Ore.

Weiss-Giltigerz, *Werner*.

Id. Wern. Pabst. b. i. s. 58. *Id. Wid.* s. 711.—Light Grey Silver-ore, *Kirw.* vol. ii. p. 119.—Weiss-Giltigerz, *Estner,* b. iii. s. 443. *Id. Emm.* b. ii. s. 195.—La Mine blanche riche, *Broch.* t. ii. p. 150.—Weiss-Gultigerz, *Reuss,* b. iv. s. 193. *Id. Lud.* b. i. s. 217. *Id. Mohs,* b. iii. s. 193. *Id. Leonhard,* Tabel. s. 55.—Argent blanc, *Brong.* t. ii. p. 255.—Weiss-Gultigerz, *Karsten,* Tabel. s. 68. *Id. Haus.* s. 74.—Plomb sulphuré antimonifere et argentifere, *Hauy,* Tabl. p. 89.—White Silver, *Aikin,* p. 22.

External Characters.

Its colour is very light lead-grey; but when it approaches to silver-glance, it inclines somewhat to black.

It occurs massive and disseminated, and always associated with lead-glance.

Internally it alternates from glimmering to glistening, and

and the lustre is metallic. The varieties that verge on silver-glance have the most lustre; those that pass into plumose antimony the least.

The fracture is even, and fine-grained uneven. When fine-grained uneven, it is passing into brittle silver-glance; when intermixed with delicate fibres, it is passing to indurated plumose antimony: Therefore, the fracture of the true white silver-ore is even.

The fragments are indeterminate angular and blunt-edged.

The streak is shining, and retains its colour.

It is soft, approaching to very soft.

It is slightly sectile.

It is easily frangible.

It is heavy.

Specific gravity, 5.322.

Chemical Characters.

Before the blowpipe, it melts, and partly evaporates, leaving a bead of impure silver, surrounded by a yellow powder.

Constituent Parts.

	Dark White Silver-ore from Himmelsfürst near Freyberg.	Light White Silver-ore from Himmelsfürst.
Lead,	41.00	48.06
Silver,	9.25	20.40
Antimony,	21.50	7.88
Iron,	1.75	2.25
Sulphur,	22.00	12.25
Alumina,	1.00	7.00
Silica,	0.75	0.25
	97.25	99.09
	Klaproth, Beit. b. i. s. 175.	Ibid. s. 172.

Geognostic

Geognostic and Geographic Situations.

It occurs in veins that traverse gneiss, along with galena or lead-glance, dark red silver-ore, brittle silver-glance, plumose antimony, arsenical and iron pyrites, black blende, brown-spar, calcareous-spar, and quartz.

It is found in considerable quantity in the mines of Himmelsfürst and Beschert Glück, near Freyberg, but rarely in other countries.

It is said to have been found in small quantity in the Hartz, and also in Bohemia.

11. Grey Silver-Ore or Carbonate of Silver.

Grausilber, *Hausmann.*

Luftsaures Silber, *Widenmann's* Min. s. 689.—L'Argent carbonaté, *Broch.* t. ii. p. 155.—Kohlensaures Silber, *Reuss*, Min. b. ii. 3. s. 376.—Argent carbonaté, *Lucas*, t. ii. p. 293.—Kohlensaures Silber, *Leonhard*, Tabel. s. 55.—Argent carbonaté, *Hauy*, Tabl. p. 76.—Grausilber, *Haus.* Handb. b. iii. s. 1008.—Carbonated Silver, *Aikin*, p. 23.

External Characters.

Its colour is ash-grey, which passes into greyish-black, and iron-black.

It occurs massive and disseminated.

Its lustre is glistening.

The fracture is uneven, inclining to earthy.

It is soft.

It becomes more shining in the streak.

It is brittle, passing into sectile.

It is heavy.

Chemical

Chemical Characters.

It is easily reduced before the blowpipe. It effervesces with nitrous acid.

Constituent Parts.

Silver, - - -	72.5
Carbonic Acid, - -	12.0
Oxide of Antimony, and a trace of Copper, -	15.5
	100.0 *Selb.*

Geognostic and Geographic Situations.

It occurs in veins that traverse granite, in the mine of Wenzeslaus, at Altwolfach in the Black Forest. In these veins, it is associated with native silver, silver-glance, and heavy-spar.

GENERAL

GENERAL OBSERVATIONS ON SILVER.

1. The most valuable silver-mines in the Old World, are situated in the Austrian dominions, consequently including those of Bohemia, Hungary, Transylvania, Salzburg, Moravia, and Austria: the next in importance are those of Russia and Saxony; and less considerable are the Hanoverian, Prussian, Bavarian, and Swedish mines. In the New World, the silver-mines of Mexico and Peru far exceed in value the whole of the European and Asiatic mines; for we are told by Humboldt, that these mines, in the space of three centuries, afforded 316,023,883 lb. Troy of pure silver *. Humboldt also states the quantity of gold and silver imported into Europe from America, between the years 1492 and 1803, at £1,166,775,322 Sterling, and gives the following table of the annual produce of the gold and silver mines of Europe, Northern Asia, and America.

ANNUAL

* Humboldt remarks, that this silver would form a solid sphere of a diameter of 91,206 feet English.

ANNUAL PRODUCE of the Gold and Silver Mines of Europe, Northern Asia, and America.

Great Political Divisions.	Gold.				Silver.				Value of Gold and Silver in Piastres.	Value of Gold and Silver in Sterling money.
	Marcs of France.	Killogr.	Value in Francs.	Value, Sterling money.	Marcs of France.	Killogr.	Value in Francs.	Value, Sterling money.		
Europe,	5,300	1,297	4,467,444	£178,697	215,200	52,670	11,704,444	£438,177	16,171,888	£646,874
Northern Asia,	2,200	538	1,853,111	74,124	88,700	21,709	4,824,222	192,966	6,677,333	267,090
America,	70,647	17,291	59,557,889	2,382,315	3,250,547	795,581	176,795,778	7,071,830	236,353,667	9,454,145
Total,	78,147	19,126	65,878,444	£2,635,136	3,554,447	869,960	193,324,444	£7,732,973	259,202,888	£10,368,109

2. The relative value of gold and silver, as will appear from the following statement, has varied considerably at different times. According to the present regulations in the British mint, a pound of standard gold is coined into $44\frac{1}{2}$ guineas: a pound weight of standard silver is coined into 62 shillings; and a guinea is current for 21 shillings. These particulars enable us to calculate the relative value of gold to silver, if we neglect the alloy in the coins; for $44\frac{1}{2}$ guineas are equivalent in value to 1869 sixpences, and 62 shillings being equal to 124 sixpences, the value of gold is to that of silver as 1869 to 124, or as $15\frac{9}{124}$ to 1. This would accurately express the relative values of the two metals, if the quantity of alloy in a pound weight of standard in each bore the same proportion to the whole, which is not the case. In a pound weight of standard gold at the British mint, one-twelfth is alloy: in a pound weight of standard silver, it is $\frac{3}{40}$; and the relative value of pure gold to pure silver, according to these regulations, and the established currency between coins of the two metals, is as $15\frac{2859}{13460}$ to 1. One of the earliest accounts of the relative value of gold and silver we possess, is that of Herodotus, who informs us, that in Persia and Greece, it was as 13 to 1. Plato, who flourished about fifty years after Herodotus, asserts, in his Hipparchus, that the value of gold in Greece was to that of silver as 12 to 1 *. Menander, who was born about the year 341 before the Christian era, estimates the value of gold to that of silver so low as 10 to 1. According to Pliny, the relative value of the two metals in Rome, was at one period as high as $14\frac{15}{28}$ to 1; but this did not continue long; for we find, in the conditions on which the Romans made peace with the Ætolians, about 189 years before the Christian era, that they coincided with the Greeks in estimating

* Platonis Opera, t. iii. p. 231. edit. H. Steph. 1578.

mating the value of gold to be to that of silver as 10 to 1. On the return of Cæsar to Rome from Gaul, he brought with him so much gold, that the value of that metal to that of silver was soon as low as 7½ to 1. We cannot say how long this last-mentioned proportion between the two precious metals continued; but we find, that in the time of Claudius, about a century after Cæsar's return from Gaul, the value of gold was considerably advanced; for under this Emperor's reign, it was thought proper, according to Tacitus * and the younger Pliny †, to limit the fee of an advocate to 10,000 sesterces, and this legal fee is stated in the Digest at 100 aurei. Now, as 10,000 sesterces were equal to 2,500 denarii, it follows, that the value of gold was to that of silver as 2,500 to 200; or as 12½ to 1. It is highly probable that this proportion continued some time after the reign of Alexander Severus, as the state of the Empire justifies such a supposition. At what period it ceased cannot be determined; but under the reign of Constantine the Great, we find, that the value of gold was much diminished, the proportion being now as 10½ to 1. Owing to the political alterations which succeeded the reign of Constantine, the value of gold was much increased: even so soon as the time of Arcadius and Honorius, about sixty years after Constantine, the proportional value of the metals was as 14⅖ to 1.

From this statement, it appears, that the lowest proportional value of the two metals in ancient times, was as 7½ to 1, and the highest as 14⅖ to 1; which latter does not differ much from that which exists at present. The various causes which gave rise to these fluctuations, are luminously detailed in Lord Liverpool's valuable "Letter to the King on the Coins of the Realm."

V. ORDER.

* Tacitus, Annalium lib. xi. cap. 7.

† C. Plinii Epist. lib. v. ep. 25.

V. ORDER.—COPPER.

This Order contains twenty-two species, viz. Native Copper, Copper-glance or Vitreous Copper-ore, Variegated Copper-ore, Copper-pyrites or Yellow Copper-ore, White Copper-ore, Grey Copper-ore, Black Copper-ore, Copper Black, Red Copper-ore, Tile-ore, Azure Copper-ore, Malachite, Brown Copper-ore or Anhydrous Carbonate of Copper, Copper-green, Ironshot Copper-green, Emerald Copper-ore, Muriate of Copper or Atacamite, Phosphate of Copper, Copper-mica or Micaceous Copper-ore, Lenticular Copper-ore, Oliven Copper-ore, and Martial Arseniate of Copper.

1. Native Copper.

Gediegen Kupfer, *Werner.*

Cuprum nativum, *Wall.* t. ii. p. 274.—Cuivre natif, *Romé de L.* t. iii. p. 305.—Gediegen Kupfer, *Werner,* Pabst. b. i. s. 62. *Id. Wid.* s. 737.—Cuivre natif, *De Born,* t. ii. p. 303.—Native Copper, *Kirw.* vol. ii. p. 128.—Gediegen Kupfer, *Estner,* b. iii. s. 459. *Id. Emm.* b. ii. s. 206.—Le Cuivre natif, *Broch.* t. ii. p. 158. *Id. Hauy,* t. iii. p. 518.-529.—Gediegen Kupfer, *Reuss,* b. iii. s. 392. *Id. Lud.* b. i. s. 219. *Id. Suck.* 2ter th. s. 168. *Id. Bert.* s. 377. *Id. Mohs,* b. iii. s. 200. *Id. Hab.* s. 106.—Cuivre natif, *Lucas,* p. 124.—Gediegen Kupfer, *Leonhard,* Tabel. s. 56.—Cuivre natif, *Brong.* t. ii. p. 211. *Id. Brard,* p. 279.—Gediegen Kupfer, *Karsten,* Tabel. s. 62. *Id. Haus.* s. 69.—Native Copper, *Kid,* vol. ii. p. 98.—Cuivre natif, *Hauy,* Tabl. p. 85.—Native Copper, *Aikin,* p. 26.

External Characters.

Its colour is copper-red, but is frequently tarnished brownish, blackish, yellowish or silver white.

It

It occurs massive, disseminated, in membranes, in plates, in angular pieces, in rolled pieces, in grains, capillary, filiform, moss-like, irregular dendritic, corroded, ramose; and crystallised in the following figures:

1. Perfect cube *, fig. 157.
2. Cube truncated on the angles, which is the middle crystal between the cube and the octahedron †, fig. 158.
3. Cube truncated on the edges, which is the middle crystal between the cube and the rhomboidal or garnet dodecahedron ‡.
4. Cube truncated on all the edges and angles, fig. 159.
5. Rhomboidal or garnet dodecahedron, fig. 160
6. Perfect octahedron, sometimes truncated on the edges ||, fig. 161.
7. Rectangular four-sided prism, flatly acuminated with four planes, which are set on the lateral planes.
8. Simple acute three-sided pyramid.
9. Simple acute six-sided prism.
10. Very low equiangular and equilateral six-sided prism, acuminated by six planes, which are set on the lateral planes §.
11. Oblique four-sided prism, acuminated by four planes, which are set on the lateral planes ¶.

The crystals are seldom middle-sized and small, usually very small and microscopic. They seldom occur singly imbedded

---

* Cuivre natif cubique, Hauy.

† Cuivre natif cubo-octaedre, Hauy.

‡ Cuivre natif cubo-dodecaedre, Hauy.

|| Cuivre natif octaedre, Hauy.

§ Cuivre natif tri-hexaedre, Hauy.

¶ The figures enumerated above, also refer to the principal crystallisations of Native Gold, Native Silver, and Silver-glance.

imbedded and superimposed, more commonly aggregated in a variety of external shapes.

The lateral planes of the crystals are sometimes smooth, sometimes drusy, and the simple six-sided pyramids are transversely streaked. The lustre of the surface of the crystals is splendent; that of the other shapes is glistening.

Internally it is intermediate between glistening and glimmering, and the lustre is metallic.

The fracture is hackly.

The fragments are indeterminate angular, and blunt-edged.

The streak is splendent, with a metallic lustre.

It is intermediate between semi-hard and soft; it is harder than silver.

It is completely malleable.

It is flexible, but not elastic.

It is difficultly frangible.

It is uncommonly heavy.

Specific gravity, 7.728, from Hungary, *Gellert.* 7.600 to 7.800, *Kirwan.* 8.5844, from Siberia, *Hauy.*

Chemical Characters.

When copper is allowed to stand for some time in ammonia, it communicates to it a blue colour: it is fusible before the blowpipe into a bead of apparently pure copper.

Constituent Parts.

Native Copper from Ekatharineburg.

Copper, - - -	99.80
Trace of Gold and Iron.	
	100.0

John, Chem. Untersuch. b. i. s. 256.

Geognostic Situation.

No metal occurs so frequently in a native state as copper, and it is often met with in large masses on the surface of the earth, particularly in uncultivated and remote regions. In the interior of the earth, it generally occurs in veins, where it is usually associated with red copper-ore, and brown ironstone, seldomer with red ironstone, copper-glance or vitreous copper-ore, copper-pyrites, malachite, and copper-green, and most rarely with oliven-ore, and its congenerous species. The rocks in which these veins are contained, are granite, gneiss, mica-slate, chlorite-slate, talc-slate, foliated granular limestone, and grey-wacke. It also occurs imbedded in masses, or in drusy cavities, in serpentine, amygdaloid, old flœtz limestone, and flœtz ironstone. The earthy minerals with which it is generally associated in the different formations, are, quartz, calcareous-spar, chlorite, and a kind of soft clay.

Geographic Situation.

Europe.—It occurs in small veins and imbedded portions in serpentine, in the Island of Yell, one of the Zetland Islands; in red sandstone, along with copper-pyrites, grey copper-ore, malachite, brown hematite, sparry ironstone, and iron-pyrites, in Mainland, the largest of the Zetland Islands. It has been long known as a mineral production of Cornwall, where it occurs in veins that traverse granite, and clay-slate, along with tinstone, red copper-ore, malachite, ironstone, common quartz, rock crystal, sometimes with chlorite, &c. It generally occurs near the surface, or only a few fathoms under it, although there are instances of its being found very deep in some of the veins. It is met with in the mines

mines named Huel-Unity, Cook's Kitchen, Mullion, Camborne, St Just, Poldory, and also in the rocks of the Lizard. It occurs in Nalsoe, one of the Faroe Islands, imbedded in amygdaloid, along with fibrous and radiated zeolite, and copper-green; in the Bear Islands in the White Sea; at Gullardsrud-schurf in Norway, in serpentine; at Friedrichs-minde, also in Norway, along with earthy azure copper-ore, and copper-green, in grey hornstone and limestone; at Guldholmen, near Moss in Norway, along with calcareous-spar, in a trap rock; at Fahlun in Sweden; in the Hartz, as at Blankenburg, where it is associated with brown-spar, and brown hematite, in veins that traverse grey-wacke-slate; in different venigenous formations that traverse gneiss, in the Saxon Erzgebirge; in beds of bituminous marl-slate at Bottendorf in Thuringia; in the Brennthal, near Mühlbach in Salzburg, in clay-slate; at Kamsdorf in the Westerwald, in beds of ironstone; at Altenkirchen, in veins that traverse grey-wacke, where it is associated with brown ironstone, malachite, red copper-ore, copper-green, copper-glance or vitreous copper-ore, and quartz; at Reichenbach, near Oberstein, in flœtz amygdaloid, along with prehnite; at St Bel, near Lyons in France; and in veins that traverse gneiss, in the Kenzigerthal in Suabia; in the mine of Maria-Taferl, at Moldowa in the Bannat, in syenite-porphyry; and in different mines in Hungary.

Asia.—In the Island of Japan, along with red copper-ore, and brown ironstone; in large masses in the Kurile Islands; in the Altain and Uralian Mountains; Kamschatka; and China.

North America.—In masses on the surface in Canada; on the banks of Copper-mine River, on the confines of the Arctic Ocean; in the mines of Ingaran, near the

 base

base of the volcano of Jorullo, in Mexico, along with copper-glance or vitreous copper-ore, and red copper-ore; in the intendancy of Valladolid; and in the province of New Mexico.

South America.—Large masses of native copper are met with on the surface of the uncultivated and thinly inhabited regions of Brazil; and Professor Vandilli informs us, that a mass weighing 2600 lb. was found in a valley near to Cachoeira, in that country. It measures 3 feet 2 inches in length, 2 feet 1 inch in breadth, and 10 inches in thickness. Its surface is rough, and covered in some places with malachite and red copper-ore. It is also met with in the upper mines of Chili.

Uses.

The copper used for economical and other purposes is obtained from the ores of copper afterwards to be described, native copper seldom occurring in any considerable quantity. Combined with zinc, it forms the useful compound called *Brass*, and with tin, *Bell-metal* or *bronze*. Is is also used in coinage, either pure, or when combined with gold or silver, to which it gives a greater degree of tenacity. Its oxide is employed in colouring glass and porcelain green; and when combined with acetic acid, it affords the well-known pigment called *Verdigris*. Great quantities of it are used for sheathing the bottom of ships intended for long voyages into warm climates, to preserve them from the attack of the *Teredo navalis*, and other destructive vermes: when covered with tin, for culinary vessels.

This metal as already mentioned, is occasionally found in great masses, dispersed over the surface of the earth in uncultivated countries: hence Werner conjectures, that it was the first metal worked by man. From its known metallic

metallic characters, this opinion may be considered as very probable, especially when supported by the account which is given of some of the native tribes of the north-western parts of America, who, though little civilised, have applied to domestic purposes the native copper with which their country abounds. It is also known, that, at a very early period, domestic utensils, and instruments of war, were made of a compound of this metal and tin: even during the Trojan war, as we learn from Homer, the combatants had no other armour but what was made of bronze, which is a mixture of copper and tin. Macrobius, who wrote in the fourth century, informs us, that when the Etruscans intended building a new city, they marked out its limits with a coulter of brass, and that priests of the Sabines were in the habit of cutting their hair with a knife of the same metal *. The Greek and Roman sculptors executed fine works of art in porphyry, granite, and hard other minerals, by means of their copper instruments. The great hardness of the ancient copper instruments, induced historians to believe, that the ancients possessed a particular secret for tempering copper, and converting it into steel. There is no doubt the axes and other ancient tools were almost as sharp as steel instruments; but it was by a mixture with tin, and not by any tempering, that they acquired their extreme hardness. Axes, and other instruments of copper, have been discovered in the tombs of the ancient Peruvians, and also in those of the early inhabitants of Mexico. These were so hard, that the sculptors of these countries executed large works in the hardest greenstone and basaltic porphyry: their jewellers

 cut

* Macrobius, Saturnalia, lib. v. cap. 19. p. 29. 512.

cut and pierced the emerald, and other precious stones, by using at the same time a metal tool and a siliceous powder. Humboldt brought with him from Lima an ancient Peruvian chisel, in which M Vauquelin found 0.94 of copper, and 0.06 of tin. This mixture was so well forged, that, by the closeness of the particles, its specific gravity was 8.815; while, according to the experiments of M. Briche, chemists never obtain this maximum of density, but by a mixture of 16 parts of tin, with 100 parts of copper. It appears that the Greeks and Romans made use of both tin and iron at the same time in the hardening of copper. Even a Gaulish axe found in France by M. Dupont de Nemours, which cuts wood like a steel axe, without breaking or yielding, contains, according to the analysis of Vauquelin, 87 of copper, 3 of iron, and 9 of tin *.

Observations.

1. Native Copper is distinguished from *Copper-nickel*, by its malleability, and inferior degree of hardness; copper-nickel being semi-hard, bordering on hard, and brittle.

2. When iron-plates are put into a solution of copper vitriol, their surfaces soon become covered with a coating or crust of malleable copper, which is called Copper of Cementation. As copper thus formed is an artificial product, it cannot be included in a system of oryctognosy.

2. Copper-

* Humboldt's New Spain.

2. Copper-glance or Vitreous Copper-ore.

Kupferglas, *Werner*.

This species is divided into three subspecies, viz. Compact Copper-glance, Foliated Copper-glance, and Malleable Copper-glance.

First Subspecies.

Compact Copper-glance or Compact Vitreous Copper-ore.

Dichtes Kupferglas, *Werner*.

Cuprum vitreum, *Wall.* Syst. Min. vol. ii. p. 277.—Dichtes Kupferglas, *Werner*, Pabst. b. i. s. 71.—Compact Vitreous Copper-ore, *Kirw.* vol. ii. p. 144.—Dichtes Kupfer Glanzerz, *Estner*, b. iii. s. 476.—Kupferglas, *Emm.* b. ii. s. 223.—Cuivre sulphuré, *Hauy*, t. iii. p. 551.–555.—Le Cuivre vitrieux compacte, *Broch.* t. ii. p. 162.—Dichtes Kupferglanz, *Reuss*, b. iii. s. 401. *Id. Lud.* b. i. s. 220. *Id. Suck.* 2ter th. s. 173. *Id. Bert.* s. 383. *Id. Mohs*, b. iii. s. 253. *Id. Leonhard*, Tabel. s. 57.—Cuivre sulphuré, *Brong.* t. ii. p. 212. *Id. Brard*, p. 289.—Gemeiner Kupferglanz, *Karsten*, Tabel. s. 62. *Id. Haus.* s. 71.—Grey Sulphuret of Copper, *Kid*, vol. ii. p. 106.—Cuivre sulphuré, *Hauy*, Tabl. p. 87.—Glance Copper, *Aikin*, p. 26.

External Characters.

Its colour is blackish lead-grey, which sometimes inclines to steel-grey, and to iron-black, and has often a tempered-steel coloured tarnish.

It occurs massive, disseminated, in membranes; and rarely crystallised in the following figures:

1. Low equiangular six-sided prism, which is either perfect or truncated on the terminal edges *.
2. Oblique four-sided prism, which is formed when two opposite planes of the preceding figure disappear.
3. Double six-sided pyramid, either perfect or truncated on the extremities †.

The crystals are small and very small, seldom middle-sized.

Externally it is shining.

Internally it is intermediate between shining and glistening, and sometimes even passes into glimmering: it is most commonly glistening, and the lustre is metallic.

The fracture is small and fine-grained uneven, which passes into small conchoidal; also into large and flat conchoidal, and sometimes into even.

The fragments are indeterminate angular, and more or less sharp-edged.

It retains its colour, and is shining in the streak.

It is soft.

It is perfectly sectile.

It is easily frangible.

It is heavy.

Specific gravity, 4.888 to 5.338, *Gellert*. 5.452, Cornish, *Kirwan*. 4.129, Hungarian, *Kirwan*. 4.8648, *Hauy*. 4.856, 4.865, *Klaproth*.

Constituent

---

* Cuivre sulphuré primitif, annulaire, et ternaire, Hauy.

† Cuivre sulphuré dodecaedre, Hauy.

Constituent Parts.

	Siberia.	Rothenburg.
Copper, - -	78.05	76.50
Iron, - - -	2.25	0.50
Sulphur, - -	18.50	22.00
Silica, - -	0.75	Loss, 1.00
	100.00	100.00
	Klaproth, Beit. b. ii. s. 279.	Ibid. b. iv. s. 39.

Second Subspecies.

Foliated Copper-glance, or Foliated Vitreous Copper-Ore.

Blättriches Kupferglanz, *Werner.*

Id. Werner, Pabst. b. i. s. 73.—Foliated Vitreous Copper-ore, *Kirw.* vol. ii. p. 146.—Blättriges Kupferglanzerz, *Estner,* b. iii. s. 477.—Blättriges Kupferglas, *Emm.* b. ii. s. 225.—Le Cuivre vitreux lamelleux, *Broch.* t. ii. p. 164.—Blättriches Kupferglanz, *Reuss,* b. iii. s. 403. *Id. Lud.* b. i. s. 222. *Id. Suck.* 2ter th. s. 178. *Id. Bert.* s. 385. *Id. Mohs,* b. ii. s. 260. *Id. Leonhard,* Tabel. s. 57.—Schuppiger Kupferglanz, *Karsten,* Tabel. s. 62. *Id. Haus.* s. 72.

External Characters.

Its colour is the same as that of the preceding subspecies.

It occurs massive and disseminated.

The principal fracture is shining; the cross fracture is glimmering, with a metallic lustre.

The

The principal fracture is pretty straight foliated, with a single cleavage; the cross fracture is fine-grained uneven.

The fragments are indeterminate angular, and blunt-edged.

It occurs always in coarse and fine granular distinct concretions.

In the remaining characters, it agrees with the preceding subspecies.

Chemical Characters of the Species.

Before the blowpipe, on charcoal, it melts very easily, and yields a globule of copper, covered with a blackish-coloured scoria. When melted with borax, it communicates to it a green colour; and when digested with ammonia, it tinges it blue.

Geognostic Situation.

It occurs in veins and beds in primitive rocks; also in beds in bituminous marl-slate, and in flœtz amygdaloid. The accompanying minerals in the primitive and transition rocks, are copper-pyrites, grey copper-ore, azure copper-ore, malachite, copper-green, and red and brown ironstone, with calcareous-spar, and quartz; in the flœtz rocks, it is associated with copper-pyrites, and variegated copper-ore.

Geographic Situation.

Europe.—It occurs in small veins, along with heavy-spar, in transition rocks, at Fassney Burn in East Lothian; also in Ayrshire; and in the Fair Isle, situated between Orkney and Zetland; at Middleton Tyas in Yorkshire;

Yorkshire; Llandidno in Caernarvonshire; Cook's Kitchen, Carvath, Tincroft, Camborne, Huel-Muttrel, and Bullen Garden, in Cornwall: in the mines of Friedrichsminde and Glittersberg in Norway, along with quartz, malachite, variegated copper-ore, &c.; also in the mines of Kongsberg; at Atwod and Sunnerskog in Sweden; in Hessia, along with grey copper-ore, white copper-ore, malachite, brown ironstone, tile-ore, variegated copper-ore, copper-pyrites, sparry ironstone, white cobalt-ore, quartz, and calcareous-spar; in Thuringia, in bituminous marl-slate, associated with copper-pyrites, and variegated copper-ore: in different mines in the Saxon Erzgebirge, where it is accompanied with various ores of copper, iron, and silver; thus, at Berggieshübel, it occurs along with copper-pyrites, iron-pyrites, compact and ochry red ironstone, native silver, and lamellar heavy-spar; and at Deutschneudorf, with quartz, lithomarge, ochry red ironstone, and copper-green: at Graupen in Bohemia, with grey copper-ore and copper-green: in talc-slate in Moravia: in amygdaloid in Deux-Ponts: at Schwaz in the Tyrol, along with fibrous malachite, slaggy ironshot copper-green, and azure copper-ore: in the Leogang, and Limberg in Bavaria: in Silesia; in the Kinzegthal in Suabia, along with copper-green, malachite, and quartz, in veins that traverse gneiss: Catalonia in Spain: in primitive limestone at Saska, in the Bannat of Temeswar; and in Hungary

Asia.—In great abundance in different mines in the Uralian Mountains.

America.—In small quantities, along with copper-green and quartz, in West Greenland.

Observations.

Observations.

1. Compact and Foliated Copper-glance are sometimes confounded with *Grey Copper-ore*, but may be readily distinguished from it, by their being sectile, whereas grey copper-ore is brittle. The red colour and red-coloured streak of *Red Copper-ore*, distinguish it at once from copper-glance; and *Silver-glance* or *Sulphureted Silver-ore*, although somewhat resembling the two first-mentioned subspecies of copper-glance in external aspect, yet may be readily distinguished from them by an obvious character, viz. Its cutting readily into slices with a knife, whereas these ores separate into small grains when we attempt to cut them.

2. The tarnished varieties of Compact and Foliated Copper-glance incline, and even sometimes pass, into Variegated Copper-ore.

3. The Frankenberg or Hessian corn-ears mentioned by authors, are sometimes aggregations of small crystals of copper-glance; sometimes, according to M. Monch, true petrifactions of a phalaris, (Phalaris pulposa ?) composed of copper-glance, white copper-ore, and grey copper-ore. They are sometimes invested with a thin cover of native silver.

4. It is rather a rare mineral, and the only country in which it has been met with in great quantity, is Siberia.

Third

Third Subspecies.

Malleable Copper-glance or Malleable Vitreous Copper-ore.

Geschmeider Kupferglanz, *Karsten.*

Id. Leonhard, Tabel. s. 57. *Id. Haus.* s. 72. *Id. Karsten,* Tabel. s. 62.

External Characters..

Its colour is steel-grey.

Internally it is strongly glimmering.

The fracture is even.

It becomes shining in the streak, and the colour is changed to lead-grey.

It is malleable.

Specific gravity, 5.099.

Geographic Situation.

It is a rare mineral, having been hitherto found only in the mines of Catharinenburg in Siberia.

Observations.

Colour, lustre, fracture, and malleability, distinguish it from the Compact and Foliated subspecies; and its inferior malleability distinguishes it from *Silver-glance* or *Sulphureted Silver-ore.*

3. Variegated

3. Variegated Copper-Ore.

Bunkupfererz, *Werner.*

Cuprum lazurcum, *Wall.* t. ii. p. 278.—Bunt Kupfererz, *Werner;* Pabst. b. i. s 73. *Id. Wid* s. 744.—Cuivre sulphuré violet, *De Born,* t. ii. p. 311.—Purple Copper-Ore, *Kirw.* vol. ii. p. 142.—Bunt Kupfererz, *Estner,* b. iii. s. 489. *Id. Emm.* b. ii. s. 228.—La Mine de Cuivre panachée ou violette, *Broch.* t. ii. p. 166.—Cuivre pyriteux hepatique, *Hauy,* t. iii. p. 536. —Buntkupfererz, *Reuss,* b. iii. s. 410. *Id. Lud.* b. i. s. 222. *Id. Suck* 2ter th s. 179. *Id. Bert.* s. 385. *Id. Mohs,* b. iii. s. 248. *Id. Leonhard,* Tabel. s. 57.—Le Cuivre pyriteux panaché, *Brong* t. ii. p. 215.—Bunt Kupfererz, *Karsten,* Tabel. s. 62. *Id. Haus.* s 74.—Variegated or Iridescent Sulphuret of Copper, *Kid,* vol. ii. p. 108—Cuivre pyriteux hepatique, *Hauy,* Tabl. p. 86.—Purple Copper, *Aikin,* p. 27.

External Characters.

Its fresh colour is intermediate between copper-red and pinchbeck-brown; it however soon acquires a tarnish, which is first reddish, then the red passes successively into violet-blue, azure-blue, and sky-blue, and lastly into green; yet several of these colours are to be observed on the same mass, so that it has a variegated aspect, and of these colours the blue is usually the predominant, and the green occurs only in spots.

It occurs massive, disseminated, in plates, in membranes; and crystallised in the following figures:

1. Cube, in which the faces are sometimes curved.
2. Cube truncated on all the angles.

Internally it is shining or glistening, and the lustre is metallic.

The

The fracture is small and rather imperfect conchoidal, which sometimes approaches to fine-grained uneven.

The fragments are indeterminate angular, and rather sharp-edged.

Neither colour nor lustre are changed in the streak.

It is soft.

It is sectile in a slight degree.

It is easily frangible.

It is heavy.

Specific gravity, from the Bannat, 4.956, *Kirwan.* From Lorraine, 4.983, *Kirwan.* 4.300, *La Metherie.* 5.467, *Wiedenman.* 5.033, *Bournon.*

Chemical Characters.

It is fusible, but not so easily as copper-glance, and with less ebullition, into a globule, which acts powerfully on the magnetic needle.

Constituent Parts.

	From Hitterdahl in Norway.	From Rudelstadt in Silesia.
Copper,	69.50	58
Sulphur,	19.00	19
Iron,	7.50	18
Oxygen,	4.00	5
	100.00	100
	Klaproth, Beit. b. ii. s. 283.	Ibid. s. 286.

Geognostic Situation.

It occurs in veins in primitive and transition rocks, particularly in gneiss, mica-slate, talc-slate, and grey-wacke. It is also met with in flœtz rocks, as in beds in bituminous marl-slate. In these repositories, it is associated

ciated with grey copper-ore, copper-pyrites, copper-glance or vitreous copper-ore, copper-green, malachite, iron-pyrites, blende, brown ironstone, quartz, common garnet, heavy-spar, calçareous-spar, tremolite, and actynolite.

Geographic Situation.

Europe.—It is found in Cook's Kitchen and Tincroft mines in Cornwall, along with grey copper-ore, copper-pyrites, &c.; in the mines of Arendal in Norway, where it is associated with copper-glance or vitreous copper-ore, copper-pyrites, and common garnets, in beds in gneiss, at Kongsberg, also in Norway, along with native silver; at Lauterberg in the Hartz, in veins that traverse grey-wacke, along with copper-pyrites, and tile-ore; and in the Fluss Mine, in the same country, along with fluor-spar, lamellar heavy-spar, calcareous-spar, and azure copper-ore; in the Saxon Erzgebirge, along with grey copper-ore, copper-glance, copper-pyrites, and different ores of silver; in bituminous marl-slate in Mansfeld and Thuringia; at Kupferberg in Silesia, in a metalliferous bed, along with copper-pyrites, azure copper-ore, malachite, tile-ore, copper-glance or vitreous copper-ore, arsenical-pyrites, iron-pyrites, lamellar heavy-spar, brown-spar, calcareous-spar, and heavy-spar; at Olonez in Russian Finland, with iron-pyrites, copper pyrites, copper-green, and quartz; at Swappawari in Lapland, in quartz and mica-slate; in Transylvania, along with amethyst; at Dognatska, with common garnet, blende, copper-pyrites, copper-green, and malachite, in calcareous-spar or quartz; and at Oravieza, along with calcareous spar and asbestous tremolite.

Asia.—At Schlangenberg, along with quartz, brown-spar, and hornstone; and in the Pochadjaschinsche mines, associated with malachite, azure copper-ore, and quartz.

America.

America.—At Coquimbo in Chili, along with copper-green and malachite.

Uses.

Copper is extracted from it, but it is not so easily reduced as copper-glance. It yields from 50 to 70 *per cent.* of copper.

Observations.

1. Its external characters and chemical composition, shew that it is a species intermediate between copper-glance and copper-pyrites.

2. It occurs equally abundant with copper-glance, but not in such great quantity as copper-pyrites.

3. This variety differs from copper-glance or vitreous copper-ore, with respect to its component parts, in containing a smaller proportion of copper, and a greater proportion of iron. The variegated colour is supposed by Klaproth to be owing to the slightly oxidated state of the metal: so in steel, and other metallic substances, the beginning of their oxidation is indicated by a similar diversity of colours. In the last-mentioned substances, indeed, the change of colour is only superficial, for the oxygen of the atmosphere can only act upon the surface of the metal: in the variegated copper-ore, the diversity of colour penetrates the whole mass, in consequence of the general distribution of the oxygen throughout the substance of the ore. As, however, the oxidation is slight, the metal is disposed to absorb a farther portion of oxygen; and the uniform brown colour is gradually

gradually produced in consequence, as often as a fresh surface is exposed to the action of the air *

4. It was formerly confounded with Copper-glance, Copper-pyrites, and Red Copper-ore; but Werner ascertained it to be a distinct species, and gave it its present name from its tarnish, which is one of the most striking features in its external aspect.

4. Copper-Pyrites or Yellow Copper-Ore.

Kupferkies, *Werner*.

Minera Cupri flava, *Wall.* t. ii. p. 282.—Mine de Cuivre jaune, *Romé de Lisle,* t. iii. p. 309.—Kupferkies, *Werner,* Pabst. b. i. s. 75. *Id. Wid.* s. 746.—Copper-Pyrites or Yellow Copper-Ore, *Kirw.* vol. ii. p. 140.—Mine de Cuivre jaune, *De Born,* t. ii. p. 313.—Kupferkies, *Estner,* b. iii. s. 494. *Id. Emm.* b. ii. s. 232.—Cuivre pyriteux, *Lam.* t. i. p. 197.—La Pyrite cuivreux, *Broch.* t. ii. p. 169.—Kupferkies, *Reuss,* b. iii. s. 415. *Id. Lud.* b. i. s. 223. *Id. Suck.* 2ter th. s. 181. *Id. Bert.* s. 386. *Id. Mohs,* b. iii. s. 239. *Id. Hab.* s. 107.—Cuivre pyriteux, *Lucas,* p. 125.—Kupferkies, *Leonhard,* Tabel. s. 57. —Cuivre pyriteux, *Brong.* t. ii. p. 213. *Id. Brard,* p. 283.—Kupferkies, *Karsten,* Tabel. s. 62. *Id. Haus.* s. 71.—Yellow Sulphuret of Copper, *Kid,* vol. ii. p. 109.—Cuivre pyriteux, *Hauy,* Tabl. p. 85.—Yellow Copper, *Aikin,* p. 28.

External Characters.

On the fresh fracture, its colour is brass-yellow, of different shades: that which contains the greatest quantity of

* Kid's Min. vol. ii. p. 108, 109.

of copper has a deep-yellow colour, approaching to gold-yellow; the poorer varieties approach to greyish brass-yellow, and steel-grey.

It is usually tarnished, either with variegated colours, as pavonine, columbine, and sometimes tempered-steel coloured, or with simple colours, as blue and black.

The tarnished colours occur sometimes on the mineral in the bosom of the earth, sometimes by the exposition of the recent fracture to the action of the air.

It occurs massive, disseminated, and in membranes; also dendritic, reniform, botryoidal, stalactitic, specular, amorphous; and crystallised as follows:

1. Tetrahedron.
   - *a.* Perfect *.
   - *b.* Truncated on the angles †.
   - *c.* Truncated on the edges ‡.
   - *d.* In which each plane is divided into three compartments ||.
   - *e.* The angles acuminated with three planes, which are set on the lateral planes.

When the angles on the base are slightly truncated, but that of the summit deeply truncated, there is formed a

2. Six-sided table, in which the terminal planes are set alternately oblique and straight on the lateral planes.
3. Octahedron.
   - *a.* Perfect.
   - *b.* Truncated on the edges, thus forming a figure intermediate

* Cuivre pyriteux primitif, Hauy.

† Cuivre pyriteux epointé, Hauy.

‡ Cuivre pyriteux cubo-tetraedre, Hauy.

|| Cuivre pyriteux dodecaedre, Hauy.

intermediate between the octahedron and the rhomboidal or garnet dodecahedron.

4. Rhomboidal or garnet dodecahedron.
5. Twin-crystal, resembling that of the spinel, which is formed by the union of two segments of deeply truncated tetrahedrons, joined base to base, so that the conjoined truncatures form three re-entering angles, and the lateral planes three salient angles.

The crystals are usually small and very small, and generally superimposed.

Externally it is intermediate between glistening and shining, and is often splendent.

Internally it is shining, which in some varieties passes into glimmering, and the lustre is metallic.

The fracture is most commonly coarse and small-grained uneven: the coarse-grained passes on the one hand into imperfect and small conchoidal, and from this into imperfect foliated: the small-grained passes into fine-grained uneven, and into even, and large and flat conchoidal. The lustre varies with the fracture; the foliated has the strongest lustre, and its colour approaches to gold-yellow: the next in intensity of lustre is the small conchoidal: and the large conchoidal and even, have the least lustre, being only glimmering.

It sometimes occurs in curved lamellar concretions.

The fragments are indeterminate angular, and rather sharp-edged.

It is intermediate between semi-hard and soft.

It is brittle.

It is easily frangible.

It is heavy.

Specific gravity, 4.160, *Gellert.* 4.080, *Kirwan.* 4.344, *Brisson.* 4.3154, *Hauy.*

Chemical

Chemical Characters.

Before the blowpipe, on charcoal, it decrepitates, emits a greenish-coloured sulphureous smoke, and melts into a black globule, which, by continuing the fire gradually, assumes the metallic lustre of copper. It imparts to borax a green tinge.

Constituent Parts.

	Cornwall.	Sainbel.
Copper, - -	30	30.0
Iron, - -	53	31.0
Sulphur, - -	12	36.5
	95	97.5
	Chenevix.	*Gueniveau.*

It sometimes also contains small portions of gold or silver.

Geognostic Situation.

It is one of the most abundant ores: it occurs in almost every kind of repository, and in all the great classes of rocks. Thus, it is met with in granite, gneiss, mica-slate, clay-slate, porphyry, syenite, trap, grey-wacke, in the first flœtz limestone, in coal formations, and also in in those of sandstone. In these rocks, it is associated with various metalliferous and earthy minerals, such as iron-pyrites, magnetic ironstone, malachite, azure copper-ore, tile-ore, red copper-ore, variegated copper-ore, copper-glance or vitreous copper-ore, galena or lead-glance, blende, cobalt-ochre, arsenical-pyrites, sparry ironstone, and sometimes native gold: the earthy minerals are calcareous-spar, fluor-spar, heavy-spar, brown-

spar,

spar, quartz, garnet, actynolite, hornblende, tremolite, &c.

Geographic Situation.

Europe.—In veins that traverse a great bed of quartz in the Clifton mine, near Tyndrum in Perthshire; in these veins, it is associated with copper-green, red cobalt-ochre, galena or lead-glance, brown and yellow blende, quartz, and heavy-spar: in a vein in red sandstone in the Mainland, the largest of the Zetland Islands, where it is accompanied with grey copper-ore, malachite, native copper, iron-pyrites, sparry ironstone, and brown ironstone: at the mines of Ecton, on the borders of Derbyshire and Staffordshire, it is contained in flœtz limestone, and is accompanied with galena or lead-glance, blende, calcareous-spar, fluor-spar, and heavy-spar: at Pary's Mountain in Anglesea, it occurs in a bed of great thickness, associated with native copper, malachite, azure copper-ore, galena or lead-glance, and calamine: in several places in Derbyshire: abundantly in the copper-mines of Cornwall, along with copper-glance, grey copper-ore, and red copper-ore. There are considerable copper-mines at Cronebane and Ballymurtach, in the county of Wicklow in Ireland, and the principal ore is copper-pyrites. This ore is met with in considerable abundance on the Continent of Europe, but the localities are so numerous, that we cannot spare room but for a few of them. It occurs in the mines of Rörras and Arendal in Norway; in that of Fahlun in Sweden; in the Hartz; the Saxon Erzgebirge; Hessia, Bohemia, Franconia, Suabia, Bavaria, Silesia, Austria, Hungary, Spain, France, and Russia.

Asia.—Siberia; and Japan.

America

America.—United States; Mexico; and Chili.

Africa.—Morocco; Abyssinia; country of the Namaquas, in Southern Africa.

Uses.

Nearly one-third of all the copper which is obtained by metallurgic operations, is extracted from this species: it is, however, a poor ore, seldom yielding above 36 pounds, more commonly only 20 pounds of copper in the hundred. Sulphur is frequently obtained from it by sublimation.

Observations.

1. It has been confounded with *Native Gold*, but it may be readily distinguished from it by its fracture, which is uneven, imperfect conchoidal, or imperfect foliated; whereas that of gold is hackly; and also by its tenacity, it being brittle, whereas gold is malleable. It is distinguished from *Iron-pyrites*, by its hardness, it being only intermediate between semi-hard and soft, whereas iron-pyrites gives fire with steel; by colour, iron-pyrites being bronze-yellow, whereas it is brass-yellow; and the crystallisations are also very different from those of iron-pyrites; in particular, it occurs in tetrahedrons, a form never observed in iron-pyrites.

2. It passes into several other species of ore, particularly into White Copper-ore, Grey Copper-ore, and Variegated Copper-ore.

3. The softer varieties of copper-pyrites contain the greatest quantity of copper, and the harder the greatest proportion of iron. Among the softer varieties, those having a tarnished surface are said to contain the greatest quantity of copper.

4. Those varieties which contain the largest proportion of sulphur, are the least affected by exposure to the air.

5. White Copper-Ore.

Weiss Kupfererz, *Werner*.

Weisslich Kies-kupfererz, *Henkel's* Kieshistorie, s. 210.—Minera Cupri alba, *Wall.* t. ii. p. 280.—Weiss Kupfererz, *Werner*, Pabst. b. i. s. 83. *Id. Wid.* s. 750.—White Copper-ore, *Kirw.* vol. ii. p. 152.—Weiss Kupfererz, *Estner*, b. iii. s. 505. *Id. Emm.* b. ii. s. 236.—Mine de Cuivre blanche arsenicale, *Lam.* t. i. p. 201.—La Mine de Cuivre blanche, *Broch.* t. ii. p. 173.—Weiss Kupererz, *Reuss*, b. iii. s. 425. *Id. Lud.* b. i. s. 224. *Id. Suck.* 2ter th. s. 184. *Id. Bert.* s. 397. *Id. Leonhard*, Tabel. s. 58. *Id. Karsten*, Tabel. s. 62. *Id. Haus.* s. 74.—White Copper, *Aikin*, p. 28.

External Characters.

Its colour is intermediate between silver-white and bronze-yellow, which sometimes inclines slightly to brass-yellow. On the fresh fracture, it soon becomes tarnished with a greyish-yellow colour.

It occurs massive and disseminated *.

Internally it is glistening, with a metallic lustre.

The fracture is small and fine-grained uneven.

It sometimes occurs in curved lamellar concretions, the surfaces of which are splendent and metallic.

The fragments are indeterminate angular, and rather sharp-edged.

It is soft, passing into semi-hard.

It

* According to Karsten and Ullman, it occurs crystallised in octahedrons.

It is brittle.
It is easily frangible.
It is heavy.
Specific gravity, 4.500, *La Metherie.*

Chemical Characters.

Before the blowpipe, it yields a white arsenical vapour, and melts into a greyish-black slag.

Constituent Parts.

Henkel, who gave the first account of this ore, informs us, that it contains 40 parts of Copper, and the remainder consists of Iron, Arsenic, and Sulphur.

Geognostic Situation.

It occurs in veins and mineral beds in primitive and transition rocks. It is usually accompanied with copper-pyrites, and copper-glance or vitreous copper-ore, seldomer with grey copper-ore, copper-green, red copper-ore, azure copper-ore, and native silver.

Geographic Situation.

Europe.—In the mine called Huel Gorland in Cornwall; in the mines Lorenz Gegentrum and Elias, near Freyberg in the Electorate of Saxony; Rudelstadt, Altenberg, and Kupferberg in Silesia; Lauterberg in the Hartz; Frankenberg in Hessia; Christophsthal, near Freüdenstadt in Wurtemberg; Strazena, behind the Creutzberg in Upper Hungary.

Asia.—Catharinenburg in Siberia.

America.—Chili.

Observations.

Observations.

1. It has been frequently confounded with Copper-pyrites, Copper-glance, Grey Copper-ore, and Arsenical-pyrites. It is, however, easily distinguished from *Copper-pyrites, Copper-glance,* and *Grey Copper-ore,* by its colour; and from *Arsenical-pyrites* by its inferior specific gravity.

2. It is one of the rarest species of copper-ore.

3. It is an intermediate species between copper-pyrites and grey copper-ore, into both of which it probably passes.

6. Grey Copper-Ore.

Fahlerz, *Werner.*

Minera Cupri grisea, *Wall.* t. ii. p. 281.—Mine d'Argent grise, et Mine de Cuivre grise, *Romé de L.* t. iii. p. 315.—Fahlerz, *Werner,* Pabst. b. i. s. 83. *Id. Wid.* s. 751.—Grey Copper-ore, *Kirw.* vol. ii. p. 146.—Fahlerz, *Estner,* b. iii. s. 509. *Id. Emm.* b. iii. s. 238.—Argent gris, *Lam.* t. i. s. 133.—Mine d'une couleur fauve, ou le Cuivre gris, ou le Fahlerz, *Broch.* t. ii. p. 175.—Cuivre gris, *Hauy,* t. iii. p. 536.-556.—Fahlerz, *Reuss,* b. iv. s. 198. *Id. Lud.* b. i. s. 224. *Id. Mohs,* b. iii. s. 231. *Id. Hab.* s. 108.—Cuivre gris, *Lucas,* p. 126.—Fahlerz, *Leonhard,* Tabel. s. 58.—Cuivre gris arsenié, *Brong.* t. ii. p. 215. *Id. Brard,* p. 286.—Fahlerz, *Karsten,* Tabel. s. 62. *Id. Haus.* s. 74.—Grey Copper-ore, *Kid,* vol. ii. p. 115.—Cuivre gris, *Hauy,* Tabl. p. 86.—Grey Copper, *Aikin,* p. 27.

External Characters.

Its colour is steel-grey, which sometimes inclines to iron-black and lead-grey.

It

It occurs rarely with a tempered-steel coloured tarnish.

It occurs massive, disseminated, seldom in membranes, and often also crystallised: its crystallisations are the following:

1. Tetrahedron, or simple three-sided pyramid *, fig. 162. which presents the following varieties:
   a. Truncated on the angles †, fig. 163.; or on the edges ‡, fig. 164.
   b. Bevelled on the edges ||, fig. 165. When the bevelling edges increase so much as to cause the original planes of the tetrahedron to disappear, a tetrahedron is formed, in which each plane is divided into three, or there is formed on each of the planes a very obtuse acumination §, fig. 166.
   c. Each of the angles of the tetrahedron very flatly acuminated with three planes, fig. 167.: sometimes the edges of the tetrahedron are bevelled at the same time ¶, and also the summits and edges of the acuminations **. When the acuminating planes increase so much that the original faces of the tetrahedron disappear, there is formed
2. The rhomboidal or garnet dodecahedron, fig. 168.
3. When the truncations on the angles of the tetrahedron

---

* Cuivre gris primitif, Hauy.

† Cuivre gris epointé, Hauy.

‡ Cuivre gris cubo-tetraedre, Hauy.

|| Cuivre gris encadré, Hauy.

§ Cuivre gris dodecaedre, Hauy.

¶ Cuivre gris apophane, Hauy; Cuivre gris progressif, Hauy.

** Cuivre gris identique, Hauy.

hedron meet, an octahedron is formed,—a figure, however, which has been scarcely observed in the species.

The crystals are small and seldom middle-sized; usually heaped on one another, sometimes also superimposed.

Their surface is shining and splendent.

Internally it is usually glistening; sometimes, however, it passes into shining, and has a metallic lustre.

The fracture is coarse and small-grained uneven; sometimes it inclines to imperfect conchoidal, and such varieties have a blackish colour, the strongest lustre, and contain the greatest proportion of silver, and the least of copper.

The fragments are indeterminate angular, and rather blunt-edged.

It is more or less semi-hard.

It gives a reddish-brown streak: some varieties do not produce any alteration of colour *.

It is brittle.

It is easily frangible.

It is heavy.

Specific gravity, 4.594, *Wiedemann*. 4.8648, *Hauy*. 4.4460 to 4.560, *Bournon*.

Chemical Characters.

Before the blowpipe, it first decrepitates, and then melts into a greyish-coloured brittle metallic globule. During fusion it disengages a white arsenical vapour: to borax it communicates a yellowish colour inclining to red. Some varieties are difficult of fusion.

Constituent

* According to Count de Bournon, those varieties that afford a reddish-brown streak, may be presumed to contain a mixture of silver and antimony, generally combined together in the state of red silver-ore.

Constituent Parts.

From Airthrie.	
Copper,	19.2
Iron,	51.0
Sulphur,	14.1
Arsenic,	15.7
	100.0
Thomson, Ed. Trans.	

	Freyberg.	Freyberg.	Freyberg.
Copper,	41.00	48.00	42.50
Iron,	22.50	25.50	27.60
Sulphur,	10.00	10.00	10.00
Arsenic,	24.10	14.00	15.60
Silver,	0.40	0.50	0.90
Antimony,			1.50
	98.00	98.00	98.00
	Klaproth, Beit. b. iv. s. 47.	Ib. s. 49.	Ib. s. 52.

Geognostic Situation.

It occurs in beds and veins, in primitive, transition, and flœtz rocks, in which it is usually accompanied with copper-pyrites, galena or lead-glance, ores of manganese, sparry ironstone, heavy-spar, calcareous-spar, fluor-spar, and quartz; seldomer with malachite, azure copper ore, and other ores of copper.

Geographic Situation.

Europe.—It occurs along with copper-pyrites in red sandstone, near Sandlodge, in the Mainland of Zetland; in small veins at Fassney Burn in East Lothian; at Airthrie, in the Ochil Hills, north-east of Stirling, in veins, along with heavy-spar, and calcareous-spar; also in Ayrshire; and at Tavistock in Devonshire; and in the copper-mines of Cornwall; at Kongsberg in Norway, along with variegated copper-ore; at Freyberg in Saxony, in veins that traverse gneiss, along with copper-pyrites, sparry ironstone, quartz, calcareous-spar, heavy-spar,

spar, and fluor-spar; in the Hartz, also with copper-pyrites, and sparry ironstone, in veins that traverse grey-wacke, and transition clay-slate; in flœtz limestone at Falkenstein in the Tyrol; at Saint Marie-aux-Mines in France; at Baigorry, in Navarre in Spain; in veins in gneiss, at Hochberg in the dukedom of Baden; in porphyry at Gablan in Silesia; in Thuringia, along with red and brown cobalt-ochre, and heavy-spar; at Saska and Oravicza, in the Bannat in Hungary; also at Kremnitz, and other parts in that kingdom; at Kapnic, Nagy-ag, and Offenbanya, in Transylvania; and in the government of Olonetz in Russia.

Asia.—Kolywan; Tobolsk, along with copper-green; and in several of the mines in the Uralian Mountains.

America.—In the mines of Guanaxuato in Mexico, in veins, along with copper-pyrites, brown-spar, calcareous-spar, amethyst, hornstone, and calcedony; and also in the mines of Zimapan, with quartz, calcareous-spar, and gypsum; and in the copper-mines of Chili.

Uses.

It is valued as an ore of copper; and when it contains silver, it is worked as an ore of that metal.

Observations.

It passes into Copper-pyrites, and Copper-glance or Vitreous Copper-ore.

7. Black

7. Black Copper-Ore.

Schwarzerz, *Werner*.

Minera Cupri grisea, (in part) *Waller*. Syst. Min. t. ii. p. 281. —Cuivre gris, (in part), *Hauy*, t. iii. p. 357.—Graugiltigerz, *Reuss*, b. ii. 3. s. 427.—Schwarzgiltigerz, *Lud.* b. i. s. 218. *Id. Suck.* 2ter th. s. 185. *Id. Bert.* s. 398. *Id. Mohs*, b. iii. s. 196.—Graugiltigerz, *Leonhard*, Tabel. s. 58.—Cuivre gris antimonié, *Brong.* t. ii. p. 216.—Graugiltigerz, *Karsten*, Tabel. s. 62.—Schwarzgiltigerz, *Haus.* s. 74.—Cuivre gris antimonifere, *Hauy*, Tabl. p. 87.—Antimonial Grey Copper, *Aikin*, 2d edit. p. 86.

External Characters.

Its colour is iron-black.

It occurs massive, disseminated; and crystallised in the following figures:

1. Tetrahedron.
   a. Perfect.
   b. Bevelled on the edges, and the angles flatly acuminated with three planes, which are set on the lateral planes.
   c. The preceding figure, in which the summits of the acuminations are truncated.
2. When the acuminations on the angles of the tetrahedron become so large that its original planes disappear, then a rhomboidal or garnet dodecahedron is formed.

The crystals are generally splendent, and are often invested with a thin crust of copper-pyrites.

Internally it is shining and splendent, and the lustre is metallic.

The

The fracture is small and imperfect conchoidal.
It is semi-hard.
It is brittle.
It is easily frangible,
It is heavy.
Specific gravity, 4.842, 4.893, *Ullmann*.

Chemical Characters.

Before the blowpipe, it decrepitates, and then melts into a black cupreous bead, giving out a white antimonial vapour.

Constituent Parts.

	Kapnik in Transylvania.	Poratsch in Hungary.	Annaberg in Saxony.	Zilla in Clausthal	St Wenzel near Wolfach.	Peru.
Copper,	37.75	39.00	40.25	37.50	25.50	27.00
Antimony,	22.00	19.50	23.00	29.00	27.00	23.50
Sulphur,	28.00	26.00	18.50	21.50	25.50	27.75
Silver, -	0.25	-	0.30	3.00	13.25	10.25
Iron, -	3.25	7.50	13.50	6.50	-	-
Lead, -	-	-	-	-	-	1.75
Arsenic,	-	-	0.75	-	-	-
Zinc, -	5.00	-	-	-	-	-
Mercury, -	-	6.25	-	-	-	-
Loss, -	3.75	1.75	3.70	2.50	1.75	2.75
	100.00	100.00	100.00	100.00	100.00	100.00

Klaproth, Beit. b. iv. s. 56. 68. 73. & 80.

From these analyses, it appears, that the essential component parts of Black Copper are, Copper, Antimony, Iron, and Sulphur, and that the Silver, Lead, Zinc, and Mercury, are only accidentally mixed.

Geognostic

Geognostic and Geographic Situations.

Europe.—It occurs in veins that traverse transition rocks at Zilla, in the Clausthal in the Hartz *; Annaberg in Saxony; in the mine of St Wenzel, at Wolfach in the Schwarzwald; at Kapnik, in red manganese-ore; at Nagyag, along with iron-pyrites, grey copper-ore, and quartz; at Kremnitz and Poratsch in Hungary; and Allemont in Dauphiny.

America.—In Peru, in veins in alpine limestone.

Use.

It is worked, both as an ore of copper and as an ore of silver.

Observations.

1. It has been confounded with *Grey Copper ore*, but it may be distinguished from it by its iron-black colour, splendent lustre, and small conchoidal fracture.

2. It is intermediate between Brittle Silver-glance and Grey Copper-ore.

3. It is the most compounded of all the ores of copper.

8. Copper-

* The crystals found at Zilla, are generally invested with a crust of copper-pyrites.

8. Copper-Black or Black Oxide of Copper.

Kupferschwärze, *Werner*.

Ochra Cupri nigra, *Wall.* t. ii. p. 291.—Kupferschwärtze, *Wern.* Pabst. b. i. s. 88. *Id. Wid.* s. 755.—Black Copper-ore, *Kirw.* vol. ii. p. 143.—Kupferschwärtze, *Emm.* b. ii. s. 244. *Id. Estner*, b. iii. s. 525.—Oxide noir de Cuivre, *Lam.* p. 312.—Le Cuivre noir, *Broch.* t. ii. p. 180.—Kupferschwärtze, *Reuss*, b. iii. s. 431. *Id. Lud.* b. i. s. 226. *Id. Suck.* 2ter th. s. 188. *Id. Bert.* s. 379. *Id. Mohs*, b. iii. s. 229. *Id. Leonhard*, Tabel. s. 58. *Id. Karsten*, Tabel. s. 62. *Id. Haus.* Handbuch. b. i. s. 243.—Black Copper, *Aikin*, p. 28.

External Characters.

Its colour is usually intermediate between bluish and brownish-black, but rather more inclining to brownish-black.

It is friable.

It occurs massive, sometimes disseminated, reniform, and sometimes thinly coating other ores of copper.

It is composed of dull dusty particles, which scarcely soil.

It is always more or less cohering.

It is heavy *.

Chemical Characters.

Before the blowpipe, it emits a sulphureous odour, melts into a slag, and communicates a green colour to borax. It forms a smalt-blue coloured solution with ammonia, the iron remaining undissolved.

Constituent

* Bournon, in his *Catalogue Mineralogique*, says, that this ore is sometimes reniform, with a fibrous fracture like hæmatite.

Constituent Parts.

It is said to be composed of Oxide of Copper and Oxide of Iron.

Geognostic Situation.

It occurs usually with copper-pyrites, malachite, copper-green, and copper-glance or vitreous copper-ore; sometimes with native copper, red copper-ore, grey copper-ore, azure copper-ore, quartz, fluor-spar, heavy-spar, and brown-spar.

Geographic Situation.

Europe.—It occurs at Carrarach and Tincroft mines in Cornwall; in veins in transition rocks in the Hartz; in the mines of Moss and Arendal in Norway; near Freyberg, in veins, along with grey copper-ore, copper-pyrites, ochry brown ironstone, and quartz; near Schwatz in the Tyrol, along with copper-pyrites, grey copper-ore, malachite, and copper-green; at Kupferberg and Rudelstadt in Silesia; Markirch in Alsace; in the Schwarzwald; and also in Hungary.

Asia.—Along with iron-pyrites at Schlangenberg; and in different mines in the Uralian Mountains.

Observations.

It appears to be formed, sometimes by the decomposition of copper-pyrites and copper-glance or vitreous copper-ore, and in other instances to be an original formation.

9. Red Copper-Ore.

Rothkupfererz, *Werner.*

This species is divided into three subspecies, viz. Compact Red Copper-ore, Foliated Red Copper-ore, and Capillary Red Copper-ore.

First Subspecies.

Compact Red Copper-Ore.

Dichtes Rothkupfererz, *Werner.*

Id. Werner, Pabst. b. i. s. 66.—Compact florid, or Cochineal-red Copper-Ore, *Kirw.* vol. ii. p. 135.—Dichtes Roth Kupfererz, *Estner,* b. iii. s. 530. *Id. Emm.* b. ii. s. 213.—Le Cuivre Oxide rouge compacte, *Broch.* t. ii. p. 181.—Dichtes Rothkupfererz, *Reuss,* b. iii. s. 433. *Id. Lud.* b. i. s. 226. *Id. Suck.* 2ter th. s. 189. *Id. Bert.* s. 380. *Id. Mohs,* b. iii. s. 213. *Id. Leonhard,* Tabel. s. 56.—Cuivre oxidulé compact, *Brong.* t. ii. p. 219.—Dichtes Rothkupfererz, *Karsten,* Tabel. s. 62. —Red Oxide of Copper, *Kid,* vol. ii. p. 101.—Cuivre oxidulé massif, *Hauy,* Tabl. p. 38.—Amorphous Red Copper, *Aikin,* p. 29.

External Characters.

Its colour is dark cochineal-red, faintly inclining to lead-grey.

It occurs massive, in membranes, stalactitic, corroded, amorphous, and also disseminated.

Internally

Internally it is glimmering, inclining to glistening, and the lustre is semi-metallic.

The fracture is even, inclining to flat conchoidal.

The fragments are indeterminate angular, and rather sharp-edged.

It is opaque.

It gives a tile-red streak, and loses thereby a little of its lustre.

It is semi-hard.

It is brittle.

It is easily frangible.

It is heavy.

Second Subspecies.

Foliated Red Copper-Ore.

Blättriches Rothkupfererz, *Werner.*

Id. Werner, Pabst. b. i. s. 66.—Foliated florid-red Copper-ore, *Kirw.* vol. ii. p. 136.—Blätteriges Rothkupfererz, *Estner,* b. iii. s. 533. *Id. Emm.* b. ii. s. 214.—Le Cuivre Oxide rouge lamelleux, *Broch.* t. ii. p. 183.—Blättriches Rothkupfererz, *Reuss,* b. iii. s. 436. *Id. Lud.* b. i. s. 226. *Id. Suck.* 2ter th. s. 189. *Id. Bert.* s. 382. *Id. Mohs,* b. iii. s. 213. *Id. Leonhard,* Tabel. s. 56.—Cuivre oxidulé crystallisé, *Brong.* t. ii. p. 219.—Blættriches Rothkupfererz, *Karsten,* Tabel. s. 62. —Cuivre oxidulé lamellaire, *Hauy,* Tabl. p. 88.

External Characters.

Its colour is dark cochineal-red, which sometimes inclines to lead-grey; but the crystals are redder, and sometimes pass into dark carmine-red.

It occurs massive, disseminated, in membranes, corroded; and crystallised in the following figures:

1. Perfect octahedron, fig. 169. Sometimes two opposite planes become so large in comparison of the others, that a *six-sided table* is formed, fig. 170.: in other instances, two opposite planes entirely disappear, when an *acute rhomboid* is formed, fig. 171.; and frequently the octahedron ends in a line, fig. 172.
2. Octahedron truncated on the angles, fig. 173. When the truncating planes become so large that the original planes of the octahedron disappear, a *cube* is formed, fig. 174.
3. Octahedron truncated on all the edges, fig. 175. When these truncating planes become so large as to cause the original planes to disappear, a *rhomboidal dodecahedron* is formed, fig. 176. When the edges of the common basis of the octahedron are deeply truncated, the figure thus formed might be described as a rectangular four-sided prism, acuminated with four planes, which are set on the lateral planes. In the same figure, the angles are also frequently truncated.
4. Octahedron, in which each of the angles is acuminated with four planes, which are set on the lateral edges, fig. 177.; sometimes the edges are truncated, and also the summits of the acuminations, fig. 178.
5. Octahedron bevelled on the edges, fig. 179: sometimes the bevelling planes become so large that each face of the octahedron appears divided into three compartments, or each plane supports a

flat

flat three-planed acumination, fig. 180. Sometimes the angles and edges of the octahedron in this figure are also truncated, and according as these bevelling and truncating planes are larger or smaller, the figure varies in appearance.

6. Octahedron, in which each angle is acuminated with four planes, which are set on the lateral planes, fig. 181. This variety is generally combined with the planes of several of the preceding ones.

7. Octahedron, in which each angle is acuminated with eight planes, two of which are set on each plane, fig. 182. This variety is always associated with some of the preceding, and crystals have been met with, exhibiting this variety combined with all the preceding ones. Fig. 183. represents such a crystal, which is marked as follows:—P, planes of the octahedron: (1.) Truncations on the angles: (2.) Truncations on the edges: (3.) Four acuminating planes on the angles, which are set on the edges of the octahedron: (4.) Bevelling planes on the edges: (5.) Four acuminating planes on the angles, which are set on the planes of the octahedron: (6.) Eight-planed acumination *.

The crystals are usually small and very small, seldom middle-sized: they occur sometimes aggregated on one another, side by side, and scalar-wise.

* These various forms, and numerous intermediate ones, are delineated in a set of plates attached to Mr Phillip's valuable Memoir on Red Copper-ore, in the first volume of the Transactions of the Geological Society of London.

The planes are smooth and splendent.

Internally it alternates from shining to glistening, and its lustre is adamantine.

The fracture is imperfect foliated, with a four-fold cleavage, parallel with the sides of an octahedron, sometimes coarse and small grained uneven.

The fragments are indeterminate angular, and rather sharp-edged.

It sometimes occurs in coarse, small, and fine granular distinct concretions, and the fine granular pass into compact.

The massive varieties are usually opaque, or very faintly translucent on the edges. The crystals are transparent and semi-transparent, and sometimes strongly translucent.

It yields a muddy tile-red streak.

It is semi-hard; it scratches calcareous-spar, but does not affect fluor-spar.

It is brittle.

It is easily frangible.

It is heavy.

Specific gravity, 3.950, *Wiedemann.* 5.600, *Phillips.* 5.691, *Lowry.*

Third

Third Subspecies.

Capillary Red Copper-ore.

Haarförmiges Kupferglas, *Werner*.

Kupferblüthe, *Werner*, Pabst. b. i. s. 68.—Fibrous red Copper-ore, *Kirw*. vol. ii. p. 137.—Kupferblüthe, *Estner*, b. iii. s. 538. *Id. Emm*. b. ii. s. 216.—Le Cuivre Oxide rouge capillaire, *Broch*. t. ii. p. 184.—Haarförmiger Rothkupfererz, *Reuss*, b. iii. s. 439. *Id. Lud*. b. i. s. 227. *Id. Suck*. 2ter th. s. 194. *Id. Bert*. s. 382. *Id. Mohs*, b. iii. s. 226. *Id. Leonhard*, Tabel. s. 56.—Cuivre oxidulé capillaire, *Brong*. t. ii. p. 219.—Haarförmiges Rothkupfererz, *Karsten*, Tabel. s. 62.—Cuivre oxydulé capillaire, *Hauy*, Tabl. p. 88.—Capillary Red Copper, *Aikin*, p. 29.

External Characters.

Its colour is most commonly carmine-red, which sometimes approaches to cochineal-red.

It occurs in small capillary crystals, also in thin tables, which are sometimes aggregated into amorphous and scopiform flakes.

It is shining, and the lustre is adamantine.

It is translucent; but its internal aspect, and the other external characters, cannot be determined, on account of the smallness of the parts of the mineral.

Chemical Characters of the Species.

It is easily reduced to the metallic state before the blowpipe: if pulverised, and thrown into nitric acid, a violent effervescence ensues, and the copper is dissolved, the

the solution at the same time acquiring a green colour; but if thrown into muriatic acid, no effervescence takes place. We can by this character distinguish red copper-ore from red silver-ore and cinnabar: red silver-ore does not effervesce in nitric acid; and cinnabar does not dissolve in it. It is soluble in ammonia, to which it communicates a blue colour.

Constituent Parts.

	Cornwall.	Siberia.
Copper,	88.5	91.0
Oxygen,	11.5	9.0
	100.0	100.0
	Chenevix, Phil. Trans. for 1801.	*Klaproth*, Beit. b. iv. s. 29.

	Foliated, Siberia.	Compact, Siberia.
Red Oxide of Copper,	97.55	99.50
Intermixed Copper,	1.45	- -
Water, - -	0.75	0.25
Oxide of Iron, -	0.25	0.25
	100.0	100.0
	John, Chem. Unter. b. i. s. 264. & 261.	

Geognostic Situation of the Species.

It occurs principally in veins that traverse primitive, and sometimes transition, rocks, rarely in flœtz rocks, and but seldom in beds, along with copper-glance or vitreous copper-ore. In the veins, it is associated with native copper, azure copper ore, malachite, copper-green, tile-ore, copper-glance or vitreous copper-ore, copper-pyrites, copper-black or black oxide of copper, oliven-ore, copper-pyrites, cube-ore or arseniate of iron, arsenical-pyrites, and brown ironstone. The vein-stones are, quartz, fluor-spar, calcareous-spar, heavy-spar, and occasionally chlorite and mica.

Geographic

Geographic Situation of the Species.

Europe.—It occurs in different veins in the mine of Huel-Gorland in Cornwall. All the veins traverse granite, and three of them, viz. the North Lode, the Great Gossan Lode, and the Muttrel Lode, afford the red copper-ore. In the North Lode, it is associated with fluor-spar. In the Great Gossan Lode, it occurs in considerable quantity, and occasionally intermixed with native copper: higher up in the same vein, there is abundance of fluor-spar, sometimes intermixed with copper-pyrites, and arsenical-pyrites. In the Muttrel Lode, the red copper-ore is occasionally accompanied with copper-glance or vitreous copper-ore, copper-black or black oxide of copper, oliven-ore, arsenical-pyrites, quartz, and fluor-spar. Native copper also occurs in considerable quantities, and generally intermixed with red copper-ore *. It is also found in the mines of Carvath and Huel-Prosper, also in Cornwall. Small portions of this ore occur, along with native copper, in the trap rocks of Nalsoe, one of the Faroe Islands; also in the mine of Aardal in Norway, and that of Garpa, in East Gothland in Sweden. It occurs but sparingly, and along with native copper, in the Rammelsberg in the Hartz; near Freyberg, along with native copper, ochry brown ironstone, lamellar heavy-spar, and quartz; at Altenkirchen, along with brown hematite, native copper, malachite, oliven-ore, and quartz; in the Zillerthal in Bavaria; at Ensiedel in Hungary, with native copper, copper and iron-pyrites; at Saska and Moldowa in the Bannat, associated with copper-glance or vitreous copper-ore, malachite, azure copper-ore, native copper, and brown ironstone.

Asia.

* Phillips, in Geological Transactions, vol. i. p. 23,—29.

Asia.—In the mines of Kolywan, along with native copper, and various ores of that metal; and in different mines in the Uralian Mountains.

America.—Chili, and Peru.

The preceding account applies to the Compact and Foliated red copper-ores: the third subspecies, the Capillary ore, occurs less frequently. In Cornwall, it is found in Huel-Gorland, St Day, and Carharrack mines. On the Continent of Europe, beautiful specimens are met with at Rheinbreitenbach in Nassau, where it is associated with ochry brown ironstone, native copper, copper-green, malàchite, copper-pyrites, white lead-ore, phosphate of copper, copper-black or black oxide of copper, and quartz, in veins that traverse grey-wacke; at Saska in the Bannat, with brown ironstone, malachite, tile-ore, native copper, foliated red copper-ore, steatite, and lithomarge; and also in the Saxon Erzgebirge, as at Freyberg and Glasshütte. It is also a production of the Siberian copper-mines.

Uses.

It is valued as an ore of copper.

Observations.

1. It is distinguished from *Copper-glance* or *Vitreous Copper-ore*, by its colour · from *Red Silver-ore*, by its crystallisations, and accompanying minerals: from *Cinnabar*, by its colour, weight, and accompanying minerals: from *Red Antimony-ore* by its colour, red antimony having a cherry-red colour.

2. Hausmann,

2. Hausmann, in his Handbuch der Mineralogie, describes in the following manner a fourth subspecies of red copper-ore.

Earthy Red Copper-ore.

Erdiges Kupferroth.

The colour is intermediate between cochineal-red and brick-red. It occurs massive, and incrusting other ores. It is fine-earthy, and dull. It is associated with malachite, tile-ore, native copper, and brown ironstone. It occurs in veins, probably also in beds, in primitive and transition rocks. It is a rare mineral, and has been hitherto found only at Rheinbreitenbach; and near Lauterberg in the Hartz.

10. Tile-Ore.

Ziegelerz, Werner.

This species is divided into two subspecies, viz. Earthy Tile-ore, and Indurated Tile-ore.

First Subspecies.

Earthy Tile-Ore.

Erdiches Ziegelerz, *Werner*.

Id. Werner, Pabst. b. i. s. 70.—Earthy Brick-red Copper-Ore, *Kirw.* vol. ii. p. 137.—Erdiges Ziegelerz, *Estner,* b. iii. s. 550. *Id. Emm.* b. ii. s. 219.—Le Ziegelerz terreuse, *Broch.* t. ii. p. 187.—Erdiges Ziegelerz, *Reuss,* b. iii. s. 443. *Id. Lud.* b. i. s. 227. *Id. Suck.* 2ter th. s. 194. *Id. Bert.* s. 382. *Id. Mohs,* b. iii. s. 226. *Id. Leonhard,* Tabel. s. 56.—Cuivre oxidulé ferrifere, *Brong.* t. ii. p. 220.—Erdiges Ziegelerz, *Karsten,* Tabel. s. 62.—Earthy Red Oxide of Copper, mixed with Brown Oxide of Iron, *Kid,* vol. ii. p. 105.—Cuivre oxydulé terreuse, *Hauy,* Tabl. p. 88.—Ferruginous Red Copper, *Aikin,* p. 29.

External Characters.

Its colour is hyacinth-red, sometimes also brownish-red, which passes into a reddish-brown, that borders on yellowish-brown.

It is intermediate between friable and solid.

It occurs massive, disseminated, and incrusting copper-pyrites.

It is composed of dull dusty particles when in crusts; but of earthy particles when it is massive.

It soils slightly.

It is almost always coherent; some varieties incline to compact.

It is heavy.

Geognostic

Geognostic Situation.

It occurs in veins, and is usually accompanied with native copper and malachite, and sometimes with red copper-ore

Geographic Situation.

It is found at Lauterberg in the Hartz; in veins in the Bannat, along with copper-pyrites, red copper-ore, grey copper-ore, ironshot copper-green, malachite, native copper, and ochry ironstone; at Falkenstein in the Tyrol, with copper-green, malachite, grey copper-ore, azure copper-ore, calcareous-spar, and quartz; and at Rezbanya, along with copper-green, malachite, and calcareous-spar.

Second Subspecies.

Indurated Tile-Ore.

Festes Ziegelerz, *Werner.*

Minera Cupri picea, *Wall.* t. ii. p. 280.—Dichtes Ziegelerz, *Werner,* Pabst. b. i. s. 70.—Indurated Brick-red Copper-ore, *Kirw.* vol. ii. p. 138.—Pecherz, *Estner,* b. iii. s. 553.—Ziegelerz, *Emm.* b. ii. s. 220.—Le Ziegelerz endurcé, *Broch.* t. ii. p. 188.—Verhärtetes Ziegelerz, *Reuss,* b. iii. s. 444. *Id. Lud.* b. i. s. 228. *Id. Suck.* 2ter th. s. 196. *Id. Bert.* s. 383. *Id. Mohs,* b. iii. s. 229. *Id. Leonhard,* Tabel. s. 56. *Id. Karsten,* Tabel. s. 62.

External Characters.

Its colour is dark hyacinth-red, brownish-red, reddish-brown, from which it passes into blackish-brown, and dark

dark steel-grey; also dark clove-brown, yellowish-brown, brownish-black.

It occurs massive, disseminated, reniform, botryoidal, and cellular.

Internally it is glimmering, passing into glistening, and is resinous.

The fracture is intermediate between even and large conchoidal, and sometimes passes into small conchoidal.

It occurs in curved lamellar concretions, which sometimes pass into granular concretions.

The fragments are indeterminate angular, and more or less sharp-edged.

The streak is feebly shining, and somewhat lighter in the colour.

It is intermediate between semi hard and soft.

It is rather brittle.

It is rather easily frangible.

Specific gravity, 3.572, the variety named Pecherz.

Chemical Characters.

Before the blowpipe it becomes black, but is very difficultly fusible. To borax it communicates a muddy green colour.

Constituent Parts.

Werner considers it to be an intimate combination of red copper-ore and brown iron-ochre. It contains from 10 to 50 *per cent.* of copper.

Geognostic Situation.

It occurs in veins, and is usually accompanied with red copper-ore, native copper, copper-pyrites, fibrous malachite, and brown iron-ochre.

Geographic

Geographic Situation.

Europe.—It occurs in veins, along with red copper-ore, native copper, copper-pyrites, and other ores, in Huel-Gorland in Cornwall; also at Llanymynich Hill in Shropshire; at Aardals copper-mine in Norway, along with compact malachite and native copper; at Lauterberg in the Hartz; Kupferberg in Silesia; Rheinbreitenbach in in Nassau, along with copper-pyrites, malachite, azure copper-ore, copper-green, &c.; Saxon Erzgebirge; and Thuringia; at Falkenstein in the Tyrol, particularly the variety named Pecherz or Pitch-ore; Iglo, Rezbanya, and the Bannat in Hungary.

Asia.—In the mines of Frolowskoi, along with red copper-ore, and brown iron-ochre.

America.—In the mine of El Rosario in Mexico, associated with copper-green and copper-pyrites.

Observations.

1. The red varieties contain the greatest quantity of copper, and the brown the greatest quantity of iron.

2. It is rather a common ore of copper, and occurs almost always where red copper-ore is found.

3. It passes sometimes, by increase of the quantity of brown iron-ochre, into brown ironstone.

4. The dark-brown variety of indurated tile-ore, on account of the resemblance of its fracture to pitch, has been denominated *Pitchore*, (Pecherz).

11. Azure Copper-Ore.

Kupferlazur, *Werner*.

This species is divided into two subspecies, viz. Earthy AzureCopper-ore, and Indurated Azure Copper-ore.

First Subspecies.

Earthy Azure Copper-Ore.

Kupferlazur, *Werner*.

Erdiche Kupferlazur, *Werner*, Pabst. b. i. s. 92. *Id. Wid.* s. 762. —Earthy Mountain-blue, *Kirw.* vol. ii. p. 129.—Erdiche Kupferlazur, *Estner*, b. iii. s. 560. *Id. Emm.* b. ii. s. 246.— L'Azur de Cuivre terreux, *Broch.* t. ii. p. 191.—Gemeine Kupferlazur, *Reuss*, b. iii. s. 449.—Erdiche Kupferlazur, *Lud.* b. i. s. 228. *Id. Suck.* 2ter th. s. 198. *Id. Bert.* s. 388. *Id. Mohs*, b. iii. s. 262. *Id. Leonhard*, Tabel. s. 58.—Gemeine Kupferlazur, *Karsten*, Tabel. s. 62. *Id. Haus.* s. 135.—Blue Carbonate of Copper, *Kid*, vol. ii. p. 117.—Cuivre carbonaté bleu terreux, *Hauy*, Tabl. p. 89.—Earthy Blue Copper, *Aikin*, p. 30.

External Characters.

Its colour is smalt-blue.

It is usually friable, seldom massive, often disseminated, and thinly coating.

It is composed of dull and fine dusty particles.

It does not soil, or at most very faintly.

It is almost always cohering.

Chemical

Chemical Characters.

It is soluble with effervescence in nitric acid. Before the blowpipe, without addition, it blackens, but does not melt: with borax, on charcoal, it effervesces, gives a metallic globule, and colours the flux green.

Geognostic Situation.

It occurs in small quantity, and is usually accompanied with malachite and copper-green. In Silesia, it is found incrusting bituminous marl-slate; in Thuringia, coating varieties of the old red sandstone; and in Siberia, disseminated in sandstone.

Geographic Situation.

Europe.—It is found in Norway; at Saalfeld, Sangerhausen, Bottendorf, and Eisleben in Thuringia; Thalitter in Hessia; Zellerfeld in the Hartz; Prausnitz, &c. in Silesia; Leogang in Salzburg; and in West Gallicia.

Asia.—Siberia.

Second Subspecies.

Indurated or Radiated Azure Copper-Ore.

Feste Kupferlazur, *Werner.*

Strahlige Kupferlazur, *Werner*, Pabst. b. i. s. 89. *Id. Wid.* s. 764.—Striated Mountain-blue, *Kirw.* vol. ii. p. 130.—Strahlige Kupferlazur, *Estner*, b. iii. s. 564. *Id. Emm.* b. ii. s. 249.—L'Azur de Cuivre rayonné, *Broch.* t. ii. p. 192.—Cuivre carbonaté bleu, *Hauy*, t iii. p. 562.—Strahlige Kupferlazur, *Reuss*, b. iii. s. 453. *Id. Lud.* b. i. s. 229. *Id. Suck.* 2ter thl. s. 203. *Id. Bert.* s. 390. *Id. Mohs*, b. iii. s. 272.—Cuivre carbonaté bleu, *Lucas*, p. 131.—Strahlige Kupferlazur, *Leonhard*, Tabel. s. 59.—Cuivre azure, *Brong.* t. ii. p. 220.—Cuivre carbonaté bleu, *Brard*, p. 293.—Strahlige Kupferlazur, *Karsten*, Tabel. s. 62.—Edler Kupferlazur, *Haus.* s. 135.—Cuivre carbonaté bleu, *Hauy*, Tabl. p. 89.—Blue Copper, *Aikin*, p. 30.

External Characters.

Its principal colour is azure-blue; it occurs also Berlin-blue, and blackish-blue, and very seldom inclines to smalt blue. These colours are of different degrees of intensity, and they all appear to possess a slight tinge of red.

It seldom occurs massive, disseminated, and as a coating; sometimes in membranes, more frequently botryoidal, small reniform, stalactitic, and cellular, but most frequently crystallised.

Its crystallisations are the following:

Rather oblique four-sided prism, rather acutely bevelled on the terminal planes, the bevelling planes set

set on the acuter lateral edges. It exhibits the following varieties:

a. The angles which the bevelling planes make with two lateral planes, or with the obtuse lateral edges, truncated.

b. The acute lateral edges bevelled, the edges of the bevelment, and the obtuse lateral edges, truncated: when these planes meet, an eight-sided prism is formed.

c. The proper edge of the bevelment on the terminal planes more or less deeply truncated; sometimes so deeply, that the figure appears as a simple oblique four-sided prism.

d. Two opposite planes, so much larger than the others, that the crystal has a tabular form.

e. The rectangular four-sided prism, or eight-sided prism, acuminated with four planes, which are set on the lateral planes.

The crystals are sometimes singly superimposed, more frequently aggregated on one another, intersecting one another; also in thin drusy vesicles, and aggregated in scopiform, globular, and budlike shapes.

The particular external shapes have always a drusy surface, but that of the planes of the crystals is streaked or smooth, as is the case with the truncating planes.

Externally the crystallised varieties are shining, but the massive and particular external shapes are dull.

Internally it is shining and glistening, and the lustre is intermediate between vitreous and resinous.

The fracture is narrow, straight, and scopiform or stellular radiated, sometimes imperfect foliated, and also small conchoidal.

The fragments of the radiated varieties are wedge-shaped; those of the foliated and conchoidal splintery.

It is usually compact. It sometimes, however, occurs in distinct concretions; the foliated varieties in small and fine granular distinct concretions; the reniform, botryoidal and stalactitic varieties in curved lamellar distinct concretions, which are bent in the direction of the external surface.

The crystals are translucent, passing into semi-transparent, and are sometimes only translucent on the edges.

The colour becomes lighter in the streak.

It is soft.

It is rather brittle.

It is easily frangible.

Specific gravity, 3.6082, *Brisson*. 3.231, *Wiedemann*. 3.400, *Bindheim*.

Chemical Characters.

Same as in the earthy azure copper-ore.

Constituent Parts.

		Siberia.	Chessy.
Copper, -	66 to 70	56.00	56.00
Carbonic Acid,	18 to 20	24.00	25.00
Oxygen, -	8 to 10	14.00	12.50
Water, -	2 to 2	6.00	6.50
		100.00	100.00
	Pelletier, in Mem. et Observ. de Chimie, t. ii. p. 20.	*Klaproth*, Beit. b. iv. s. 33.	*Vauquelin*, Ann. du Mus. t. xx. p. 3.

Geognostic

Geognostic Situation.

This mineral occurs in veins that traverse primitive, transition, and flœtz rocks; also in beds, but in smaller quantity, and less frequently. Thus, at Catharinenburg in Bohemia, and in the electorate of Triers, it occurs in gneiss; at Kleingabel, near Pries in Lower Hungary, in mica-slate, passing to clay-slate; at Zellerfeld in the Hartz, in grey-wacke and grey-wacke-slate; in the limestone mountains of the Tyrol; in the first flœtz limestone of Krakau in Lower Saxony; in the sandstone of the Uralian Mountains; in the old red sandstone of Thuringia, along with copper-green, forming what is called sand-ore, *(sanderz)*; and also in rocks of the coal formation.

There are several formations of this species: one of considerable antiquity, which contains malachite, and brown ironstone, also red copper-ore, and tile-ore, and probably grey copper-ore and copper-pyrites; another newer, in which it is associated with white and green lead-ores; and a third, in which the ore is associated with earthy cobalt-ochre, and straight lamellar heavy-spar.

At Leadhills, in Lanarkshire, it is accompanied with galena or lead-glance, ochre of manganese, lead-earth, sparry ironstone, calamine, ochry brown ironstone, brown hematite, iron-pyrites, green lead-ore, white lead-ore, and lead-vitriol or sulphate of lead.

In Hungary, it is associated with copper-pyrites, malachite, copper-green, grey copper-ore, and iron-ochre; in the Bannat, which produces very beautiful specimens, it occurs along with fibrous malachite, earthy tile-ore, compact red copper-ore, iron-ochre, copper-green, and asbestous actynolite; near Laak in Upper Carniola, with

 quartz

quartz and malachite; in the district of Kamsdorf in Saxony, it is accompanied with yellow and brown iron-ochre, ironshot copper-green, and other ores of copper; at Saalfeld in Thuringia, with straight lamellar heavy-spar, grey copper-ore, malachite, ironshot copper-green, tile-ore, and iron-ochre; at Kupferberg in Silesia, with brown-spar and malachite: the Siberian, which rivals that of the Bannat in beauty, is accompanied with copper-green, malachite, tile-ore, green and white lead-ores, brown ironstone, heavy-spar, and quartz.

Geographic Situation.

Europe.—It occurs at Leadhills and Wanlockhead in Lanarkshire; Huel-Virgin and Carharrack in Cornwall. On the Continent, it is met with in the iron-mines at Arendal in Norway; in the government of Olnetz in Russia; in the Hartz; Thalitter in Hessia; Moschellandsberg in Deux-Ponts; in Salzburg; Schwatz in the Tyrol; West Gallicia; Corsica; and Spain.

Asia.—Kamtschatka, and Kolywan; and in many mines in the Uralian Mountains.

America.—In veins in granite in Chili.

Uses.

This species is not only used as an ore of copper, but also as a blue colour, (called Mountain-blue), of which there is a manufactory at Schwatz in the Tyrol.

Observations.

1. The *Armenian Stone* of the ancients, which was brought from Armenia, is a limestone impregnated with earthy

earthy azure copper-ore, and in which copper and iron pyrites are sometimes disseminated *.

2. The Κύανος αὐτοφυής of Theophrastus, the *Cyanos* of Pliny, appear to include this mineral and some others. It is the *Cæruleum montanum* of Wallerius, and the *Azur de Cuivre*, or *Fleurs de Cuivre bleues*, of Romé de Lisle.

12. Velvet Copper-Ore.

Kupfersammterz, *Werner*.

Kupfersammterz, *Karsten*, Tabel. s. 62.

External Characters.

Its colour is intermediate between smalt-blue and sky-blue, and sometimes passes into sky-blue.

It occurs in very small and delicate capillary crystals, which generally form a velvety crust, and are seldom aggregated in balls.

Externally and internally the lustre is glistening and pearly.

It is very soft.

It is sectile.

Geognostic and Geographic Situations.

It is a very rare mineral, and has hitherto been found only at Oravicza in the Bannat, along with malachite and brown ironstone †.

Observations.

---

* Vid. A. Boetius de Boot, Gemmarum et Lapidum Historia, lib. ii. cap. 144.

† This species is so rare, that 50 dollars are given for single specimens of it.

Observations.

According to Werner, it forms the connecting link between Azure Copper-ore and Malachite.

13. Malachite.

Malachit, *Werner.*

This species is divided into four subspecies, viz. Fibrous Malachite, Compact Malachite, Foliated Malachite, and Earthy Malachite.

First Subspecies.

Fibrous Malachite.

Fasricher Malachit, *Werner.*

Ærugo nativa crystallisata, *Wall.* t. ii. p. 287.—Fasriche Malachit, *Werner*, Pabst. b. i. s. 92. *Id. Wid.* s. 768.—Fibrous Malachite, *Kirw.* vol. ii. p. 131.—Fasriger Malachit, *Estner*, b. iii. s. 577. *Id. Emm.* b. ii. s. 254.—Cuivre vert soyeux, *Hauy*, t. iii. p. 571.-575.—La Malachite fibreuse, *Broch.* t. ii. p. 197.—Cuivre carbonaté vert, *Hauy*, t. iii. p. 571.—Fasriger Malachit, *Reuss*, b. iii. s. 461. *Id. Lud.* b. i. s. 230.—Malachit Kupfer, *Suck.* 2ter th. s. 203. *Id. Bert.* s. 390. *Id. Mohs*, b. iii. s. 272.—Cuivre carbonaté vert, *Lucas*, p. 132.—Fasriger Malachit, *Leonhard*, Tabel. s. 59.—Cuivre Malachit soyeux, *Brong.* t. ii. p. 222.—Cuivre carbonaté, *Brard*, p. 296.—Fasriger Malachit, *Karsten*, Tabel. s. 62.—Edler Malachit, *Haus.* s. 135.—Green Carbonate of Copper, *Kid*, vol. ii. p. 119.—Cuivre carbonaté vert aciculaire soyeux, *Hauy*, Tabl. p. 90.—Malachite, *Aikin*, p. 30.

External Characters.

Its most common colour is perfect emerald-green, sometimes

sometimes inclining to grass-green, and sometimes passing to dark leek-green.

It is seldom massive, sometimes disseminated, tuberose, stalactitic, reniform, botryoidal fruticose, most frequently as a coating, and crystallised. The following crystallisations have been observed:

1. Rather oblique four-sided prism, bevelled on the extremities, the bevelling planes set on the obtuse lateral edges *.
2. The preceding figure truncated on the obtuse lateral edges, which thus forms a six-sided prism, in which the bevelling planes are set on two opposite lateral planes.
3. Acute angular three-sided prism, in which the terminal planes are set on, either straight or oblique.

The crystals are generally short, capillary, and acicular. When very short, they form velvety drusy pellicles; and when longer, they are scopiformly aggregated.

Internally it is intermediate between glistening and glimmering, and the lustre is pearly or silky.

The fracture is extremely delicate, and usually scopiform or stellular fibrous; sometimes it is coarse fibrous. In some varieties, the coarse fibrous borders on, and even passes into, narrow radiated.

The fragments are wedge-shaped and splintery.

It occurs in large and coarse, sometimes longish, granular distinct concretions, which sometimes pass into thick and short wedge-shaped distinct concretions.

The crystals are translucent, but the massive varieties only translucent on the edges.

It

* According to Bournon, the lateral planes of the prism meet under angles of 103° and 77°; whereas those in the oblique fonr-sided prism of azure copper-ore are said to meet under angles of 116° and 56°.

It is soft, and very soft.

The colour of the streak is somewhat lighter.

It is brittle, inclining to sectile.

Specific gravity, 3.5718, *Brisson.*

Chemical Characters.

Before the blowpipe, it decrepitates and becomes black, and is partly infusible, partly reduced to a black slag. It melts with borax, to which it communicates a dark yellowish-green colour, and readily affords with it a bead of copper. It effervesces with acids, and forms a blue-coloured solution with ammonia.

Constituent Parts.

Siberia.		Arragon.		Chessy.	
Copper,	58.00	Copper,	56.8	Copper,	56.10
Carbonic Acid,	18.00	Carbonic Acid,	27.0	Carbonic Acid,	21.25
Oxygen,	12.50	Oxygen,	14.2	Oxygen,	14.00
Water,	11.50	Sand, -	1.0	Water, -	8.75
		Lime, -	1.0		
	100.00		100.0		100.00
Klaproth, Beit. b. ii. s. 290.		Ann. Mus. t. xx. p. 7.		Ibid. p. 8.	

Geognostic Situation.

It occurs principally in veins, and these generally contain, besides this ore, red copper-ore, tile-ore, brown ironstone, azure copper-ore, copper-pyrites, copper-glance or vitreous copper-ore, along with calcareous-spar and quartz. These veins traverse primitive, transition, and flœtz rocks.

Geographic Situation.

Europe.—It occurs at Sandlodge in Mainland, one of the Zetland Islands, in veins that traverse red sandstone, in in which it is associated with grey copper-ore, copper-pyrites, and brown ironstone; at Llandidno in Ca rnarvonshire; in the mines of Arendal in Norway, along with magnetic ironstone, copper-pyrites, and grey copper-ore; Fahlun and Sahlberg in Sweden; at Lauterberg in the Hartz, with copper-pyrites, compact brown ironstone, and tile-ore; at Frankenberg in Hessia, in flœtz limestone, along with radiated azure copper-ore; at Thalitter in Hesse-Darmstadt, with copper-green, azure copper-ore, variegated copper-ore, and mineral pitch; at Kaisersteinmel in Nassau, in veins in clay-slate and grey-wacke, along with native copper, copper-pyrites, copper-glance or vitreous copper-ore, fibrous and foliated oliven-ore, red copper-ore, iron-pyrites, brown ironstone, and quartz; in the copper-mine of Malscheid, also in Nassau, with black and white lead ore, and galena or lead-glance, blende, copper and iron pyrites, copper-black or black oxide of copper, and quartz; Kupferberg in Silesia; on the Buchberg, near Landshut in Lusatia, in basalt, and along with calcareous-spar; in the Bannat, along with red copper-ore, azure copper-ore, tile-ore, and brown ironstone; and in the government of Olnetz in Russia.

Asia.—In the mines of Kolywan, along with tile-ore, red copper-ore, brown ironstone, copper-glance or vitreous copper-ore, copper-green, azure copper-ore, white lead-ore, brown-spar, ironshot quartz, and hornstone; and in several of the mines situated in the Uralian Mountains.

America.—In Maryland and Pennsylvania in the United States; and in the copper mines of Chili, in South America.

Second

Second Subspecies.

Compact Malachite.

Dichter Malachit, *Werner.*

Molochites, *Plin.* Hist. Nat. xxxvii. 8. s. 36.—Ærugo nativa fissilis, stalactitica, solida, *Wall.* t. ii. p. 287.—Dichter Malachit, *Werner,* Pabst. b. i. s. 94. *Id. Wid.* s. 770.—Compact Malachite, *Kirw.* vol. ii. p. 132.—Dichter Malachit, *Estner,* b. iii. s. 586. *Id. Emm.* b. ii. s. 256.—Cuivre carbonaté vert concretionné, *Hauy,* t. iii. p. 571.—La Malachite compacte, *Broch.* t. ii. p. 199.—Dichter Malachit, *Reuss,* b. iii. s. 467. *Id. Lud.* b. i. s. 230. *Id. Suck.* 2ter th. s. 203. *Id. Bert.* s. 390. *Id. Mohs,* b. iii. s. 272. *Id. Leonhard,* Tabel. s. 69. —Cuivre Malachite concretionné, *Brong.* t. ii. p. 223.—Dichter Malachit, *Karsten,* Tabel. s. 62.—Gemeiner Malachite, *Haus.* s. 135.—Cuivre carbonaté vert compacte, *Hauy,* Tabl. p. 90.—Massive Malachite, *Aikin,* p. 31.

External Characters.

Its colour is intermediate between emerald-green and verdigris-green: sometimes it passes into verdigris-green and blackish-green, and occasionally inclines to grass-green and mountain-green. The colours are often disposed in concentric delineations, and are varied with dark coloured dendritic markings.

It occurs massive, disseminated, and in membranes; most frequently reniform and botryoidal, frequently tuberose, stalactitic, fruticose, and also cellular, amorphous, in supposititious crystals, from azure copper-ore and red copper-ore; and crystallised in rather oblique four-sided prisms, which are set on the lateral planes.

Externally

Externally it is rough and dull.

Internally it passes, according to the kind of fracture, from glistening through glimmering to dull, but it is most commonly glimmering, and the lustre is silky.

The fracture is sometimes extremely delicate and scopiform fibrous, which passes into even, and this into flat and small conchoidal, which latter sometimes inclines to small-grained uneven.

The fragments are indeterminate angular, and rather sharp-edged.

It occurs almost always in thin lamellar distinct concretions, which are bent in the direction of the external surface; also in large, coarse and small granular distinct concretions.

The surface of the concretions is rough and dull, and apparently covered with a thin greenish-white film. The intensity of the colour is different in the individual lamellar distinct concretions, and hence this subspecies has usually a striped aspect.

It is opaque, or very faintly translucent on the edges.

It is soft, passing into semi-hard.

It is rather brittle.

It is easily frangible.

It retains its colour in the streak, only becomes somewhat paler.

It is not particularly heavy.

Specific gravity, 3.500, 3.994, *Muschenbröck*. 3.653, *Kirwan*. 3.6412, *Brisson*.

Its chemical characters and constituent parts are nearly the same with the preceding subspecies.

Geognostic

Geognostic Situation.

It occurs in veins, which traverse primitive, transition, and flœtz rocks.

Geographic Situation.

Europe.—In the copper-mines of Huel-Carpenter and Huel-Husband, in Cornwall; in the copper-mines of Aardal in Norway; in small quantity in veins in Lauterberg in the Hartz, along with copper-pyrites, fibrous malachite, copper-green, tile-ore, quartz, heavy-spar, and calcareous-spar; in many of the mines in the Saxon Erzgebirge, but in none of them in great quantity; also in several copper-mines in Silesia, and Hessia; at Schwatz in the Tyrol, in limestone; Herrengrund, Kasemarkt, Schmolnitz, and other places in Hungary; but most abundantly, and in greatest variety, in the mines of Moldawa, Oravicza and Saska, in the Bannat.

Asia.—In the mines of Kolywan, Gamashersk, Turja, &c. in Siberia, where the most beautiful and largest specimens of this mineral are met with, along with tile-ore, red copper-ore, brown ironstone, copper-glance or vitreous copper-ore, azure copper-ore, copper-green, white lead-ore, brown-spar, ironshot quartz, hornstone, &c. It is also met with in different parts of China.

Africa.—In the land of the Namaquas, in Southern Africa.

America.—In Maryland, Pennsylvania, Mexico, and St Christopher's, in North America: and in Chili in South America.

Use.

Uses.

Independent of its use as an ore of copper, it is also when pure employed as a green pigment. The compact varieties receive a beautiful polish, and present an agreeable colour, and hence are used in jewellery.

Third Subspecies.

Foliated Malachite.

Blättriger Malachit, *Leonhard.*

Id. Leonhard, in Leonhard & Selb's Mineralogische Studien, b. i. s. 3.

External Characters.

Its colour is leek-green, varying in intensity, and sometimes approaching to grass-green.

Its external shape is massive, disseminated, and crystallised in rather oblique four-sided tables.

The crystals are small and very small, and are attached by their lateral planes, so that they form small druses.

It is shining and splendent, and is either pearly or vitreous.

The fracture is foliated, with a simple cleavage.

The fragments are tabular.

It occurs in small angulo-granular concretions.

It is translucent.

It is semi-hard, approaching to soft.

It is brittle, but in a low degree.

The streak is greenish-white.

Chemical Characters.

It effervesces briskly with nitric acid.

Constituent Parts.

It would appear, from a letter of Bucholz to Leonhard, that this mineral agrees with the preceding subspecies of malachite in composition; but Hauy is rather inclined to consider it as a mixture of Carbonate of Copper and Phosphate of Copper.

Geognostic and Geographic Situations.

It has been hitherto found only at Rheinbreitenbach, associated with phosphate of copper, copper-green, tile-ore, compact and fibrous malachite, and ochry brown ironstone.

Fourth Subspecies.

Earthy Malachite.

Erdiger Malachit, *Hausmann.*

Χρυσοκολλα, *Theophr.* 46, 47.—Chrysocolla, *Plin.* Hist. Nat. xxxiii. 26. 30.—*G. Agricola,* de Natura Fossilium.—*Lehman's* Ueb. iii. 173.—*Beckmann,* in Aristot. Mirab. p. 124.—*Schwarze,* De quodam Pseudo-Smaragdorum apud veteres genere, Gorl. 1803.—Cuivre carbonaté vert pulverulent, *Hauy,* t. iii. p. 573.—Cuivre Malachite chrysocollé, *Brong.* t. ii. p. 223.

External Characters.

Its colours are emerald-green and verdigris-green.

It

It occurs massive, disseminated, in crusts, membranes, and sometimes stalactitic.

It is dull.

The fracture is earthy.

It is very soft, bordering on friable.

Geognostic and Geographic Situations.

It occurs along with the other subspecies of malachite at Sandlodge in Mainland, Zetland; also at Lauterberg in the Hartz; in Mansfeldt; Saxony; Silesia; Hungary; the Bannat; and Siberia.

Uses.

The ancients appear to have employed this mineral, along with silver and gold, in the soldering of gold: hence the name *Chrysocolla* given to it by Theophrastes and Pliny.

Observations on the Species.

1. The name of the species is derived from the word μαλάχη, "*malva*," the "*marsh-mallow*," the colour of which malachite resembles. The Greek word is sometimes corruptly written μολόχη, whence Pliny has derived the term *molochites*: "Non translucet molochites, spissius virens et crassius quàm smaragdus, a colore *malvæ* nomine accepto."

2. The Compact and Fibrous subspecies are distinguished from *Uran-Mica*, by fracture, and by their effervescence in nitric acid; and the Foliated subspecies from the same mineral also by its effervescence during solution in nitric acid: from *Green Lead-ore*, by the deep green colour of its powder, the powder of the green lead-ore

 being

being yellowish-green: from *Atacamite* or *Muriate of Copper*, by its effervescence during solution in nitric acid: from the *Arseniates of Copper*, by the same circumstance, besides which, the arseniates of copper emit a strong smell of garlic under the action of the blowpipe, which malachite does not.

3. The finest specimens of European malachite, are those of the Tyrol and the Bannat; but they do not equal, either in size or magnificence, the Siberian. M. Patrin observed at St Petersburgh a plate of malachite about 32 inches long by 17 inches broad, which was valued at 20,000 livres.

4. The bones and teeth of animals which are coloured with malachite, azure copper-ore, or phosphate of copper, are so hard as to receive a high polish. These are named *Turquoises*, because the first specimens of this kind were brought from Turkey: they are also found in Persia, and in France, and are esteemed as ornamental articles. These turquoises, however, must not confounded with the Turquois of Blumenbach, which is an earthy mineral found in small kidneys in beds of clay in Eastern Persia, and composed, according to Dr John, of 73 of alumina, 18 of water, 4.50 of oxide of copper, and 4 of oxide of iron. We still want an accurate account of this mineral.

14. Brown

14. Brown Copper-Ore or Anhydrous Carbonate of Copper.

Analysis of a new species of Copper-ore by *Dr Thomson*, Phil. Trans. for 1814.

External Characters.

Its colour, when pure, is dark blackish-brown; but it is very generally intermixed with malachite and red copper-ore, so that the colour appears a mixture of green, red, and brown, sometimes one and sometimes another prevailing. Small green veins of malachite likewise traverse it in different directions.

It occurs massive, with numerous imbedded small rock-crystals.

Its lustre is glimmering and resinous.

The fracture is small conchoidal, and sometimes inclining to foliated.

It is soft, being easily scratched by the knife.

It is sectile.

The streak is reddish-brown.

Specific gravity, 2.620, *Thomson.*

Chemical Characters.

It effervesces in acids, and dissolves, letting fall a red powder. The solution is green or blue, according to the acid, indicating that it consists chiefly of copper.

 Constituent

Constituent Parts.

Carbonic Acid, - -	16.70
Per-oxide of Copper, -	60.75
Per-oxide of Iron, - -	19.50
Silica, - - - -	2.10
Loss, - - - -	0.95
	100.00

Thomson, in Phil. Trans. for 1814.

Geognostic Situation.

It appears to occur in nests in primitive rocks, which are of greenstone, or some similar rock of the primitive trap series subordinate to mica-slate. It is associated with malachite.

Geographic Situation.

In the peninsula of Hindostan, near the eastern border of the Mysore country.

Observations.

This mineral was discovered by Dr Benjamin Heyné, about the year 1800, in the Mysore country, and was first described and analysed by Dr Thomson in 1813.

15. Copper-

15. Copper-Green.

Kupfergrün, *Werner.*

Id. Werner, Pabst. b. i. s. 96. *Id. Wid.* s. 772.—Mountain Green, *Kirw.* vol. ii. p. 134.—Kupfergrün, *Estner,* b. iii. s. 595. *Id. Emm.* b. ii. s. 260.—Le vert de Cuivre, ou la Chrysocolle, *Broch.* t. ii. p. 203.—Kupfergrün, *Reuss,* b. iii. s. 477. *Id. Lud.* b. i. s. 231. *Id. Suck.* 2ter th. s. 210. *Id. Bert.* s. 393. *Id. Mohs,* b. iii. s. 287. *Id. Leonhard,* Tabel. s. 59.—Kiesel Malachit, *Haus.* Handbuch, b. iii. s. 1028.—Chrysocolla, *Aikin,* p. 91. 2d edit.

External Characters.

Its principal colour is verdigris-green, of different degrees of intensity, which in some varieties passes into emerald-green, and in others inclines to leek-green.

It occurs massive, disseminated, and coating or incrusting malachite, and sometimes stalactitic, small reniform, and small botryoidal.

Internally it is shining, passing into glistening, and the lustre is resinous.

The fracture is small conchoidal.

The fragments are indeterminate angular, and more or less sharp-edged.

It is more or less translucent on the edges, and translucent in thin pieces.

It is soft and very soft.

It is easily frangible, and rather brittle.

Specific gravity, 2.371, *Ullmann,* in Leonhard's Taschenbuch, b. viii. s. 504.

Chemical Characters.

Before the blowpipe, it becomes first black, then brown, but is infusible: on the addition of borax, it melts rapidly, and effervesces, tinging the flame green, and is reduced to the metallic state. In diluted muriatic acid, it effervesces slightly; the oxide of copper dissolves, and there remains behind a nearly colourless and often semi-gelatinous mass of silica, of the same size as the original specimen.—*Aikin.*

Constituent Parts.

Copper, - -	40.00	42.00
Oxygen, - -	10.00	7.63
Carbonic Acid, -	7.00	3.00
Water, - -	17.00	17.50
Silica, - -	26.00	28.37
Sulphate of Lime, -		1.50
	100.00	100.00
	Klaproth, Beit. b. iv. s. 36.	*John*, Chem. Unters. b. ii. s. 260.

Geognostic Situation.

It is met with in the same geognostic situations as malachite, and is usually associated with copper-pyrites, tile-ore, grey copper-ore, malachite, brown ironstone, and other ores.

Geographic Situation.

Europe.—It occurs in Cornwall, along with oliven-ore, and also in the vale of Newlands, near Keswick. It is found at Zinwald, along with tinstone, wolfram, tungsten, copper, iron, and arsenical pyrites, copper-glance or vitreous copper-ore, galena or lead-glance, fluor-spar, and quartz; at Spitz

Spitz in Austria, with tile-ore; Falkenstein in the Tyrol, in compact limestone, with grey copper-ore, and earthy tile-ore; in the Bannat, where it formerly occurred in great beauty, along with azure copper-ore, malachite, copper-black or black oxide of copper, foliated red copper, tile-ore, and copper-pyrites, in clay-porphyry; at Herrngründe in Lower Hungary, along with grey copper-ore, malachite, and selenite, in grey-wacke; in Upper Hungary, associated with tile-ore, compact malachite, compact red copper-ore, and white lead ore.

Asia.—In Siberia, along with azure copper-ore, compact and foliated red copper-ore, native copper, copper-black or black oxide of copper, malachite, tile-ore, iron-ochre, and brown-spar, in sandstone; and in China, along with tile-ore.

America.—In Mexico, along with red copper-ore, malachite, and azure copper-ore.

Observations.

1. It is distinguished from *Malachite*, by colour, lustre, fracture, and translucency; its feeble effervescence with acids also distinguishes it from that mineral; and it is distinguished from certain varieties of *Steatite*, by its brittleness, and easy frangibility.

2. John, in his Chem. Unters. b. ii. s. 252, gives the following account of a mineral allied to Copper-green, and which he considers to be a new species:—

Siliceous Copper-ore.

Kieselkupfer, *John.*

Its colours are asparagus-green, and celandine-green, inclining to sky-blue.

It occurs in crusts.

It is dull, or faintly glistening, and resinous.

The fracture is even or earthy.

It

It is opaque, and rarely translucent on the edges.

It is soft.

Constituent Parts.—Copper, 37.8. Oxygen, 8. Water, 21.8. Silica, 29; and Sulphate of Lime, 3.

Observations.—It is nearly allied in chemical composition to the Emerald Copper-ore, if we are to rely on the analysis of that mineral by Lowitz.

16. Ironshot Copper-Green.

Eisenschüssiges Kupfergrün, *Werner*.

It is divided into two subspecies, viz. Earthy Ironshot Copper-green, aud Slaggy Ironshot Copper-green.

First Subspecies.

Earthy Ironshot Copper-Green.

Erdiches eisenschüssiges Kupfergrün, *Werner*

Id. Werner, Pabst. b. i. s. 96. *Id. Wid.* s. 773.—Earthy Ironshot Mountain-green, *Kirw.* vol. ii. p. 151.—Erdiches eisenschüssiches Kupfergrün, *Estner*, b. iii. s. 605. *Id. Emm.* b. ii. s. 262.—Le vert de Cuivre ferrugineux terreux, *Broch.* t. ii. p. 205.—Erdiges eisenschussig-Kupfergrün, *Reuss*, b. iii. s. 482. *Id. Lud.* b. i. s. 232. *Id. Suck.* 2ter th. s. 210. *Id. Bert.* s. 393. *Id. Mohs*, b. iii. s. 290. *Id. Leonhard*, Tabel. s. 59.—Cuivre Malachite ferrugineux terreux, *Brong.* t. ii. p. 223.—Erdiges eissenschüssiges Kupfergrün, *Karsten*, Tabel. s. 62.—Eeissenschüssiger erdiger Malachit, *Haus.* s. 135.—Cuivre hydraté siliciferc compacte, *Vauquelin*, Journ. de Mines, t. xxxiii. p. 341.

External Characters.

Its colour is olive-green, which sometimes passes into pistachio-green.

It

It occurs massive, disseminated, and in crusts.

It is dull.

The fracture is earthy.

It soils slightly.

The fragments are indeterminate angular, and blunt-edged.

It is opaque.

It retains its colour in the streak, only it becomes somewhat paler.

It is very soft, passing into friable.

It is rather brittle.

It feels meagre.

It is easily frangible.

It is rather heavy.

Second Subspecies.

Slaggy Ironshot Copper-Green.

Schlackiges eisenschüssiges Kupfergrün, *Werner.*

Id. Werner, Pabst. b. i. s. 97. *Id. Wid.* s. 775.—Glassy Ironshot Mountain-green, *Kirw.* vol. ii. p. 152.—Schlackiges eisenschüssiges Kupfergrün, *Estner,* b. iii. s. 606. *Id. Emm.* b. ii. s. 263.—Le Vert de Cuivre ferrugineux scoriacé, *Broch.* t. ii. p. 206.—Schlackiges eisenschüssiges Kupfergrün, *Reuss,* b. iii. s. 483. *Id. Lud.* b. i. s. 232. *Id. Suck.* 2ter th. s. 211. *Id. Bert.* s. 394. *Id. Mohs,* b. iii. s. 290. *Id. Leonhard,* Tabel. s. 62.—Cuivre Malachite ferrugineux resineuse, *Brong.* t. ii. p. 224.—Schlackiges eisenschüssiges Kupfergrün, *Karsten,* Tabel. s. 62.—Muschlicher Pharmakochalzit, *Haus.* s. 136. —Cuivre hydraté silicifere resinite, *Vauquelin,* Journ. des Mines, t. xxxiii. p. 341.

External Characters.

Its colours are dark olive and pistachio green, which latter

latter passes into dark blackish-green, verging on greenish-black.

It occurs massive and disseminated.

Internally it is shining and glistening, and the lustre is resinous.

The fracture is small conchoidal.

The fragments are indeterminate angular, and more or less sharp-edged.

It is opaque.

It becomes paler in the streak.

It is soft, verging on very soft.

It is rather brittle.

It is easily frangible.

It is rather heavy.

Constituent Parts.

According to Vauquelin, both subspecies of this mineral are compounds of Oxide of Copper, Silica, and Water.

Geognostic Situation.

Both subspecies usually occur together, and they frequently pass into each other. They are usually accompanied with copper-green, azure copper-ore, and malachite; frequently also with grey copper-ore, foliated copper-glance, tile-ore, ochry and compact brown ironstone, compact red copper-ore, quartz, and straight lamellar heavy-spar.

Geographic Situation.

Europe.—It occurs in Cornwall, along with oliven-ore; at Saalfeldt in Thuringia, it is associated with malachite, azure copper-ore, copper-green, copper-pyrites, grey copper-ore, per-ore, yellow and brown cobalt ochre, red cobalt-ochre, and straight lamellar heavy-spar; at Lauterberg in the Hartz, along with azure copper-ore, malachite, and grey copper-ore; at Schwatz in the Tyrol, along with foliated copper-glance, copper-green, fibrous malachite, and azure copper-ore; at Saska in the Bannat, along with copper-green and red copper-ore.

Asia.—In the Gumashevsk mines in Siberia, associated with compact and ochry brown ironstone, tile-ore, malachite, azure copper-ore, grey copper-ore, red copper-ore, white and yellow lead-ores, and native silver, with quartz; also in other mines in Siberia.

America.—Chili.

Observations.

Mr Kirwan suspects, from its olive-green colour, that it may contain arsenic acid. Heergen says, that he convinced himself of the presence of this acid in many varieties, and proposes to consider it as a subspecies of oliven-ore *; but the experiments of Vauquelin already mentioned, shew that it cannot be considered as a phosphate of copper.

17. Emerald

* Heergen, Descripcion y annuncio de varios Minerales del Regno de Chile; and in the Anales de ciencias naturales, mes. d. Julio 1801, n. 11. t. 4. p. 198.; also in Von Moll's Annalen der Berg et Huttenkünde, 1r B. 2te Lieferung, s. 150.

17. Emerald Copper-Ore.

Kupfer-Schmaragd, *Werner*.

Emeraudine, *Lam.* t. ii. p. 230.—Dioptase, *Hauy,* t. iii. s. 136. *Id. Broch.* t. ii. p. 511.—Achirite, *Hermann,* in Nov. Act. Petrop. xiii. 339.—Kupferschmaragd, *Reuss,* b. iii. s. 472. *Id. Lud.* b. i. s. 233. *Id. Mohs,* b. iii. s. 297. *Id. Leonhard,* Tabel. s. 69.—Cuivre dioptase, *Brong.* t. ii. p. 225.—Dioptase, *Brard,* p. 161. *Id. Karsten,* Tabel. s. 62. *Id. Haus.* s. 136.—Cuivre dioptase, *Hauy,* Tabl. p. 91.—Emerald Copper, *Aikin,* p. 37.

External Characters.

Its colour is emerald-green.

It occurs crystallised in six-sided prisms, which are rather acutely acuminated on both extremities by three planes, which are set on the alternate lateral edges.

The lateral planes are smooth.

Externally and internally it is shining, and the lustre is intermediate between vitreous and pearly.

It has a threefold cleavage, and the folia are parallel to the faces of an obtuse rhomboid, of which, however, only one is very distinct.

It is translucent, passing to semi-transparent.

It scratches glass feebly.

It is brittle.

Specific gravity, 2.850, *La Metherie.* 3.300, *Hauy.*

Chemical Characters.

It becomes of a chesnut-brown colour before the blowpipe, and tinges the flame green, but is infusible; with borax it gives a bead or globule of copper.

Constituent

Constituent Parts.

Oxide of Copper,	28.57	Oxide of Copper,	55
Carbonate of Lime,	42.85	Silica, - -	33
Silica, - -	28.57	Water, - -	12
	99.99		100
	Vauquelin, in Hauy, t. iii. p. 137.		*Lowitz*, in Nova Acta Petrop. xiii.

Geognostic and Geographic Situations.

It is found, according to Hermann, in the land of the Kirguise, 125 leagues from the Russian frontier, where it is associated with fibrous and compact malachite, calcareous-spar, and limestone.

Observations.

1. Vauquelin's analysis of this mineral was made with a very small quantity, not many grains; whereas that of Lowitz was made with a considerable quantity: hence it is probable that the latter analysis is the most correct. In this view, the Siliceous Copper-ore of John, mentioned at p. 169. will prove to be a subspecies of Emerald Copper-ore.

2. It was brought to Petersburgh about twenty-seven years ago by General Bogdanof, who obtained it from the discoverer, Achir Mahmed, a Bucharian merchant.

18. Muriate

18. Muriate of Copper or Atacamite.

Salzkupfererz, *Werner.*

This species is divided into two subspecies, viz. Compact and Arenaceous.

First Subspecies.

Compact Muriate of Copper.

Festes Salzkupfererz, *Werner.*

Cuivre muriaté massif, *Brong.* t. ii. p. 228.—Gemeines Salzkupfererz, *Karsten,* Tabel. s. 64.—Cuivre muriaté, *Hauy,* Tabl. p. 89.—Blättricher & Strahliger Smaragdochalzit, *Haus.* Handbuch, b. iii. s. 1039.—Muriate of Copper, *Aikin,* p. 34.

External Characters.

Its colours are emerald-green, leek-green, olive-green, and blackish-green.

It occurs massive, disseminated, in crusts, or investing; and in short needle-shaped crystals, of the following forms:

1. Oblique four-sided prism, bevelled on the extremities, the bevelling planes set on the acute lateral edges.
2. The preceding figure, in which the acuter lateral edges are deeply truncated, thus forming a six-sided prism.

Internally it is shining and glistening, inclining to resinous.

The

The fracture is radiated, which passes on the one side into fibrous, on the other into foliated.

The fragments are indeterminate angular.

It sometimes occurs in small granular distinct concretions.

It is translucent on the edges, or translucent.

It is soft.

It is brittle.

It is easily frangible.

Specific gravity, 4.4.

Chemical Characters.

It tinges the flame of the blowpipe of a bright green and blue, muriatic acid rises in vapours, and a bead of copper remains on the charcoal. It is soluble in nitric acid without effervescence.

Constituent Parts.

Oxide of Copper,	73.0	76.595
Water, - -	16.9	12.767
Muriatic Acid, -	10.1	10.638
	100.0	100.000
	Klaproth, Beit. b. iii. s. 200.	*Proust*, in Journ. de Phys. t. 50. p. 63.

Geognostic and Geographic Situations.

It occurs in veins at Remolinos in Chili, along with red copper-ore, malachite, brown ironstone, selenite, rock-crystal, and calcedony; and in Peru, along with silver-glance, corneous silver-ore, and calcareous-spar.

It is said also to occur in Nassau *; and a variety of it is met with, incrusting the lavas of Vesuvius, particularly those of the years 1804 and 1805 †.

Observations.

This mineral was first brought from Chili to Europe, by an eminent mineral-dealer, and zealous and liberal promoter of the study of mineralogy, Mr Heuland of London.

Second Subspecies.

Arenaceous Muriate of Copper, or Copper-Sand.

Kupfersand, *Werner.*

Cuivre muriaté pulverulent, *Hauy,* t. iii. s. 561. *Id. Brong.* t. ii. p. 229.—Sandiges Salzkupfer, *Karsten,* Tabel. s. 64—Cuivre muriaté pulverulent, *Hauy,* Tabl. p. 89.—Sandiger Smaragdochalzit, *Haus.* Handbuch, b. iii. s. 1040.—Muriate of Copper in form of sand, *Aikin,* p. 34.

External Characters.

Its colour is emerald-green, inclining to grass-green.

It occurs in scaly particles, which are shining and glistening.

It does not soil.

It is translucent.

Constituent

* According to the report of Professor Ullmann.

† Dr Thompson of Naples.

Constituent Parts.

Oxide of Copper, -	63	70.5
Water, - - -	12	18.1
Muriatic Acid, -	10	11.4
Carbonate of Iron, -	1	
Mixed Siliceous Sand,	11	
	97	100.0
	La Rochefoucault, *Berthollet*, and *Fourcroy*, Mem. de l'Acad. 1786, p 158.	*Proust*, Journ de Phys t. 50. p. 63.

Geognostic and Geographic Situations.

It is found in the sand of the river Lipes, 200 leagues beyond Copiapu, in the desart of Atacama, which separates Chili from Peru.

Observations.

It was brought from South America by the traveller Dombey.

19. Phosphate of Copper.

Phosphorkupfererz, *Werner.*

Phosphorsaures Kupfer, *Karsten,* in d. N. Schriften der Berlin, Ges. Natf. Fr. b. iii. s. 304.—Cuivre phosphaté, *Broch.* t. ii. p. 544.—Phosphorsaures Kupfer, *Leonhard,* Tabel. s. 61 — Cuivre phospaté, *Brong.* t. ii. p. 227.—Phosphor Kupfer, *Karsten,* Tabel. s. 64 —Cuivre phosphaté, *Hauy,* Tabl p. 92. —Pseudo-malachit, *Haus.* Handb. b. iii. s. 1035.—Phosphat of Copper, *Aikin,* p. 34.

This

This species is divided into three subspecies, viz. Foliated Phosphate of Copper, Fibrous Phosphate of Copper, and Compact Phosphate of Copper.

First Subspecies.

Foliated Phosphate of Copper.

Cuivre phosphaté rhomboidal, *Hauy,* Tabl. p. 92.—Blättriches pseudo-malachite, *Haus.* Handbuch, b. iii. s. 1036.

External Characters.

Externally the colour is greyish-black: internally between emerald-green and verdigris-green; also leek-green and olive-green.

It occurs crystallised in the following figures, viz.

1. Octahedron.
2. Cuneiform or lengthened octahedron.
3. Rhomboid, with small curvilinear faces, and which is sometimes truncated on the edges and angles.

The crystals are small and very small, and form drusy coverings.

Externally smooth; and sometimes incrusted with copper-green.

Internally it is shining and splendent, and vitreous, inclining to pearly.

The longitudinal fracture is imperfect and small conchoidal; the cross fracture uneven, passing into small conchoidal.

The fragments are indeterminate angular, and not particularly sharp-edged.

It is translucent, passing into semi-transparent.

It

It is semi-hard.
It is greyish-white in the streak.
It is brittle.
It is easily frangible.
Specific gravity, 3.5142, *Kopp.*

Chemical Characters.

It is insoluble in water; but dissolves in nitric acid, without effervescence. On the first impression of heat, it fuses into a brownish globule, which, by the further action of the blowpipe, extends on the surface of the charcoal, and acquires a reddish-grey metallic colour. The brownish globule, on cooling, crystallises into three sided and six sided facets.

Constituent Parts.

According to Bucholz, it is a compound of Copper and Phosphoric Acid *.

Geognostic and Geographic Situations.

It is found at Liebethen, in the neighbourhood of Neusohl in Hungary, in an ironshot quartz, along with copper-green and malachite, in a rock apparently of clay-slate: also at Virneberg, near Rheinbreitenbach on the Rhine, in veins, along with capillary red copper-ore, native copper, tile-ore, malachite, and other ores of copper, and associated with quartz.

* Vid. Leonhard and Selb's Studien, b. i. s. 89.

Second Subspecies.

Fibrous Phosphate of Copper.

Cuivre phosphaté mamelonné fibreux, *Hauy,* Tabl. p. 92.—Phosphorsaures Kupfer, *Leonhard,* in d. Schriften d. Wetterauischen Gesellsch. b. i. s. 83.—Fasriger pseudo-malachit, *Haus.* Handbuch, b. iii. s. 1037.

External Characters.

Its colour is emerald-green, inclining to verdigris-green.

It occurs massive, disseminated, in crusts, small botryoidal, and imperfectly crystallised.

Internally it alternates from shining to glimmering, and the lustre is intermediate between resinous and pearly.

The fracture is diverging fibrous, sometimes passing into radiated.

It sometimes occurs in lamellar concretions.

It is faintly translucent on the edges; seldom translucent.

Geographic Situation.

It occurs at the Virneberg, near Rheinbreitenbach on the Rhine.

Third

Third Subspecies.

Compact Phosphate of Copper.

Cuivre phosphaté compacte, *Hauy,* Tabl. p. 92.—Dichtes phosphorsauer Kupfer, *Jordan's* Reisebem. 217.—Dichtes Pseudomalachit, *Haus.* Handbuch, b. iii. s. 1037.

External Characters.

Its colour is emerald-green, inclining to verdigris-green.

It occurs massive, disseminated, in crusts or investing, reniform, and globular.

Internally it is glimmering or dull, and the lustre is resinous.

The fracture is flat conchoidal, inclining to even, small splintery, and sometimes very faintly fibrous.

It sometimes occurs in indistinct lamellar concretions.

Constituent Parts.

Oxide of Copper,	-	68.13
Phosphoric Acid,	-	30.95
		99.08

Klaproth, Beit. b. iii. s. 206.

Geognostic and Geographic Situations.

It is met with at the Virneberg, near Rheinbreitenbach on the Rhine, in veins in grey-wacke, along with quartz, calcedony, hornstone, red copper-ore, native copper, malachite, and azure copper-ore; also as Liebethen, near Neusohl in Hungary, in ironshot-quartz.

 20. Copper-Mica.

20. Copper-Mica or Micaceous Copper-Ore.

Kupferglimmer, *Werner*.

Blättriges Olivenerz, *Karsten,* Journ. de Phys. an 10. p. 348. —Arseniate of Copper in hexaedral laminæ, with inclined sides, *Bournon,* Phil. Trans. part i. 1801.—Blättriches Olivenerz, *Reuss,* b. iii. s. 504.—Kupferglimmer, *Mohs,* b. iii. s. 294. —Cuivre arseniaté lamelliforme, *Brong.* t. ii. p. 230.—Kupferglimmer, *Karsten,* Tabel. s. 64.—Cuivre arseniaté lamelliforme, *Hauy,* Tabl. p. 90.—Hexaedral Arseniate of Copper, *Aikin,* p. 32.

External Characters.

Its colour is emerald-green, which in some varieties inclines to verdigris-green.

It occurs massive, disseminated, seldom crystallized in very thin six-sided tables, which are bevelled on the terminal planes.

Externally it is smooth and splendent.

Internally it is splendent, and the lustre is pearly, inclining to metallic.

The fracture is perfect foliated, with a single cleavage; the cross fracture small-grained uneven, inclining to conchoidal.

The fragments are indeterminate angular and tabular.

It occurs in coarse and small granular distinct concretions.

The massive varieties are translucent; the crystallised transparent.

It is soft; scratches gypsum slightly, but does not affect calcareous-spar.

It

It is sectile.
It is rather brittle.
Specific gravity, 2.5488.

Chemical Characters.

It decrepitates before the blowpipe; and passes, first to the state of a black spongy scoria, after which it melts into a black globule, of a slightly vitreous appearance.

Constituent Parts.

Oxide of Copper, -	39	58
Arsenic Acid, -	43	21
Water, - - -	17	21
	99	100
	Vauquelin, Journ. des Mines, N. 55. p. 562.	*Chenevix*, Phil. Tr. for 1801, p. 201.

Geognostic and Geographic Situations.

It has been hitherto found only in veins in the copper-mines in Cornwall, where it is accompanied with red copper-ore, copper-pyrites, copper-glance or vitreous copper-ore, variegated copper-ore, copper-black or black oxide of copper, compact and fibrous malachite, ironshot copper-green, azure copper-ore, indurated tile-ore, oliven-ore, and brown iron-ochre.

21. Lenticular

21. Lenticular Copper-Ore.

Linsenerz, *Werner*.

Arseniate of Copper, in the form of an obtuse octahedron, *Bournon*, Phil. Trans. part i. 1801.—Linsenerz, *Mohs*, b. iii. s. 292. —Cuivre arseniaté obtus, *Brong*. t. ii. p. 230.—Linsenerz, *Karsten*, Tabel. s. 64.—Cuivre arseniaté primitif, *Hauy*, Tabl. p. 90.—Octahedral Arseniate of Copper, *Aikin*, p. 32.

External Characters.

Its colour is sky-blue, which sometimes inclines to Berlin-blue; also verdigris-green, grass-green, apple-green, and bluish-white.

It occurs crystallised, in small and very small, and very obtuse octahedrons, that sometimes terminate in a line.

Externally it is shining; internally it is glistening and vitreous.

The fracture is foliated, in the direction of the planes.

It is translucent; seldom semi-transparent.

It is semi-hard; it is harder than calcareous-spar, but not so hard as fluor-spar.

It is rather brittle.

It is uncommonly easily frangible.

Specific gravity, 2.8819, *Bournon*.

Constituent Parts.

Oxide of Copper, - -	49
Arsenic Acid, - -	14
Water, - - -	35
	98

Chenevix, Phil. Trans. for 1801.

Geognostic

Geographic Situation.

It has been hitherto found only in Cornwall, where it is associated with the preceding subspecies, and many other ores of copper.

22. Oliven-Ore or Olive Copper-Ore.

Olivenerz, *Werner.*

This species is divided into four subspecies, viz. Prismatic Oliven-ore, Trihedral Oliven-ore, Fibrous Oliven-ore, and Earthy Oliven-ore.

First Subspecies.

Prismatic Oliven-Ore.

Blättriches Olivenerz, *Werner.*

Arseniate of Copper, in the form of an acute octahedron, *Bournon,* Phil. Trans. part i. for 1801.—Cuivre arseniaté aigue, *Brong.* t. ii. p. 231.—Dichtes Olivenerz, *Karsten,* Tabel. s. 64.—Cuivre arseniaté octaedre aigue, *Hauy,* Tabl. p. 91.—Cuivre arseniaté en prisme tetraedre rhomboidal, *Bournon,* Catalogue Mineralogique, p. 254.—Gemeines Oliven Kupfer, *Haus.* Handbuch, b. iii. s. 1045.—Prismatic Arseniate, *Aikin,* p. 32.

External Characters.

Its colour is perfect olive-green, which passes on the one side into dark leek-green, and blackish-green; on the other into pale leek-green, siskin-green, and sulphur-yellow.

It seldom occurs massive, usually in drusy crusts; and in small crystals, which present the following varieties of form:

1. Rhomboidal

1. Rhomboidal four-sided prism, in which the lateral planes meet under angles of 96° and 84°, bevelled on the extremities, the bevelling planes set on the acute edges.
2. The preceding figure, in which the obtuser edges are more or less deeply truncated.

The planes of the crystals are smooth, shining, and splendent.

Internally it is shining or glistening, and the lustre is vitreous, inclining to resinous.

The fracture is imperfect foliated ; the cross fracture is small conchoidal.

The fragments are indeterminate angular.

The massive varieties occur in coarse and small granular distinct concretions.

It is translucent, but the crystals are sometimes transparent.

It yields a straw-yellow coloured streak.

It scratches fluor-spar and heavy-spar, but does not affect glass.

It is sectile.

Specific gravity, 4.280, *Bournon.*

Chemcial Characters.

Before the blowpipe, it first boils, and then gives a hard reddish-brown scoria.

Constituent Parts.

Oxide of Copper, -	60.0
Arsenic Acid, - -	39.7
	99.7

Chenevix, Phil. Trans. 1801.

Geognostic

Geognostic and Geographic Situations.

It has been hitherto found only in the copper-mines of Cornwall.

Second Subspecies.

Trihedral Oliven-Ore.

Nadelförmiges Oliven-Kupfer, *Hausmann.*

Arseniate of Copper, in the form of a trihedral prism, *Bournon,* Phil. Trans. 1801.—Cuivre arseniaté trihedre, *Brong.* t. ii. p. 231.—Dichtes Olivenerz, *Karsten,* Tabel. s. 64.—Cuivre arseniaté prismatique triangulaire, *Hauy,* Tabl. p. 91.—Cuivre arseniaté en prisme triedre, *Bournon,* Catalogue Mineralogique, p. 257.—Nadelformiges Olivenkupfer, *Haus.* b. iii. s. 1046.—Trihedral Arseniate, *Aikin,* p. 33.

External Characters.

Its colour is deep celandine-green; but owing to its ready oxidation, it has generally a blackish superficial tarnish.

It occurs massive; and crystallised in the following figures:

1. Three-sided prism, which is sometimes truncated on one of the edges: when the truncations become deep, there is formed a
2. Four-sided prism, which is flat when two opposite planes become very broad.
3. Acute rhomboid, sometimes truncated on the diagonally opposite angles.
4. Irregular octahedron.

5. When

5. When two elongated tetrahedral prisms are joined together by one of the sides of the prism, a twin-crystal is formed, which may be viewed as a tetrahedral prism of 60° and 120°. When two opposite and lateral edges of this figure are deeply truncated, the prism becomes six-sided.

The lustre is intermediate between vitreous and resinous.

The fracture is foliated in the direction of the planes of the three-sided prism, but is uneven in every other direction.

It is transparent or semi-transparent.

It scratches calcareous-spar with difficulty, but does not affect fluor-spar or heavy-spar.

Specific gravity, 4.280.

Chemical Characters.

Before the blowpipe, it flows like water, and in cooling crystallises in small rhomboidal plates of a brown colour.

Constituent Parts.

Oxide of Copper, -	54	50.62
Arsenic Acid, - -	30	45.00
Water, - - -	16	3.50
	100	99.12
	Chenevix, in Phil. Trans for 1801.	*Klaproth*, Beit. b. iii. s. 192.

Geognostic and Geographic Situations.

It occurs in the Cornish mines along with the other subspecies.

Third

Third Subspecies.

Fibrous Oliven-Ore.

Fasriges Olivenerz, *Werner.*

Hæmatitiform and Amianthiform Arseniate, *Bournon,* Phil. Trans. for 1801.—Cuivre arseniaté capillaire et mamelonné, *Brong.* t. ii. p. 231, 232.—Fasriges Olivenerz, *Karsten,* Tabel. s. 64.—Cuivre arseniaté mamelonné fibreux, *Hauy,* Tabl. p. 91.—Cuivre arseniaté en petites masses habituellement fibreuses et mamelonnées, *Bournon,* Catalogue Mineralogique, p. 259.—Fasriges Oliven Kupfer, *Haus.* Handbuch, b. iii. s. 1047.—Hæmatitic and Amianthiform Arseniate, *Aikin,* p. 83.

External Characters.

Its colour is olive-green of different degrees of intensity. The darker varieties border on blackish-green, the lighter pass into pistachio-green, liver-brown, wood-brown, and greenish-white.

The colours are sometimes arranged in spotted and striped delineations.

It occurs massive, reniform, and crystallised in capillary crystals.

The crystals are small and very small, and externally shining.

Internally the massive varieties are glistening or glimmering, with an adamantine lustre.

The fracture is delicate and scopiform fibrous.

The fragments are indeterminate angular, and wedge-shaped.

It

It occurs sometimes in coarse granular concretions, sometimes in curved lamellar concretions, which traverse the former; and such varieties have a strong resemblance to brown hematite and Cornish tin-ore.

It is translucent on the edges.

It scratches calcareous-spar with great difficulty.

It is rather brittle.

The fibres are sometimes flexible *.

Specific gravity, between 4.100 and 4.200, *Bournon.*

Constituent Parts.

	Amianthiform.	Hæmatitiform.
Oxide of Copper, -	50	50
Arsenic Acid, -	29	29
Water, - - -	21	21
	100	100

Chenevix, in Phil. Trans. for 1801.

Geographic Situation.

It has been hitherto found only in Cornwall, where it is associated with the other arseniates of copper, and various ores of copper.

Fourth

* The fibres are sometimes so delicate, so short, and so confusedly grouped together, that the whole appears like a dusty cottony mass, the true nature of which is discoverable only by the lens. At other times, this variety appears in thin laminæ, rather flexible, sometimes scarcely perceptible to the naked eye, sometimes tolerably large, and perfectly like Amianthus papyraceus.—*Bournon*, Phil. Trans for 1801, part i. p. 180.

Fourth Subspecies.

Earthy Oliven-Ore.

Cuivre arseniaté terreux, *Hauy,* Tabl. p. 91.—Erdiches Olivenkupfer, *Haus.* Handbuch, b. iii. s. 1049.

External Characters.

Its colours are olive-green, verdigris-green, and siskin-green.

It occurs massive, disseminated, and in crusts.

It is dull.

The fracture is fine earthy.

It sometimes occurs in concentric lamellar distinct concretions.

It is opaque.

It is soft and very soft.

Geognostic and Geographic Situations.

It occurs along with the other subspecies of oliven-ore in the copper-mines of Cornwall.

Observations.

It is now about thirty-five years since the Arseniate of Copper was discovered in Cornwall: it was first found, either in Carrarach mine, in the parish of Gwennap, or in Tincroft mine, in the parish of Allogan, and some years afterwards in the mine of Huel-Gorland.

23. Martial Arseniate of Copper.

Cupreous Arseniate of Iron, *Bournon,* Phil. Trans. for 1801, part i. p. 191.—Cuivre arseniaté ferrifere, *Brong.* t. ii. p. 232. —Strahlenkupfer, *Karsten,* Tabel. s. 64.—Cuivre arseniaté ferrifere, *Hauy,* Tabl. p. 91.—Strahlenkupfer, *Haus.* Handbuch, b. iii. s. 1050.—Martial Arseniate of Copper, *Aikin,* p. 33.

External Characters.

Its colour is pale sky-blue, or intermediate between sky-blue and celandine-green.

It occurs massive, flat-reniform, and crystallised in compressed oblique rhomboidal four-sided prisms, acuminated with four planes; sometimes the acute edges are truncated, when the prism appears six-sided, or all the lateral edges are truncated, when it appears eight-sided.

The crystals are generally very small, and grouped into rose-like or globular forms.

Internally the lustre is shining and pearly, inclining to vitreous.

The fracture is scopiform radiated.

It is transparent.

It scratches calcareous-spar with considerable facility, but does not affect fluor-spar or heavy-spar.

Specific gravity, 3.400.

Constituent

Constituent Parts.

Oxide of Copper, -	22.5
Oxide of Iron, - -	27.5
Arsenic Acid, - -	33 5
Water, - - -	12.0
Silica, - - - -	3.0
	98.5

Chenevix, in Phil. Trans. for 1801, part i. p. 220.

Geognostic and Geographic Situations.

This mineral is found in Muttrell mine, immediately contiguous to Huel-Gorland mine, also in Tincroft and Carrarach mines, in Cornwall, where it is associated with oliven-ore, grey copper-ore, copper-pyrites or yellow copper-ore, copper-glance or vitreous copper-ore, brown ironstone, arsenical pyrites, tinstone, and quartz; also at St Leonhard, in Haute-Vienne in France.

COPPER-MINES.

The most considerable copper-mines in the world, are the English: the next in importance are those of Russia, Austria, Sweden, and Westphalia as it was in the year 1808; and the least considerable are those of Denmark, France, Saxony, Prussia, and Spain.

TABLE of the ANNUAL QUANTITY OF COPPER raised from the Earth in different Countries.

	In Quintals,—the quintal valued at 100 lb.
1. England, - - - - - -	200,000
2. Russia, - - - - - - - -	67,000
3. Austria, including Bohemia, Gallicia, Hungary, Transylvania, Stiria, Carinthia, Carniola, Salzburg, Moravia, and Austria, - - - -	60,000
4. Sweden, - - - - - -	22,000
5. Kingdom of Westphalia in 1808, -	17,229
6. States of Denmark, - - - - -	8,500
7. Bavaria, including the Tyrol, - -	3,000
8. France, - - - - - - -	2,500
9. Saxony, in 1808, - - - - -	1,320
10. Prussia, in her state of abasement after the Treaty of Tilsit, - - - - -	337
11. Spanish European mines, - - - -	309
Total,	382,186

VI. ORDER

VI. ORDER.—IRON.

This Order contains nineteen species, viz. Native Iron, Iron-pyrites, Radiated Pyrites, Hepatic Pyrites, Magnetic Pyrites, Magnetic Ironstone or ore, Specular Iron-ore or Iron-glance, Red Ironstone, Brown Ironstone, Umber, Black Ironstone, Sparry Ironstone, Clay Ironstone, Bog Iron-ore, Pitchy Iron-ore, Blue Iron-ore, Chromate of Iron, Cube-ore or Arseniate of Iron, and Pyrosmalite.

1. Native Iron.

Gediegen Eisen, *Werner.*

This species is divided into two subspecies, Terrestrial Native Iron, and Meteoric Native Iron.

First Subspecies.

Terrestrial Native Iron.

Tellureisen, *Werner.*

Gediegen Eisen, *Charpentier,* Mineralogische Geographie von Sachsen, s. 343.—Fossiles gediegen Eisen, *Klaproth,* Beit. b. iv. s. 102.—Fer natif amorphe, *Hauy,* Tab. p. 93.—Massive Native Iron, *Aikin,* p. 96.

External Characters.

Its colour is steel-grey.

 It

It occurs massive, in plates, and in leaves.

Internally it is glistening, and the lustre is metallic.

The fracture is hackly.

It is opaque.

It is malleable; but not in so high a degree as meteoric iron.

It is hard.

It is magnetic.

Constituent Parts.

From the mine named Johannes, near Great Kamsdorf in Saxony.

Iron, - - -	92.50
Lead, - - -	6.00
Copper, - - -	1.50
	100.00

Klaproth, Beit. b. iv. s. 106.

Geognostic and Geographic Situations.

It is said to have been found associated with brown ironstone, sparry ironstone, and heavy-spar, at Kamsdorf *; along with clay and hematite at Eibenstock †; with brown ironstone and quartz, in a vein in the mountain of Oulle, in the vicinity of Grenoble ‡; at Miedziana-Gora in Poland ||; in the scoriæ of the volcanic mountain of Graveneire, in the department of Puy de Dôme §; imbedded

* Charpentier, Mineralog. Geographie v. Sachsen, s. 343.

† Werner's Pabst. b. i. s. 130.

‡ Schreiber, in Journal de Physique, Juillet 1792.

|| Journal de Physique, t. 65. p. 128.

§ Mossier, in Lucas's Tableau, t. 2. p. 367.

imbedded in American iron-pyrites *; and it is said in the island of Bourbon †.

Observations.

Lucas mentions a pseudo-volcanic *meteoric steel*, found near the village of Bouiche, in the department of the Allier in France. It was discovered by M. Mossier, in the form of small globules, imbedded in minerals which had been scorified by the fire of a coal-mine, formerly in a state of inflammation.

Second Subspecies.

Meteoric Native Iron.

Meteoreisen, *Karsten.*

Plin. Hist. Nat. xxxiv. 14. (41. ed. Bip. v. 260.) & ii. 56. (ed. Bip. s. 166.) ‡.—Meteoreisen, *Klaproth,* Beit. b. iv. s. 99. 101. *Id. Leonhard,* Tabel. s. 62. *Id. Karsten,* Tabel. s. 64. Fer natif meteorique, *Hauy,* Tabl. p. 93.—Meteoreisen, *Haus.* Handbuch, b. i. s. 114.—Fer natif meteorique, *Lucas,* t. ii. p. 358.—Meteoric Native Iron, *Aikin,* p. 95. 2d edit.

External Characters.

Its colour is pale steel-grey, which inclines to silver-white,

* Proust, Journal de Physique, t. 61. p. 272.

† Brong. t. 2. p. 148.

‡ Differentia ferri numerosa: Prima, in genere *terræ cœlive:* Item, ferro (pluisse) in Lucanis, anno antequam M. Crassus a Parthis interemtus est, omnesque cum eo Lucani milites, quorum magnus numerus in exercitu erat. *Effigies quæ pluit, spongiarum fere similis fuit.*

white, like platina. It is generally covered with a thin brownish crust of oxide of iron.

It occurs ramose, imperfect globular, and disseminated in meteoric stones.

Its surface is smooth and glistening.

Internally it is intermediate between glimmering and glistening, and the lustre is metallic.

The fracture is hackly.

The fragments are blunt-edged.

It yields a splendent streak.

It is intermediate between soft and semi-hard.

It is malleable.

It is flexible, but not elastic.

It is difficultly frangible.

Specific gravity, 7.573, *Karsten.*

Constituent Parts.

	Agram.	Mexico.
Iron, - -	96.5	96.75
Nickel, -	3.5	3.25
	100.0	100.00

Klaproth, Beit. b. iv. s. 101, 102.

According to Mr Howard, the native iron found in Siberia, South America, and Senegal, contains a portion of Nickel. The American contains 0.10; the Siberian 0.17; and the Senegambian 0.5 and 0.6.

Geographic Situation.

This subspecies of iron appears to be formed in the atmosphere by some process hitherto unknown to us. It is precipitated towards the surface of the earth in masses of greater or lesser magnitude, and which generally appear

pear to proceed from fire-balls. The fall of masses of iron from the heavens, has been known from a very early period, and instances of it have even occurred in our own times, as will appear from the following enumeration:

1. About 56 years before the Christian era, a mass of spongy iron fell from the atmosphere in Lucania *.

2. In the year 648, a glowing mass, like a fiery anvil, fell from the air at Constantinople. This appears to have been a mass of iron.

3. Avicenna speaks of a mass of iron weighing 50 pounds, which fell from the air near Lurgea; and Averrhoes of a mass of iron, estimated to weigh 100 pounds, which fell at Cordova in Spain, and of which swords were made.

4. In the year 1164, during the feast of Pentecost, a shower of iron fell in Missnia †.

5. A great mass of iron fell from the air, in a forest near to Neuhof, between Leipsic and Grimma, between the years 1540 and 1550 ‡.

6. In the year 1559, five stones or masses of iron fell near Miskoz in Transylvania ||.

7. In the years 1560 and 1570, many masses of iron fell in different places in Piedmont.

8. In the year 1603, a mass of metal, probably iron, fell in Bohemia.

9. In the year 1652, a mass of iron, weighing 5 pounds, fell in India, about 100 leagues south-east of Lahore.

10. There

* Plin. Hist. Nat. ii. p. 56.

† Georg. Fabric. Rer. Misnic. lib. i. p. 32.

‡ Albini, Meissnische Berg-Chronik, p. 139.

|| Nic. Isthuansii, Hist. Hungar. l. xx. fol. 394.

10. There is preserved in the town of Ellbogen in Bohemia, a mass of iron weighing 200 pounds, which appears to have fallen from the air in the year 1647; and it is said about the same time a ball or mass of iron fell from the air on board a ship in the open sea, and killed two men.

11. On the 12th of January 1683, a stone or mass of iron fell near Castrovillari in Calabria *.

12. In the bishopric of Agram in Croatia, on the afternoon of the 26th of May 1751, a fire-ball burst with a loud explosion, and two masses of iron fell from it; the one fragment, which weighed 71 pounds, sunk a considerable depth into the earth; and the other, which was 16 pounds weight, fell on the surface of a meadow, at the distance of 2000 paces from the former. The largest fragment is still preserved in the Imperial cabinet in Vienna.

Besides these undoubted instances of meteoric iron, others less certain are mentioned by authors; and of these the most remarkable are the following.

1. Professor Pallas, many years ago, discovered a mass of native iron, about 1600 pounds weight, lying on the surface of a hill between Krasnojark and Abakunsk. It is considered as a holy relic by the natives, who believed that it fell from heaven. It is enveloped in a slight brownish-coloured crust, and the vesicular cavities are filled with a mineral of the nature of olivine. It has all the characters, both external and chemical, of meteoric iron: hence it is generally supposed to have had a similar origin.

2. Goldberry, in his journey through Western Africa, in the years 1805–7, found a mass of native iron in the Great Desart

* Mercati, Metallotheca Vaticani, cap. xix. p. 248.

Desart of Sahra. Fragments of it were brought to Europe by Colonel O'Hara, and were analysed by Mr Howard, who found it composed of 96 parts of Iron and 4 of Nickel.

3. Barrow mentions a mass of iron he met with on the banks of the Great Fish River, in Caffraria, in Southern Africa. Chladni is of opinion that it is meteoric; but Barrow considers it as an artificial mass.

4. Several masses of native iron have been met with in Mexico. A mass found at Zacatecas about fifteen years ago, according to Humboldt, still weighed nearly 2000 pounds.

5. Bougainville, the French circumnavigator, discovered an enormous mass of native iron on the banks of the river La Plata in South America. It is calculated to weigh about 100,000 pounds. It has not been analysed.

6. Many years ago, a mass of native iron, calculated to weigh about 30 tons, was discovered in the district of St Jago del Estro, in South America. It lies in the middle of a great plain, and no rock or mountain within an hundred miles of it. Proust ascertained that it contained nickel; and Howard found it composed of 90 parts of iron and 10 of nickel in 100 parts. This fact, with its general aspect, strongly favours the idea of its meteoric origin.

7. Dr Bruce, in his interesting American Mineralogical Journal, mentions a mass of iron, weighing 3000 pounds, which is said to have been found near the Red River. It is 3 feet 4 inches in length, and 2 feet 2 inches in breadth. Its specific gravity is 7.400. According to Professor Silliman and Colonel Gibbs, it contains nickel as a constituent part.

2. Iron-

2. Iron-Pyrites.

Schwefelkies, *Werner*.

This species is divided into three subspecies, viz. Common Iron-pyrites, Capillary or Hair Pyrites, and Cellular Pyrites.

First Subspecies.

Common Iron-Pyrites.

Gemeiner Schwefelkies, *Werner*.

Pyrites colore aureo, *Plin.* Hist. Nat. xxxvi. (ed Bip. v. 371.)* —Sulphureus et Marcasita, *Waller.* Syst. Min. t. ii. p. 126. —Gemeiner Schwefelkies, *Werner,* Pabst. b. i. s. 130. *Id. Wid.* s. 794.—Common Sulphur Pyrites, *Kirw.* vol. ii. p. 76. —Gemeiner Schwefelkies, *Emm* b. ii. s. 289.—Fer sulphuré, *Hauy,* t. iv. p. 65.-97.—La Pyrite martiale commune, *Broch.* t. ii. p. 221.—Gemeiner Schwefelkies, *Reuss,* b. iv. s. 14. *Id. Lud.* b. i. s. 236.—Eeisenkies, *Suck.* 2ter th. s. 234.—Gemeiner Schwefelkies, *Bert.* s. 412. *Id. Mohs,* b. iii. s. 322. *Id. Hab* s. 110.—Fer sulphuré, *Lucas,* p. 139.—Gemeiner Schwefelkies, *Leonhard,* Tabel. s. 62.—Fer sulphuré crystallisé, *Brong.* t. ii. p. 151.—Gemeiner Schwefelkies, *Karsten,* Tabel. s. 64. *Id. Haus.* s. 72.—Sulphuret of Iron, *Kid,* vol. ii. p. 184.—Fer sulphuré, *Hauy,* Tabl. p. 96.—Common Pyrites, *Aikin,* p. 35.

External Characters.

Its colour is perfect bronze-yellow; seldom tarnished, reddish, and brownish.

It

* The πυριμαχος or πυρομαχος of the Greeks is not our iron-pyrites, as is maintained by Henckel and Wallerius.—Vid. Beckmann, in his Notes to Aristot. lib. de Mirab. auscult. p. 96.

It occurs most commonly massive, disseminated, and in membranes; frequently also crystallised. Its crystallisations are as follows:

1. *Cube*, in which the faces are either straight, or spherical-convex or spherical-concave *, fig. 184. It is the most common crystallisation of this species.
2. Cube, truncated on its edges, in such a manner that each truncation is more inclined towards one face than the other, and each face supports two opposite truncating planes †, fig. 185.
3. When the truncating planes of the preceding figure become so large that the original planes disappear, the *pentagonal dodecahedron* is formed ‡, fig. 186.
4. Cube truncated on all the angles ||, fig. 187.
5. When the truncations in the preceding figure become so large as to obliterate the original planes of the cube, an *octahedron* is formed §, fig. 188.
6. Octahedron bevelled on all the edges; the bevelment once broken, fig. 189.
7. Octahedron, in which the angles are acuminated with four planes, which are set on the lateral planes, and the summits of the truncations truncated, fig. 190.
8. Octahedron bevelled on all the angles ¶, fig. 191.

9. When

* Fer sulphuré primitif, Hauy.

† Fer sulphuré cubo-dodecaedre, Hauy.

‡ Fer sulphuré dodecaedre, Hauy.

|| Fer sulphuré cubo-octaedre, Hauy.

§ Fer sulphuré octaedre, Hauy.

¶ Fer sulphuré icosaedre, variet. *a.* Hauy.

9. When the bevelling planes of the octahedron become large, the figure passes into the *icosahedron* *, fig. 192.
10. Cube, in which each angle is acuminated with three planes, which are set on the lateral edges †, fig. 193. Sometimes the acuminating planes become so large, that the original faces of the cube appear as small rhombs ‡, fig. 194.
11. Cube, in which each angle is acuminated with three planes, which are set on the lateral planes: sometimes the acuminating planes become so large, that the original faces of the cube entirely disappear, when there is formed
12. The *leucite crystallisation*, or very acute double eight-sided pyramid, in which the lateral planes of the one are set on those of the other, and both extremities are acuminated with four planes, which are set on the alternate lateral edges ||.

The cube is middle-sized and small. The icosahedron and dodecahedron only small.

The crystals are seldom single, particularly the cube, which is variously aggregated.

The surface of the crystals is sometimes smooth, sometimes alternately streaked, and the lustre extends from specular-splendent to glistening.

Internally it is usually shining and glistening, and the lustre is metallic.

The

* Fer sulphuré icosaedre, Hauy.

† Fer sulphuré quadriepointé, Hauy.

‡ Fer sulphuré triacontaedre, Hauy.

|| Fer sulphuré trapezoidal, Hauy.

The fracture is coarse, small and fine-grained uneven *.

The fragments are indeterminate angular, and rather sharp-edged.

It sometimes occurs in fine granular distinct concretions.

It is hard.

It is brittle.

It is rather difficultly frangible.

When rubbed, or struck with steel, it emits a strong sulphureous smell.

Specific gravity,—

Dodecahedral pyrites,	4.830,	*Hatchett.*
Pyrites in smooth-planed cubes,	4.831,	*Id.*
Pyrites from Freyberg,	4.682,	*Gellert.*
———— Cornwall,	4.789,	*Kirwan.*
Cubic pyrites,	4.600,	*Brisson.*
Id.	4.7016,	

Chemical Characters.

Before the blowpipe it emits a strong sulphureous odour, and burns with a bluish flame. It afterwards changes into a brownish-coloured globule, which is attractable by the magnet, and by continuance of the heat, passes into a blackish slag, which communicates a dirty-green colour to borax.

Constituent

* Hausmann mentions a variety of common iron-pyrites with perfect, large, and flat conchoidal fracture, and nearly splendent lustre, and hence proposes to subdivide this subspecies into two kinds, viz. *Uneven* and *Conchoidal* common iron-pyrites.—Vid. Leonhard's Tasehenbuch, b. viii. s. 444.

Constituent Parts.

	Dodecahedral Pyrites.	Pyrites in striated Cubes.	Pyrites in smooth Cubes.
Sulphur,	52.15	52.50	52.70
Iron,	47.85	47.50	47.30
	100	100	100

Hatchett, Phil. Trans for 1804.

Some varieties, particularly those in striated cubes and dodecahedrons, contain a portion of gold, and hence have been named Auriferous Pyrites : other varieties contain silver.

Geognostic Situation.

It occurs in beds, in primitive, transition, and flœtz mountains ; also disseminated through various rocks, as granite, gneiss, mica-slate, clay-slate, primitive greenstone, porphyry, grey-wacke, sandstone, slate-clay, limestone, &c. In the veins in primitive mountains, it is associated with galena or lead-glance, copper-pyrites, arsenical-pyrites, blende, frequently with native gold, seldomer with ores of silver : in transition mountains, with galena or lead-glance, blende, copper-pyrites, sparry ironstone, calcareous and fluor spar ; and in flœtz rocks, with ores of lead, copper, and zinc, and quartz, calcareous-spar, and fluor-spar.

It is worthy of remark, that quartz is one of the most constant attendants of common iron-pyrites.

Geographic Situation.

This mineral is so universally distributed, that it is not necessary to enter here into any geographical details in regard to it.

Use.

Uses.

It is never worked as an ore of iron ; it is principally valued on account of the sulphur which can be obtained from it by sublimation, and the iron-vitriol which it affords by exposure to the air, either with or without previous roasting. It was formerly cut into ornaments, but they are now out of use ; and at one period it was used in place of flint in gun-locks.

Observations.

1. It is distinguished from *Pale-yellow Native Gold*, by its brittleness, the gold being malleable: from *Copper-pyrites*, by colour, crystallisation, fracture, and hardness: from *Arsenical pyrites*, by colour.

2. Iron-pyrites sometimes contains gold, silver, and copper; and it is remarked, that these metals generally occur in the striated cubes, the pentagonal dodecahedron, and the intermediate varieties of form between the cube and the dodecahedron.

3. Marcasite, according to Henckel, is a word used by the Arabians to express any substance in an imperfectly metallic state, and not easily reduced, and he supposes it may be derived from the Hebrew word " *marach*," " flavescere." Apothecaries at one time applied this name to bismuth, antimony, and several other metals. Miners apply the name to the crystallised varieties of iron-pyrites.

4. Polished pieces of a yellow-coloured mineral have been found in the graves of the early inhabitants of Peru, and are known by the name *Piedra de los Incas*. All of them appear to be either large crystals of iron-pyrites, or massive pieces of the same mineral more or less polished, and which appear to have been used as mirrors.

 5. The

5. The most complete treatise on the natural and economical history of Pyrites, is that of Henckel, of which an English translation was published in London in 1757.

Second Subspecies.

Capillary or Hair Pyrites.

Haarkies, *Werner.*

Id. Werner, Pabst. b. i. p. 143.—Capillary Pyrites, *Kirw.* vol. ii. p. 79.—Haarkies, *Emm.* b. ii. s. 297.—Fer sulphuré capillaire, *Hauy,* t. iv. p. 89.—La Pyrite capillaire, *Broch.* t. ii. p. 227.—Haarkies, *Reuss,* b. iv. s. 33. *Id. Lud.* b. i. s. 239. *Id. Mohs,* b. iii. s. 350. *Id. Leonhard,* Tabel. s. 63. *Id. Karsten,* Tabel. s. 64.—Haarförmiger Eisenstein, *Haus.* s. 72. —Fer sulphuré fibreux-entrelacé, *Hauy,* Tabl. p. 97.

External Characters.

Its colour is dark bronze-yellow, which sometimes passes into steel-grey.

It occurs in very delicate capillary crystals, which are either promiscuous or scopiform aggregated.

The lustre is metallic.

It is brittle.

It is slightly flexible.

Geognostic and Geographic Situations.

It is the rarest subspecies of iron-pyrites. It is usually accompanied with grey copper-ore, galena or lead-glance, quartz, calcareous-spar, and fluor-spar. It occurs at at Zellerfeld and Andreasberg in the Hartz, along with zeolite, cross-stone, and native silver.

Observations.

The capillary pyrites of Johanngeorgenstadt has been analysed by Klaproth, who finds it to be native nickel: it must therefore be removed to the Nickel Order.

Third Subspecies.

Cellular Pyrites.

Zellkies, *Werner.*

Zellkies, 2. Unterart, *Reuss,* b iv. s. 34. *Id. Mohs,* b. iii. s. 347. *Id. Leonhard,* Tabel. s. 63. *Id. Karsten,* Tabel. s. 64.

External Characters.

The colour is bronze-yellow, which inclines very much to steel-grey, and slightly to green.

By exposure it acquires a grey tarnish.

It occurs massive, but most frequently cellular, and of this it exhibits the hexagonal, polygonal, and indeterminate cellular varieties.

The surface of the cells is drusy.

Internally it is strongly glimmering; seldom, and only when it passes into common pyrites, glistening.

The fracture is even and flat conchoidal, seldom passing into fine-grained uneven.

The fragments are indeterminate angular, and pretty sharp-edged.

In other characters it agrees with the foregoing subspecies.

 Geognostic

Geognostic and Geographic Situations.

It occurs in veins at Johanngeorgenstadt in the electorate of Saxony, where it is accompanied with hepatic pyrites, common pyrites, galena or lead-glance, sparry ironstone, iron-ochre, brown-spar, heavy-spar, fluor-spar, and quartz.

Observations.

It is the least liable to decomposition of all the subspecies of pyrites.

3. Radiated Pyrites.

Strahlkies, *Werner.*

Globuli pyritacei, *Wall.* Syst. Min. t. ii. p. 129.—Strahlkies, *Werner,* Pabst. b. i. s. 136. *Id. Wid.* s. 797.—Striated Pyrites, *Kirw.* vol. ii. p. 78.—Strahlkies, *Emm.* b. ii. s. 293.—Fer sulphuré radié, *Hauy,* t. iv. p. 89.—La Pyrite rayonnée, *Broch.* t. ii. p. 225.—Strahlkies, *Reuss,* b. iv. s. 25. *Id. Lud.* b. i. s. 337. *Id. Mohs,* b. iii. s. 337. *Id. Hab.* s. 111. *Id. Leonhard,* Tabel. s. 63.—Fer sulphuré radié, *Brong.* t. ii. p. 152.—Strahliger Schwefelkies, *Karsten,* Tabel. s. 64. *Id. Haus.* s. 74.—Fer sulphuré aciculaire-radié, *Hauy,* Tabl. p. 97.—Fer sulphuré prismatique rhomboidal, *Bournon,* Cat. Min. p. 301.—White Pyrites, *Aikin,* p. 35.

External Characters.

Its colour is bronze-yellow, (usually paler than in common pyrites), which sometimes inclines to steel-grey; and the surface is often tarnished.

It occurs massive, but most commonly dendritic, reniform,

form, stalactitic, globular, botryoidal, fruticose, tuberose; and also crystallised in the following figures:

1. Oblique four-sided prism, flatly bevelled on the extremities, the bevelling planes set on the acute lateral edges: or it may be viewed as a very flat octahedron, which is more or less elongated, fig. 195.
2. The preceding figure, in which the edge formed by the meeting of the two bevelling planes is deeply truncated, fig. 196.: sometimes the obtuse edges are deeply truncated, fig. 197.
3. N° 1. in which the angles formed by the meeting of the bevelling and lateral planes on the obtuse edges are truncated.
4. N° 1. bevelled on the acute lateral edges.
5. Broad wedge-shaped octahedron, in which the smaller lateral faces are sometimes curved.

It also occurs in twin and triple crystals, of various complicated forms *.

The crystals are sometimes aggregated in rows, and the flat wedge-shaped octahedrons are so arranged as to resemble the comb of a cock: hence this variety has been named *Cock's-comb Pyrites* †.

The crystals are usually small and very small; externally shining and glistening, which inclines to glimmering.

The fracture, is narrow and scopiform or stellular radiated, which sometimes passes into fibrous, and inclines to compact.

The fragments are wedge-shaped.

 It

* Vid. Bournon's Cat. Min. 304.—313.

† This is the Kammkies of Werner.

It occurs in coarse and large granular distinct concretions, and in each concretion the radii proceed from a central point; also in thin and curved lamellar distinct concretions, which are bent in the direction of the external surface, and traverse the granular concretions; and lastly, in thin columnar distinct concretions.

It is hard.

It is brittle.

It is very easily frangible, and breaks more easily in the direction of the lamellar than the granular concretions.

When rubbed, or struck with steel, it emits a sulphureous odour.

It is heavy.

Specific gravity, from 4.698 to 4.775, *Hatchett.* 4.729, *Wiedemann.*

Constituent Parts.

Sulphur,	53.60	54.34
Iron,	46.40	45.66
	100	100

Hatchett, Phil. Trans. for 1804.

Geognostic Situation.

It is much rarer than common iron-pyrites; occurs principally in small variously shaped masses, in chalk, clay, &c. and in veins, which contain lead and silver ores.

Geographic Situation.

It is found in Cornwall, Isle of Sheppey, Kent, Derbyshire, and other places in England; Freyberg, Gersdorf, Schneeberg, Annaberg, Johanngeorgenstadt, in Saxony; Bohemia;

Bohemia; Zellerfeld and Goslar in the Hartz; Arendal in Norway; the islands of Morn and Seeland in Denmark; Schlangenberg in Siberia.

Observations.

It decomposes more readily than common iron-pyrites, particularly when it is exposed to a varying temperature in damp places, and then its surface becomes covered with greyish-white capillary iron-vitriol. In other instances the sulphur is volatilised, and the mineral remains unaltered in form, but in the state of an oxide of iron, forming what Hauy calls *cristaux epigenes.*

" Mr Proust is of opinion, that the pyrites which contain the smallest quantity of sulphur, are those which are most liable to vitriolisation; and, on the contrary, that those which contain the largest proportion, are the least affected by the air or weather. This opinion of the learned Professor, by no means accords with such observations as I have been able to make; for the cubic, dodecahedral, and other regularly crystallised pyrites, are liable to oxidisement, so as to become what are called Hepatic Iron-ores, but not to vitriolisation; whilst the radiated pyrites (at least those of this country) are by much the most subject to the latter effect: and therefore, as the result of the preceding analyses shew that the crystallised pyrites contain less sulphur than the radiated pyrites, I might be induced to adopt the contrary opinion. But I am inclined to attribute the effect of vitriolisation, observed in some of the pyrites, not so much to the proportiou, as to the state of the sulphur in the compound; for I much suspect, that a predisposition to vitriolisation in these pyrites, is produced by a small portion of oxygen being previously combined with a part, or with the gene-

 ral

ral mass of the sulphur, at the time of the original formation of these substances, so that the state of the sulphur is tending to that of oxide, and thus the accession of a farther addition of oxygen becomes facilitated."—*Hatchett*, Phil. Trans. 1804.

4. Hepatic or Liver-Pyrites *.

Leberkies, *Werner*.

Pyrites fuscus, *Wall.* t. ii. p. 133.—Pyrite hepatique, *Romé de L.* t. iii. p. 265.—Leberkies, *Werner*, Pabst. b. i. s. 139. *Id*, *Wid.* s. 800.—Hepatic Pyrites, *Kirw.* vol. ii. p. 83.—Leberkies, *Emm.* b. ii. s. 298.—La Pyrite hepatique, *Broch.* t. ii, p. 228.—Leberkies, *Reuss*, b. iv. s. 20. *Id. Lud.* b. i. s. 238. *Id. Mohs*, b. iii. s. 349.—Graugelber Eisenkies, *Hab.* s. 112. —Leberkies, *Leonhard*, Tabel. s. 63. *Id. Karsten*, Tabel. s. 64.—Dichter Schwefelkies, *Haus.* s. 72.—Liver Pyrites, *Aikin*, p. 36.

External Characters.

Its colour is very pale brass-yellow, which inclines more or less to steel-grey.

It changes its colour on the fresh fracture, and becomes brown, or acquires a columbine tarnish.

It occurs massive, disseminated, globular, tuberose, reniform, stalactitic, and straight and small cellular. Also crystallised in the following figures:

1. Regular six-sided prism.
2. Six-sided pyramid, truncated on the extremity.

Internally

* The name of the species is from the brown colour it exhibits on the fracture surface.

Internally it is usually glimmering, seldom approaching to glistening, and the lustre is metallic.

The fracture is even, which sometimes passes into small-grained uneven, sometimes into flat conchoidal.

The fragments are indeterminate angular, and sharp-edged.

It occurs in distinct concretions.

Specific gravity, 4.834, *Karsten.*

Geognostic Situation.

It occurs only in veins in primitive rocks, and is usually accompanied with red silver-ore, native silver, galena or lead-glance, common pyrites, black and brown blende, sparry ironstone, iron-ochre, seldomer with cobalt-glance, red cobalt-ochre, cinnabar, and grey antimony ore: the accompanying vein-stones are, quartz, heavy-spar, brown-spar, fluor-spar, and calcareous-spar.

Geographic Situation.

It is found in Derbyshire; Freyberg, Johanngeorgenstadt in Saxony; Wolfstein in the Palatinate; Salzburg; Goslar in the Hartz; Hungary; Transylvania; Bohemia; Iceland; Norway; Sweden; and Siberia.

Observations.

It is very nearly allied to Magnetic Pyrites: it is probably only a subspecies of that mineral.

4. Magnetic

5. Magnetic Pyrites.

Magnetkies, *Werner*.

Id. Werner, Pabst. b. i. s. 144. *Id. Wid.* s. 792.—Magnetic Pyrites, *Kirw.* vol. ii. p. 79.—Magnet-kies, *Emm.* b. ii. s. 286. —La Pyrite magnetique, *Broch.* t. ii. p. 232.—Magnetkies, *Reuss,* b. iv. s. 35. *Id. Lud.* b. i. s. 239. *Id. Suck.* 2ter th. s. 245. *Id. Bert.* s. 416. *Id. Mohs,* b. iii. s. 352. *Id. Hab.* s. 112. *Id. Leonhard,* Tabel. s. 63.—Fer sulphuré magnetique, *Brong.* t. ii. p. 155.—Fer sulphuré ferrifere, *Hauy,* Tabl. p. 98.—Magnetic Pyrites, *Aikin,* p. 35.

This species is divided into two subspecies, viz. Compact Magnetic Pyrites, and Foliated Magnetic Pyrites.

First Subspecies.

Compact Magnetic Pyrites.

Dichtes Magnetkies, *Hausmann*.

Gemeiner Magnetkies, *Karsten,* Tabel. s. 64.—Dichter Magnetkies, *Haus.* s. 73. *Id. Haus.* Handbuch, b. i. s. 144.

External Characters.

Its colour is intermediate between bronze-yellow and copper-red, and sometimes inclines to pinchbeck-brown. On exposure to the air, it gradually loses its lustre, and acquires a brownish tarnish.

It occurs only massive and disseminated.

Internally it is shining and glistening, and the lustre is metallic.

The

The fracture is fine and coarse grained uneven, which sometimes passes into imperfect conchoidal.

The fragments are indeterminate angular, and blunt-edged.

It is semi-hard in a high degree.

It is brittle.

It is easily frangible.

It is heavy.

Specific gravity, 4.518, *Hatchett.*

It affects the magnetic needle.

Chemical Characters.

Before the blowpipe, it emits a feeble sulphureous smell, and melts easily into a greyish-black globule, attractable by the magnet.

Constituent Parts.

According to Hatchett, it contains in 100 parts,—

Sulphur, - - -	36.50
Iron, - - -	63.50
	100

Geognostic Situation.

This mineral occurs principally in primitive mountains, in beds, in gneiss, mica-slate, and primitive limestone, associated with iron-pyrites, copper-pyrites, arsenical pyrites, magnetic ironstone, galena or lead-glance, blende, quartz, garnet, actynolite, common hornblende, and rarely with tinstone. It is also found, either massive or disseminated, in transition greenstone and clay-slate.

Geographic

Geographic Situation.

Europe.—It occurs in the Criffle, Windy-Shoulder, and other hills in Galloway; and at the base of the mountain called Moel Elion in Caernarvonshire. On the Continent of Europe, it is met with at Gillebeck and Kongsberg in Norway; Andreasberg and Treseberg in the Hartz; Breitenbrun in the Saxon Erzgebirge; Kupferberg in Silesia; Bodenmais in Bavaria; and in the Muhlbachthal in Salzburg.

Asia.—In the mines of Catharinenburg in Siberia.

America.—Zacatecas in Mexico.

Uses.

It is used for the same purposes as common pyrites.

Second Subspecies.

Foliated Magnetic Pyrites.

Blättricher Magnetkies, *Hausmann.*

Blättriger Magnetkies, *Karsten,* Tabel. s. 64. *Id. Leonhard,* Tabel. s. 63. *Id. Haus.* Handbuch, b. i. s. 145.

External Characters.

Its colour is intermediate between bronze-yellow and copper-red.

It occurs massive, disseminated; and crystallised in the following figures:

1. Regular six-sided prism; sometimes truncated on the lateral edges, sometimes on the terminal edges, and occasionally on the terminal angles.

2. Six-

2. Six-sided pyramid, truncated on the extremity.

Internally it is splendent, and the lustre is metallic.

The principal fracture is foliated; specular-foliated in one direction, less perfect in two other directions: the cross fracture is small conchoidal, inclining to uneven.

It occurs in coarse granular distinct concretions.

In other characters it agrees with the preceding subspecies.

Geognostic and Geographic Situations.

It occurs in the mines of Andreasberg; and at Bodenmais in Bavaria.

Observations.

1. It was formerly conjectured, that the iron in this species was less oxidised than that in common iron-pyrites, and in this way its magnetic property was accounted for. Mr Hatchett has shewn, however, that iron, when combined naturally or artificially with 36.50 or 37 of sulphur, is not only still capable of receiving the magnetic fluid, but is also rendered capable of retaining it, so as to become in every respect a permanent magnet; and the same, he thinks, may, in a great measure, be inferred respecting iron, which has been artificially combined with 45.50 *per cent.* of sulphur.

2. Mr Hatchett has also shewn that magnetic pyrites agrees in chemical properties with artificial sulphuret of iron or pyrites.

6. Magnetic

6. Magnetic Ironstone.

Magneteisenstein, *Werner.*

This species is divided into three subspecies, viz. Common Magnetic Ironstone, Iron-sand or Arenaceous Magnetic Ironstone, and Earthy Magnetic Ironstone.

First Subspecies.

Common Magnetic Ironstone.

Gemeiner Magneteisenstein, *Werner.*

Ferrum mineralisatum crystallisatum, et Ferrum mineralisatum, minera ferrum trahente et polos mundi ostendente, *Wall.* t. ii. p. 234.-235.—Æthiops martial natif, *Romé de L.* t. iii. p. 176. —Magnetischer Eisenstein, *Werner,* Pabst. b. i. s. 144. *Id. Wid.* s. 787.—Common Magnetic Ironstone, *Kirw.* vol. ii. p. 158.—Gemeiner magnetischer Eisenstein, *Emm.* b. ii. s. 278. —Fer oxydulé, *Hauy,* t. iv. p. 10.-38.—Le Fer magnetique commun, *Broch.* t. ii. p. 235.—Gemeiner Magneteisenstein, *Reuss,* b. iv. s. 38. *Id. Lud.* b. i. s. 240. *Id. Suck.* 2ter th. s. 247. *Id. Bert.* s. 401. *Id. Mohs,* b. iii. s. 355. *Id. Hab.* s. 113.—Fer oxydulé, *Lucas,* p. 136.—Gemeiner Magneteisenstein, *Leonhard,* Tabel. s. 63.—Fer oxydulé, *Brong.* t. ii. p. 156. *Id. Brard,* s. 310.—Gemeiner & Blättricher Magneteisenstein, *Karsten,* Tabel. s. 64.—Blättricher, körniger & dichter Magneteisenstein, *Haus.* s. 105.—Magnetic Iron-ore, *Kid,* vol. ii. p. 165.—Fer oxydulé, *Hauy,* Tabl. p. 93.—Magnetic Iron-ore, *Aikin,* p. 36.

External Characters.

Its colour is iron-black, sometimes inclining to steel-grey, and in some varieties it is variously tarnished.

Besides

Besides massive and disseminated, it occurs also crystallised.

Its crystallisations are the following:

1. Octahedron *, fig. 198. which sometimes ends in a line.
   - *a.* Truncated on the edges †, fig. 199.
   - *b.* Bevelled on the edges, fig. 200.
   - *c.* Truncated on the angles.
   - *d.* Cuneiform ‡.
2. Garnet or rhomboidal dodecahedron ||, fig. 201.
3. Rectangular four-sided prism, acuminated with four planes, which are set on the lateral edges, fig. 202.
4. Cube, either perfect, or more or less deeply truncated on the angles, fig. 203.

The planes of the garnet dodecahedron are streaked in the direction of the larger diagonal of the rhomb; the planes of the four-sided prism are transversely streaked, and those of the octahedron are smooth.

The crystals are usually imbedded, aggregated on one another, and are small and middle-sized.

Externally it is shining and glistening.

Internally it is intermediate between shining and glistening, which sometimes (as in the foliated) passes into splendent, or into feebly glimmering (as in the even); and the lustre is metallic.

The fracture is small and coarse-grained uneven, which sometimes approches to even, sometimes to imperfect and small

---

* Fer oxydulé primitif, Hauy.

† Fer oxydulé emarginé, Hauy.

‡ Fer oxydulé cuneiforme, Hauy.

|| Fer oxydulé dodecaedre, Hauy.

small conchoidal; and some varieties are straight or curved foliated, with a fourfold cleavage.

The fragments are indeterminate angular, and rather sharp-edged.

It occurs unseparated, or in distinct concretions, which are occasionally so loose that they can be separated by the finger; also in straight and curved lamellar concretions.

The colour of the streak is black.

It is intermediate between semi-hard and hard, and some varieties pass into hard.

It is brittle.

It is difficultly frangible, but by the action of the weather it readily becomes more easily frangible.

It is heavy, inclining to uncommonly heavy.

Specific gravity, Compact, 4.587,—5.283: Foliated, 5.400, *Ullmann*. From Danemora, 4.9364, *La Metherie*. 4.094, *Kirwan*.

It is highly magnetic, with polarity.

Chemical Characters.

Before the blowpipe it becomes brown, and does not melt: it communicates to glass of borax a dark-green colour.

Constituent Parts.

Peroxide of Iron,	-	71.86
Protoxide of Iron,	-	28.14
		100.00 *Berzelius.*

Dr Thomson analysed a specimen of this ore from Greenland, and which he found to contain, besides the Iron, a small portion of Titanium.—Wern. Memoirs, vol. ii. part i. p. 55.

Geognostic

Geognostic Situation.

It occurs principally in beds, often of great magnitude, in primitive rocks, as gneiss, mica-slate, chlorite-slate, clay-slate, greenstone associated with hornblende, augite, actynolite, asbestus, epidote, garnet, felspar, calcareous-spar, fluor-spar, quartz, iron-pyrites, copper-pyrites, magnetic pyrites, arsenical pyrites, blende, galena or lead-glance, and other ores and minerals: also disseminated in granite, chlorite-slate, serpentine, gabbro, &c.; less frequently in beds and nests in transition rocks, as in transition porphyry.

Geographic Situation.

Europe.—It occurs in serpentine, in Unst, one of the Zetland Islands; St Just in Cornwall; and Tavistock in Devonshire. In the iron-mines of Arendal in Norway, it occurs in beds in gneiss: these beds are short, but vary in thickness from four to sixty feet; they are frequently intermixed with the gneiss at their line of junction with it; cotemporaneous wedges of the gneiss also occur dispersed through the ironstone, and sometimes an uninterrupted transition is to be observed from the ironstone beds into the gneiss, in which they are contained. In these interesting repositories, the ironstone is associated with a great variety of different minerals: of these the most frequent are, granular garnet, augite, hornblende, epidote or pistacite, calcareous-spar, and the three constituents of gneiss. The garnet and augite are the most abundant, are generally in a granular form, and so intimately intermixed with the ironstone, that an inattentive observer might confound them together. The minerals of less frequent occurrence in these beds, are the following:

sphene or rutilite, the variety of common garnet named colophonite, apatite, scapolite, sahlite, actynolite, glassy tremolite, chlorite, common shorl, zeolite, iron-pyrites; and still rarer minerals are, prehnite, cubicite, rutile, sparry ironstone, molybdena, copper-pyrites, blende, azure copper-ore, copper-green, and graphite. These minerals are either intermixed with the ironstone in an irregular manner, or they are disposed in cotemporaneous veins included in it, or that shoot from the mass of the bed into the bounding strata of gneiss, just as is the case with cotemporaneous veins of granite shooting from massive granite into the adjacent strata. The remarkable hill named Taberg, in Smoland in Sweden, is a great mass of primitive greenstone, richly impregnated with magnetic ironstone, and resting on gneiss *; and in the island of Utö, also in Sweden, there are extensive mines of magnetic ironstone, in which the ironstone occurs as a wedge-shaped bed in gneiss, about 120 feet thick, and nearly half a mile long; but the most considerable of the Swedish iron-mines are those of Dannemora, in which the magnetic ironstone occurs as a bed several hundred feet thick, in gneiss, and is associated with tremolite, chlorite, asbestus, actynolite, and, what is worthy of particular notice, mineral pitch. There are also great beds of magnetic ironstone, at Gellivara in Luleo Lappmark, Luossavara, Kensivara, and Junossuwando. At Breitenbrunn, in the Saxon Erzgebirge, it is associated with common garnet, common hornblende, amianthus, actynolite, fluor-spar, iron-pyrites, magnetic pyrites, arsenical pyrites, blende, and tinstone; at Geier, with magnetic pyrites, galena or lead-glance, and actynolite; at Kupferberg in Silesia, along with

* Vid. Thomson's Travels in Sweden, and Hausmann's Travels in the same country, for descriptions of the Taberg.

with copper and iron pyrites; at Presnitz, also in Silesia, in beds in gneiss; in Bavaria; Franconia; Lusatia; Hartz; and Thuringia; at Cogne in Piedmont, there is a bed of this ore, about seventy-five feet thick, which is contained in a great bed of serpentine, subordinate to mica-slate; it is also found in Corsica, Sardinia, Switzerland, Spain, France, Hungary, and Transylvania.

Asia.—It is found in the Mysore country in Hindostan *; Nertschinsk, Parmien, and other places in Siberia; Siam; China; the Philippine Islands; and New Holland.

North America.—It occurs in West Greenland, in New Spain; immense beds of it extend, with little interruption, from Canada to the neighbourhood of New-York. Colonel Gibbs describes a bed of this ore as occurring at the Franconia Iron-works in New Hampshire †.

South America.—It occurs in Chili.

Uses.

When pure, it affords excellent bar-iron, but indifferent cast-iron; and as it is easily fusible, it requires but little flux. When it happens to have intermixed copper or iron pyrites, it affords a red-shot iron. Careful roasting of ore thus mixed, diminishes the bad effects of the sulphur, which is evidently the cause of the deterioration of the iron. In Sweden, particularly at Dannemora, the ore is quite pure, and affords excellent bar-iron, which is imported into Great Britain, for the purpose of steel-making.

* Dr Ainslie's Materia Medica of Hindoostan, and Artizans and Agriculturists Nomenclature, p. 55. 4to, printed at Madras in 1813.

† Bruce's Mineralogical Journal, p. 5. and 6.

Second Subspecies.

Iron-Sand or Arenaceous Magnetic Ironstone.

Eisensand, *Werner.*

Id. Werner, Pabst. b. i. s. 147. *Id. Wid.* s. 790.—Magnetic Sand, *Kirw.* vol. ii. p. 161.—Eisensand, *Emm.* b. ii. s. 284. —Le Fer magnetique sablonneux, *Broch.* t. ii. p. 241.—Sandiger Magneteisenstein, *Reuss,* b. iv. s. 48.—Eisensand, *Lud.* b. i. s. 241. *Id. Suck.* 2ter th. s. 252. *Id. Bert.* s. 402. *Id. Mohs,* b. iii. s. 363.—Magnetischer Eisensand, *Hab.* s. 144.—Sandiger Magneteisenstein, *Leonhard,* Tabel. s. 64. —Fer oxydulé sablonneux, *Brong.* t. ii. p. 157.—Fer oxydulé arenacé, *Brard,* p. 311.—Sandiger Magneteisenstein, *Karsten,* Tabel. s. 64.—Körniger Magneteisenstein, *Haus.* s. 105.— —Fer oxydulé titanifere, *Hauy,* Tabl. p. 94.—Sandy Magnetic Iron-ore, *Aikin,* p. 37.

External Characters.

Its colour is very dark iron-black.

It occurs in grains, which are sometimes angular, sometimes roundish ; and also in octahedral crystals.

The grains and crystals are small and very small.

The grains have a feeble glimmering, and rough surface.

Internally it is intermediate between shining and splendent, and the lustre is metallic.

The fracture is perfect conchoidal, and very rarely imperfect foliated.

The fragments are indeterminate angular, and sharp-edged.

It

It is semi-hard.

It is brittle.

It is easily frangible.

It is heavy.

Specific gravity, 4.600, *Kirwan*. 4.76, *Thomson*.

It is strongly attracted by the magnet.

Constituent Parts.

	Niedermenich.	Teneriffe.	Puy.	Shore of the Baltic.	River Dee, Aberdeenshire.
Oxide of Iron,	79.0	79.2	82.0	85.50	85.3
Oxide of Titanium,	15.9	14.8	12.6	14.00	9.5
Oxide of Chrome,	a trace	a trace	..	..	...
Oxide of Manganese,	2.6	1.6	4.5	0.50	...
Arsenic, -	..	..	..	..	1.0
Silica and Alumina,	1.0	0.8	0.6	..	1.5
	98.5	96.4	99.7	100.0	97.3
	Cordier, Journal de Mines, N. 124. p. 249.			*Klaproth*, Beit. b. v. s. 210.	*Thomson*, Tr. R. Soc. Edin. May 1807.

Geognostic Situation.

It occurs imbedded in basalt, clinkstone, and wacke, and loose in the beds of rivers, and in the sands of coasts and plains.

Geographic Situation.

Europe.—Imbedded in flœtz-trap rocks in Fifeshire; and Island of Skye; in the river Dee in Aberdeenshire, in a sand composed of quartz, felspar, and mica; and also in Argyleshire; at Hunstanton, Norfolk, and Arklow, near Wicklow, with native gold. On the Continent, it is met with in Norway; in flœtz-trap rocks at Hohen-

 stein;

stein; and in loose sand with hyacinth, iserine, nigrine, hornblende, and augite, in the province of Meissen in Saxony; in sand on the banks of the Elbe, at Schandau, near Pirna; and at Sebnitz, in the same district, along with small grains of hyacinth, and nigrine; in the clinkstone porphyry of the Milleschau, in Bohemia; and at Treblitz and Podsedlitz, also in Bohemia, intermixed with rolled pieces of basalt, pyrope, sapphire, and hyacinth; at Greifeswald, on the shore of the Baltic; at Puzzoli, near Naples, in the sand of the shore, along with pieces of pumice, lava, hornblende, and oliven; also in the Island of Ischia; at Messina in Sicily; and in the Island of Milo in the Archipelago; in the Tyrol; France; Piedmont; and Hungary.

Asia.—On the shores of Lake Baikal in Siberia.

America.—In the Islands of St Domingo and Guadaloupe; West Greenland; Virginia; and Cayenne.

Uses.

It is, although rarely, smelted as an ore of iron. In the Tyrol, near Naples, and in Virginia, it is smelted in considerable quantity; and, owing to its purity, affords most excellent bar-iron.

Third

Third Subspecies.

Earthy Magnetic Ironstone.

Ochriger Magneteisenstein, *Hausmann.*

Eisenschwärze, *Schumacher,* Verzeichniss der Dän. Nord. Mineralien, s. 135. *Id. Reuss,* b. iv. s. 53.—Eisenmulm, *Leonhard,* Tabel. s. 69.—Fer oxydulé fuligineux, *Hauy,* Tabl. p. 94.—Erdiger Magneteisenstein, *Haus.* Handbuch, b. i. s. 249.

External Characters.

Its colour is bluish-black.

It occurs in blunt-edged rolled pieces, in which the surface is sometimes vesicular.

Internally it is dull, or feebly glimmering on spots.

The fracture is fine-grained uneven, passing into earthy.

It is opaque.

It is soft.

It yields a black shining streak.

It soils.

It is sectile.

It is easily frangible.

It emits a faint clayey smell when breathed on.

Specific gravity, 2.200, *Schumacher.*

Geognostic and Geographic Situations.

It occurs in the iron-mines of Arendal, and in the mine of Eiserfeld in Siegen. It appears to be common magnetic ironstone in a state of decomposition.

Observations on the Species.

1. Common Magnetic Ironstone and Iron-sand are distinguished from *Iron-glance*, with which it is often confounded, by the colour of the streak, which is black; whereas that of iron-glance is cherry-red; by being powerfully magnetic, whereas iron-glance is scarcely affected by the magnet; and the crystallisations of magnetic ironstone are different from those of iron-glance.

2. Werner was the first who observed that this species of ironstone is not magnetic when at a depth in the earth, but that it acquires this property after exposure to the influence of the atmosphere.

3. The *Titaneisen* of Reuss and others, appears to be but a variety of common magnetic ironstone; and Hausmann says, that Karsten's fibrous magnetic ironstone is asbestous actynolite intimately intermixed with granular magnetic ironstone, and is therefore not a distinct subspecies.

7. Specular Iron-ore or Iron-glance.

Eisenglanz, *Werner*.

This species is divided into two subspecies, viz. Specular Iron-ore or Common Iron-glance, and Iron-mica or Micaceous Iron-glance.

First

First Subspecies.

Specular Iron-ore, or Common Iron-glance.

Gemeiner Eisenglanz, *Werner.*

Minera Ferri grisea, *Wall.* t. ii. p. 239.—Minera Ferri cœrulescens, *Ibid.* p. 241.—Mine de Fer grise ou speculaire, *Romé de Lisle,* t. iii. p. 186.—Fer speculaire, *De Born,* t. ii. p. 265.—Gemeiner Eisenglanz, *Werner,* Pabst. b. i. s. 147. *Id. Wid.* s. 802.—Specular Iron-ore, *Kirw.* vol. ii. p. 162.—Gemeiner Eisenglanz, *Emm.* b. ii. s. 301.—Fer speculaire, *Lam.* t. i. p. 220. to 225.—Fer oligiste, *Hauy,* t. iv. p. 38.-56.—Le Fer speculaire commun, *Broch.* t ii. p. 242.—Gemeiner Eisenglanz, *Reuss,* b. iv. s. 61. *Id. Lud.* b. i. s. 242. *Id. Suck.* 2ter th. s. 257. *Id. Bert.* s. 403. *Id. Mohs,* b. iii. s. 367. *Id. Hab.* s. 115. *Id. Leonhard,* Tabel. s. 64.—Fer oligiste compacte, *Brong.* t, ii. p. 160.—Gemeiner, körniger, & schiefriger Eisenglanz, *Karsten,* Tabel. s. 64.—Dichter, körniger, blättriger Blutstein, *Haus.* s. 105, 106.—Specular Iron-ore, *Kid,* vol. ii. p. 168 —Fer oligiste, *Hauy,* Tabl. p. 94.—Iron-glance, *Aikin,* p. 37.

External Characters.

Its most common colour is dark steel-grey, but which sometimes, in those varieties that approach to red iron-stone, falls into brownish-red. It seldom (and almost only in the crystallised varieties) passes into iron-black. It occurs very frequently tarnished on the external surface, or on that of the distinct concretions The tarnish is either that of tempered-steel, or is pavonine or columbine.

It

It occurs massive, disseminated, and frequently crystallised.

The following are its crystallisations:

1. Flat double three-sided pyramid, in which the lateral planes of the one are set on the lateral edges of the other. The planes are streaked in the direction of the larger diagonals. It is the fundamental crystal of this species *, fig. 204.
2. Double three-sided pyramid, in which the angles on the common base are truncated, and the truncating planes obliquely set on the lateral edges, so that three of the planes incline towards one summit and three towards the other †, fig. 205.
3. The preceding figure, in which the angles formed by the meeting of the truncating planes are more or less deeply bevelled ‡, fig. 206, 207.
4. Sometimes the truncations on the angles of the base become so large that they nearly touch each other, and then a *rhomboid*, nearly passing into the *cube*, is formed, in which the remains of the pyramidal planes become truncations on the lateral edges. When the pyramidal planes entirely disappear, the rhomboid ||, fig. 208. is formed.
5. The preceding rhomboid viewed as a nearly rectangular double three-sided pyramid, deeply truncated on the summits, gives a very oblique octahedron,

* Fer oligiste binaire, Hauy.

† Fer oligiste bi-rhomboidal, Hauy.

‡ Fer oligiste binoternaire, Hauy. This is the most common crystallisation of the Elba Iron-glance.

|| Fer oligiste primitif, Hauy.

tahedron *, fig. 209. This octahedron is sometimes truncated on the angles †, fig. 210.

6. When the truncating planes on the angles of the octahedron become of equal size with the planes of the octahedron, there is formed an equiangular six-sided table, in which the terminal planes are flatly bevelled ‡, fig. 211.

7. When the summits of the double three-sided pyramid are deeply truncated, there is formed a *six-sided table*, in which the terminal planes are set alternately oblique and straight on the lateral planes; and the lateral planes are sometimes straight and sometimes spherical-convex.

8. The lens, which is the double three-sided pyramid, having its summits and lateral edges rounded ||.

9. When the bevelling planes of N° 3. become so large as to obliterate the other planes, there is formed an acute double six-sided pyramid, in which the lateral planes of the one rest on the lateral planes of the other, and the pyramid is deeply truncated on the extremities §, fig. 212. Sometimes the alternate lateral edges are truncated ¶, fig. 213.; and sometimes the angles on the common base **, fig. 214.

10. Low

* Fer oligiste basé, Hauy.

† Fer oligiste imitatif, Hauy.

‡ Fer oligiste imitatif, var. Hauy.

|| Fer oligiste lenticulaire, Hauy.

§ Fer oligiste trapezien, Hauy.

¶ Fer oligiste uniternaire, Hauy.

** Fer oligiste progressif, Hauy.

10. Low equiangular six-sided prism, in which the terminal edges, and sometimes the angles, are slightly truncated.

The crystals are small and very small, seldom middle-sized.

The planes of the crystals are sometimes smooth, sometimes streaked.

Externally it alternates from splendent to glistening, but is most commonly splendent, and the lustre is metallic.

Internally it is generally glistening, but sometimes passes into shining and splendent, and the lustre is metallic.

The fracture is sometimes compact, sometimes foliated; the compact is coarse, small and fine-grained uneven, which sometimes passes into imperfect and small conchoidal; and in some varieties the fracture is perfect and large conchoidal, and in others slaty. The foliated has a fourfold rectangular cleavage, and the folia are triply streaked. The conchoidal and foliated varieties have the strongest, the uneven the weakest lustre.

The fragments are octahedral or pyramidal; sometimes also indeterminate angular, and rather blunt-edged.

It is usually unseparated *, yet sometimes it occurs in distinct concretions, which are large, coarse, and small granular, sometimes imperfect wedge-shaped columnar, and more frequently thick, sometimes straight, sometimes curved lamellar. The surface of the concretions is sometimes

* The terms *compact* and *unseparated* are by some considered as synonymous, which, however, is not the case. *Compact* refers to fracture-surface; *unseparated* to the mass itself; so that a mineral may be compact, and also in distinct concretions.

sometimes smooth, sometimes transversely streaked, and is shining, inclining to glistening.

It yields a cherry-red streak.

It is hard.

It is brittle.

It is usually rather difficultly frangible.

It is heavy.

Specific gravity, 5.158, *Gellert*. 4.6770 to 5.0116, *Brisson*. 4.793 to 5.2180, *Kirwan*.

Physical Characters.

When pulverised, it is magnetic in a slight degree, but it does not, like magnetic ironstone, attract filings of iron.

Chemical Characters.

Before the blowpipe, without addition, it is infusible; melted with borax, it gives a dirty yellow-coloured scoria.

Constituent Parts.

From Zocka.	
Oxide of Iron,	88.00
Oxide of Manganese,	0.75
Iron-pyrites,	8.25
Silica, - -	0.50
Magnesia, -	0.125
Loss, - -	2.53

Brocchi, Trattato Mineralogico e Chemico sulle Miniere de Turro del Departemento del Mella, vol. ii. p. 42.

From Grengesberget.	
Reddish-brown Oxide of Iron, -	94.38
Phosphate of Lime,	2.75
Magnesia, -	0.16
Mineral Oil? -	1.25
Loss by heating,	0.50
	98.94

Hisinger, Afhandlingar, iii. p. 32, 33.

Geognostic

Geognostic Situation.

It generally occurs in beds in primitive mountains, which are sometimes so large as to form mountain-masses. In these beds, it is associated with magnetic ironstone, red ironstone, iron-pyrites, copper-pyrites, arsenical pyrites, quartz, hornstone, and calcareous-spar. It also occurs in veins that traverse granite, gneiss, mica-slate, clay-slate, and grey-wacke, in which it is accompanied with red and brown ironstone, iron-pyrites, tinstone, quartz, lithomarge, brown-spar, fluor-spar, felspar, epidote or pistacite, and asbestus. It rarely occurs in vesicular cavities and fissures of volcanic rocks, and in veins in some sandstone and flœtz-trap rocks.

Geographic Situation.

Europe.—It occurs, along with red and brown ironstone, at Cumberhead in Lanarkshire; in the iron-mines of Norberg in Westmannland; at Wika in Dalecarlia; Langbanshytta in Wärmeland; at Bitsberg; also in the mountains of Haukiwara, near Luossawara in Lapland; in Norway; and the government of Olnetz in Russia. It is found in many of the iron-mines in the Saxon Erzgebirge, and generally associated with red ironstone, as is also the case in the iron-mines of Franconia, Bavaria, and Hessia. In Bohemia, it occurs in beds in mica-slate; in Silesia, in mica-slate, subordinate to the gneiss, which rests immediately on the central granite, and also in the hornblende formation which rests on the gneiss. The mountains of Switzerland do not afford much iron-glance; small portions of it are met with in St Gothard, and in mica-slate at the foot of the Great St Bernard. Although it is not a frequent ore in France, it is mentioned

tioned as occurring along with red ironstone at Framont; in small quantity at Markirch in Alsace; and in Dauphiny; and in the Puy de Dôme, and Volvic. In the Island of Corsica, it is associated with brown ironstone; and in the Island of Stromboli, in lava; but of all the islands in the Mediterranean, Elba is that which affords iron-glance in the greatest abundance. There, it is associated with brown ironstone, and quartz; and the mines, which are of great extent, have been worked for upwards of 3000 years. It is also one of the mineral productions of Salzburg, and Hungary.

Asia.—It occurs in the mines of Beresowskoi, in the Uralian Mountains; and also in the Mysore country in Hindostan *.

America.—In the United States; and in the mines of Sombrerete in Mexico; and in Chili in South America.

Uses.

When it occurs in quantity, it is smelted as an ore of iron. It affords an excellent malleable iron, which is, however, harder than that obtained from magnetic ironstone. It affords also good cast-iron, but which is not so much valued as that obtained from other ores of iron.

Observations.

1. It passes, on the one hand, into Common Magnetic Ironstone; and on the other into Red Ironstone.

2. It is easily distinguished from *Magnetic Ironstone* by its streak; magnetic ironstone yielding a black, whilst iron-glance affords a cherry-red streak.

Second

* Ainslie's Materia Medica, p. 55.

Second Subspecies.

Iron-Mica or Micaceous Iron-Glance.

Eisenglimmer, *Werner*.

Mica ferrea, *Wall.* t. ii. p. 242.—Mine de Fer micacée grise, *Romé de L.* t. iii. p. 205.—Eisenglimmer, *Werner*, Pabst. b. i. s. 152. *Id. Wid.* s. 805.—Micaceous Iron-ore, *Kirw.* vol. ii. p. 184.—Mine de Fer micacée grise, *Lam.* t. i. p. 241.—Eisenglimmer, *Emm.* b. ii. s. 306.—Fer oligiste ecailleux, *Hauy*, t. iv. p. 45.—Le Fer micacé, *Broch.* t. ii. p. 247.—Schuppiger Eisenglanz, *Reuss*, b. iv. s. 71.—Eisenglimmer, *Lud.* b. i. s. 243. *Id. Suck.* 2ter th. s. 262. *Id. Bert.* s. 405. *Id. Mohs*, b. iii. s. 378.—Schuppiger Eisenglanz, *Leonhard*, Tabel. s. 64.—Fer oligiste ecailleux, *Brong.* t. ii. p. 162.—Schuppiger Eisenglanz, *Karsten*, Tabel. s. 64.—Schuppiger Blutstein, *Haus.* s. 106.—Fer ecailleux, *Hauy*, Tabl. p. 95.—Micaceous Ironglance, *Aikin*, p. 38.

External Characters.

Its colour is iron-black, of different degrees of intensity: thin plates or folia, when held between the eye and the light, appear blood-red.

It occurs most commonly massive and disseminated: also crystallised in small thin six-sided tables, in which the terminal planes are set alternately oblique and straight on the lateral planes.

These tables sometimes intersect each other, so as to form cells.

The surface of the crystals is smooth and splendent.

Internally it is splendent, which in some varieties passes into shining, and the lustre is metallic.

The

The fracture is perfect and curved foliated, with a single cleavage.

The fragments are sometimes indeterminate angular, sometimes tabular.

The massive variety occurs in distinct concretions, which are large, coarse, small and fine granular.

It is slightly translucent on the edges; but translucent in thin plates.

Its streak is cherry-red.

It is semi-hard, approaching to soft when it passes to red scaly iron-ore.

It is brittle.

It is uncommonly easily frangible.

It is heavy.

Specific gravity, 4.500 to 5.070, *Kirwan.*

Physical Characters.

It slightly affects the magnet.

Constituent Parts.

According to Bucholz, this subspecies consists entirely of Peroxide of Iron.—*Gehlen's* Journal, 2d series, b. iii. s. 104.

Geognostic Situation.

It generally occurs in veins or in beds in newer primitive rocks, as mica-slate, and clay-slate; and in these repositories it is usually associated with red and brown ironstone, and iron-pyrites, and sometimes with copper-pyrites, sparry ironstone, calcareous-spar, fluor-spar, and quartz.

Geographic Situation.

Europe.—It occurs in veins in primitive rocks near Dunkeld, and in Benmore in Perthshire; also in Fitful-head, and other places in Mainland, the largest of the Zetland Islands. In England, it is met with at Tavistock in Devonshire; Eskdale in Cumberland; near Bristol; and in Caernarvonshire. The iron-mines in Norway and Sweden, afford small quantities of this ore; and it is also met with in the iron-mines of Olnetz in Russia, and in those in Saxony, Bohemia, Lusatia, Silesia, Franconia, Suabia, Bavaria, France, Island of Elba, and Hungary.

Asia.—In the mines of Catharinenburg in Siberia.

America.—Chili.

Uses.

It melts better than common iron-glance, but requires a greater addition of limestone. The iron which it affords is sometimes cold-short, but is well fitted for cast-ware.

Observations.

1. It passes into Red Scaly Iron-ore.
2. It affords from 70 to 80 *per cent.* of iron.
3. Iron-glance occurs usually with quartz; whereas magnetic ironstone is frequently accompanied with limestone.
4. It is the *Eisenmann* of older mineralogists.

8. Red

8. Red Ironstone.

Rotheisenstein, *Werner*.

This species is divided into four subspecies, viz. Scaly Red Iron-ore, Ochry Red Ironstone, Compact Red Ironstone, and Red Hematite.

First Subspecies.

Scaly Red Iron-ore or Red Iron-froth.

Rother Eisenrahm, *Werner*.

Hæmatites micaceus, *Wall.* t. ii. p. 248.—Rother Eisenrahm, *Werner,* Pabst. b. i. s. 153. *Id. Wid.* s. 807.—Red scaly Iron-ore, *Kirw.* vol. ii. p. 172.—Rother Eisenrahm, *Emm.* b. ii. s. 308.—Fer oxidé rouge luisant, *Hauy,* t. iv. p. 106.—Le Eisenrahm rouge, *Broch.* t. ii. p. 249.—Rother Eisenrahm, *Reuss,* b. iv. s. 76. *Id. Lud.* b. i. s. 244. *Id. Suck.* 2ter th. s. 264. *Id. Bert.* s. 406. *Id. Mohs,* b. iii. s. 385.—Schuppiger Rotheisenstein, *Hab.* s. 116.—Rother Eisenrahm, *Leonhard,* Tabel. s. 65.—Fer oxidé rouge luisant, *Brong.* t. ii. p. 164.—Schuppiger Rotheisenstein, *Karsten,* Tabel. s. 66.—Schaumiger Blutstein, *Haus.* s. 106.—Fer oligiste luisant, *Hauy,* Tabl. p. 95.—Red scaly Iron-ore, *Aikin,* p. 38.

External Characters.

Its colour is intermediate between cherry-red and brownish-red, and sometimes passes into steel-grey.

It occurs sometimes massive, sometimes coating, and disseminated, and is composed of scaly particles, which are generally slightly cohering, and glimmering, bordering on glistening, with a semi-metallic lustre.

The scales sometimes occur in four, six, and eight-sided tables.

It soils strongly.

It feels greasy.

Chemical Characters.

It is infusible before the blowpipe without addition, but it communicates to borax an olive and asparagus green colour.

Constituent Parts.

Iron, - -	66.00
Oxygen, - -	28.50
Silica, - - -	4.25
Alumina, - -	1.25
	100 *Henry.*

Bucholz found it to be a pure red oxide, mixed with a little quartz-sand.—*Gehlen's* Journal, 2d series, b. iii. p. 106.

Geognostic Situation.

It occurs in veins in primitive rocks, sometimes also in transition and in flœtz rocks. It is usually accompanied with compact and ochry red ironstone, red hematite, iron-mica, sometimes also magnetic ironstone, sparry ironstone, quartz, heavy-spar, and brown-spar.

Geographic Situation.

Europe.—It is found at Ulverstone, and several other places on the borders of Lancashire; in the mine called Oerve-

Oerve-Aase in Norway, along with iron-mica or micaceous iron-glance; Iberg and Blankenberg in the Hartz, with compact red ironstone; Schmalkalden in Hessia, with brown ironstone; Schneeberg, with iron-mica; Ehrenfriedersdorf, with magnetic ironstone: Eibenstock, with ochry red ironstone; Berggieshübel, with common iron-glance; Suhl in Henneberg; and in Silesia, and Hungary.

America.—Chili.

Use.

At Suhl, in the dutchy of Henneberg, where it occurs in very considerable quantity, it is melted, and yields good iron.

Second Subspecies.

Ochry-red Ironstone or Red Ochre.

Ochriger Rotheisenstein, *Werner.*

Ochra Ferri rubra, *Wall.* t. ii. p. 259.—Rotheisenokker, *Wid.* s. 813.—Red Ochre, *Kirw.* vol. ii. p. 171.—Roth Eeisenokker, *Emm.* b. ii. s. 317.—Fer oxidé rouge grossier, *Haüy,* t. iv. p. 106, 107.—L'Ocre de Fer rouge, *Broch.* t. ii. p. 256.—Ochriger Rotheisenstein, *Reuss,* b. iv. s. 83. *Id. Lud.* b. i. s. 246. *Id. Suck.* 2ter th. s. 269. *Id. Bert.* s. 408. *Id. Mohs,* b. iii. s. 386. *Id. Leonhard,* Tabel. s. 65.—Fer oxidé rouge ocreux, *Brong.* t. ii. p. 166.—Ochriger Rotheisenstein, *Karsten,* Tabel. s. 66.—Ochriger Blutstein, *Haus.* s. 106.—Red Ochre, *Aikin,* p. 39.

External Characters.

Its colour is light blood-red, which inclines to brownish-red.

It is usually friable, but in some varieties it approaches and even passes into solid, and occurs as a coating on the other ores of iron; also disseminated, and sometimes massive.

It consists of dull dusty particles, which are very faintly glimmering.

It soils more or less strongly.

It feels more meagre than greasy.

Its streak is blood-red.

It is easily frangible.

Specific gravity, 2.947, *Wiedemann*. 3.00, *Aikin*.

Geognostic and Geographic Situations.

It occurs in veins, and is almost always accompanied with compact red ironstone, and red hematite, and sometimes sparry ironstone, but it is seldom quite pure, being usually mixed with other species of iron-ore.

Its geographic situation is nearly the same as that of the other species of red ironstone. It occurs particularly abundant in the Irrgang, near Platte in Bohemia.

Use.

It melts more easily than any of the other ores of iron, and affords excellent malleable iron.

Third

Third Subspecies.

Compact Red Ironstone.

Dichter Rotheisenstein, *Werner.*

Hæmatites ruber solidus, *Wall.* t. ii. p. 246.—Dichter Rotheisenstein, *Werner,* Pabst. b. i. s. 154. *Id. Wid.* s. 807.—Hématite compacte rouge, *De Born,* t. ii. p. 267.—Compact red Ironstone, *Kirw.* vol. ii. p. 170.—Dichter Rotheisenstein, *Emm.* b. ii. s. 310.—La Mine de Fer rouge compacte, *Broch.* t. ii. p. 251.—Dichter Rotheisenstein, *Reuss,* b. iv. s. 79. *Id. Lud.* b. i. s. 244. *Id. Suck.* 2ter th. s. 265. *Id. Bert.* s. 406. *Id. Mohs,* b. iii. s. 386. *Id. Hab.* s. 116. *Id. Leonhard,* Tabel. s. 65.—Fer oxidé rouge compact, *Brong.* t. ii. s. 165. —Dichter Rotheisenstein, *Karsten,* Tabel. s. 66.—Gemeiner Blutstein, *Haus.* s. 106.—Fer oligiste compacte, *Hauy,* Tabl. p. 95.—Compact Iron-glance, *Aikin,* p. 39.

External Characters.

Its colour is intermediate between dark steel-grey and brownish-red; sometimes, however, inclining more to the one, sometimes more to the other.

It occurs most commonly massive, sometimes also disseminated, specular, with impressions, in pyramidal supposititious crystals; and seldom crystallised in the following figures:

1. Cube.
2. Cube truncated on all the angles.
3. Cube deeply truncated on two diagonally opposite angles.

The crystals are middle-sized, small, and sometimes very intimately grown together.

The surface of the true crystals is smooth; that of the supposititious crystals is rough.

Externally the true crystals are glistening; the specular external shape splendent; the supposititious crystals dull; and the lustre is metallic.

Internally it is commonly only glimmering; but the slaty borders on glistening, and the lustre is semi-metallic.

The fracture is usually even, from which, although but seldom, it passes into coarse-grained uneven and into large conchoidal, and is sometimes slaty.

The fragments are indeterminate angular, and more or less sharp-edged.

It yields a blood-red streak.

It is generally intermediate between hard and semi-hard; sometimes, however, it passes from hard into semi-hard, and nearly into soft.

It is more or less easily frangible.

Specific gravity, 3.423, *Kirwan*. Cubic from Siberia, 3.760, *Kirwan*. From Lancashire, 3.5731, *Brisson*. 3.439, *Karsten*. Cubic, 3.961, *Bournon*.

Physical Character.

When pure, it does not affect the magnet.

Chemical Characters.

It becomes darker before the blowpipe, but is infusible either alone or with glass of borax, to which, however, it communicates an olive-green colour.

Constituent

Constituent Parts.

Oxide of Iron, - -	70.50
Oxygen, - - -	29.50
	100.00

Bucholz, in Gehlen's Journ, b. iii. s. 158.

Geognostic Situation.

It occurs in beds and veins in gneiss, clay-slate, grey-wacke, and various flœtz rocks, usually associated with red hematite and ochry red ironstone, quartz, hornstone, red jasper, and sometimes with red iron-flint, heavy-spar, and calcareous-spar. In some mines it is accompanied with specular iron-ore or common iron-glance, or with uran-mica.

Geographic Situation.

Europe.—It occurs in considerable quantity at Ulverstone in Lancashire; in the mine called Oevre-Aase in Norway; at Leerbach, Elbingerode, Andreasberg in the Hartz; Konigsberg, near Giessen in Hessia; at Schellerhau near Altenberg, Schneeberg, Johanngeorgenstadt, Eibenstock, Suhl, and Saalfeld, in Saxony; Rudelstadt in Silesia; in several iron-mines in Bohemia, Franconia, Bavaria, Salzburg, Spain, and France.

Uses.

As it affords good cast-iron, and also bar-iron, it is often smelted at iron-works.

Observations.

It passes on the one side into Specular Iron-ore or Common Iron-glance; on the other into Clay Ironstone, and sometimes also into Common Jasper.

Fourth

Fourth Subspecies.

Red Hematite or Fibrous Red Ironstone.

Rother Glaskopf, *Werner.*

Hæmatites ruber, *Wall.* t. ii. p. 247.—Rother Glaskopf, *Werner,* Pabst. b. i. s. 156. *Id. Wid.* s. 811.—Hématite rouge, *De Born,* t. ii. p. 288.—Red Hematites, *Kirw.* vol. ii. p. 168.—Rother Glaskopf, *Emm.* b. ii. s. 313.—L'Hématite rouge, *Broch.* t. ii. p. 254.—Fer oxidé hématite rouge, *Hauy,* t. iv. p. 105. 109. 111, 112.—Fasriger Rotheisenstein, *Reuss,* b. iv. s. 85.—Rother Glaskopf, *Lud.* b. i. s. 245.—Fasriger Rotheisenstein, *Mohs,* b. iii. s. 387. *Id. Hab.* s. 117. *Id. Leonhard,* Tabel. s. 65.—Fer oxidé rouge hematite, *Brong.* t. ii. p. 164.—Fasriger Rotheisenstein, *Karsten,* Tabel. s. 66.—Fasriger Blutstein, *Haus.* s. 106.—Fibrous Hematite, *Kid,* vol. ii. p. 171.—Fer oligiste concretionné, *Hauy,* Tabl. p. 95. —Red Hematite, *Aikin,* p. 38.

External Characters.

Its colour is usually intermediate between brownish-red and dark steel-grey. Some varieties incline to blood-red, others to dark steel-grey, and others to bluish.

It occurs most frequently massive and reniform; also botryoidal, stalactitic, and globular.

The external surface is generally rough and glimmering, seldom smooth and shining.

Internally it is usually glistening, which sometimes passes into glimmering, and the lustre is semi-metallic.

The fracture is always fibrous, and is straight, delicate and stellular, or scopiform.

The

The fragments are commonly cuneiform, seldom, as in the coarse fibrous, splintery.

It generally occurs in distinct concretions, which are large, small or fine angulo-granular, and traversed by others which are curved lamellar: more rarely it occurs in cuneiform prismatic concretions. The surface of the concretions is either smooth or streaked, and the colour inclines to iron-black, with a shining and metallic lustre.

The streak is always blood-red.

It is hard, passing into semi-hard.

It is brittle.

It is rather easily frangible.

It is heavy, inclining to uncommonly heavy.

Specific gravity, 4.740, *Gellert.* 5.005, *Kirwan.* 4.8983, *Brisson.* 4.840, *Wiedemann.* 5.025, *Ullmann.*

Constituent Parts.

Oxide of Iron, - -	90	94
Trace of Oxide of Manganese,	-	..
Silica, - - - -	2	2
Lime, - - - -	1	a trace
Water, - - - -	3	2
	96	98

Daubuisson, Ann. de Chimie, Sept. 1810.

Geognostic Situation.

It occurs in every situation where the compact subspecies is found, and like it in veins, beds, and lying masses *(liegende stöcke)* that approach in magnitude to mountain-masses, principally in primitive mountains, but also in transition and flœtz mountains. The different subspecies

subspecies frequently occur together, both in beds and in veins: in veins, it is the compact and ochry that predominate; the hematite occurs principally in drusy cavities, the walls of which are incrusted with the scaly subspecies.

Geographic Situation.

Europe.—It occurs in veins that traverse sandstone at Cumberhead in Lanarkshire; in veins in flœtz greenstone at Salisbury Craigs, near Edinburgh; at Ulverstone in Lancashire; in Cumberland; and also in Devonshire; and near Bristol in Gloucestershire. It is found in considerable quantity in Saxony, from Berggieshübel to Voightland; in Bohemia, but not so abundantly as in Saxony; Bareuth; Wolfstein in the Palatinate; Silesia; Lauterberg, Walkenried, Andreasberg, Wernigerode in the Hartz; and Salzburg.

Asia.—In Siberia.

America.—Mexico.

Uses.

It affords excellent malleable and cast iron; and when ground, it is also used for polishing tin, silver, and gold vessels, and for colouring iron brown.

Observations.

The name *Hematite*, which is derived from the Greek, αἱμα, *sanguis*, was given to this ore of iron from its red colour.

Observations

Observations on the Species in general.

1. Respecting its geographic situation, it is to be observed, that it occurs in great quantity in the kingdom of Saxony, less abundantly in the east side of the Hartz, and Bohemia; not so abundantly in the Fichtelgebirge, and in considerable quantity in Norway, Sweden, Poland, Hungary, and Russia. In England, it occurs particularly abundant in Lancashire.

2. As a test of the goodness of the iron obtained from this species of ore, it may be mentioned, that plate-iron and wire are prepared from it.

3. It is one of the most common species of ironstone.

9. Brown Ironstone.

Braun Eisenstein, *Werner.*

This species is divided into four subspecies, Scaly Brown Iron-ore, Ochry Brown Ironstone, Compact Brown Ironstone, and Brown Hematite.

First

First Subspecies.

Scaly Brown Iron-ore or Brown Iron-froth.

Brauner Eisenrahm, *Werner*.

Id. Werner, Pabst. b. i. s. 159. *Id. Wid.* s. 814.—Brown scaly Iron-ore, *Kirw.* vol. ii. p. 166.—Brauner Eisenrahm, *Emm.* b. ii. s. 318.—Le Eisenrahm brun, *Broch.* t. ii. p. 258.—Brauner Eisenrahm, *Reuss,* b. iv. s. 90. *Id. Lud.* b. i. s. 247. *Id. Suck.* 2ter th. s. 270. *Id. Bert.* s. 409. *Id. Mohs,* b. iii. s. 391. *Id. Leonhard,* Tabel. s. 65.—Schuppiger Brauneisenstein, *Karsten,* Tabel. s. 66.—Scaly Brown Iron-ore, *Aikin,* p. 39.

External Characters.

Its colour is intermediate between steel-grey and clove-brown, sometimes inclining more to the one, sometimes more to the other.

It is intermediate between friable and solid; some varieties are completely friable.

It occurs massive, coating, spumous, sometimes also fruticose, and irregular dendritic.

It is composed of scaly particles, which are intermediate between shining and glistening, with a metallic lustre.

The fracture of the massive varieties is glimmering, with a semi-metallic lustre.

It soils strongly.

It is more or less cohering, sometimes verging on solid.

It feels greasy.

It is sometimes light, sometimes swimming.

Chemical

Chemical Characters.

It blackens before the blowpipe, but does not melt, and gives to glass of borax an olive-green colour.

Geognostic Situation.

It occurs almost always lining drusy cavities in brown hematite. These cavities occur more frequently in hematite which is found in veins, than that which is found in beds.

Geographic Situation.

Europe.—It is found near Sandlodge in Mainland, one of the Zetland Islands; Schmalkalden in Hessia; Clausthal in the Hartz; Bredgangs Mine in Norway; Schmottseifen in Silesia; Grosskamsdorf, and Voightsberg in Saxony; Nebra and Naila in Bareuth; Lautereck in the Palatinate; Rathhausberg in Salzburg; Carinthia; Carniola; Stiria.

America.—Chili.

Observations.

At Kamsdorf, it is known under the names *Eisennann* and *Eisenblüthe.*

Second

Second Subspecies.

Ochry Brown Ironstone.

Ockriger Brauneisenstein, *Werner.*

Ochra ferri flava? *Wall.* t. ii. p. 258.—Ochra Ferri fusca, *Ibid.* p. 344.—Braune Eisenokker, *Wid.* s. 819.—Brown Iron-Ochre, *Kirw.* vol. ii. p. 167.—Braune Eisenokker, *Emm.* b. ii. s. 327.—Fer oxydé rubigineux pulverulent, *Hauy,* t. iv. p. 108. d.—L'Ocre de Fer brune, *Broch.* t. ii. p. 263.—Ockriger Brauneisenstein, *Reuss,* b. iii. s. 96. *Id. Lud.* b. i. s. 248. *Id. Suck.* 2ter th. s. 275. *Id. Bert.* s. 412. *Id. Mohs,* b. iii. s. 394. *Id. Hab* s. 119. *Id. Leonhard,* Tabel. s. 65. —Fer oxidé brun ocreux, *Brong.* t. ii. p. 172.—Ochriger Brauneisenstein, *Karsten,* Tabel. s. 66. *Id. Haus.* s. 108.—Fer oxydé pulverulent? *Hauy,* Tabl. p. 98.—Ochrey Brown Ironstone, *Aikin,* p. 40.

External Characters.

Its colour is very light yellowish-brown, which in some varieties inclines to ochre-yellow, in others to clove-brown.

It occurs massive and disseminated.

It is intermediate between solid and friable.

It is composed of dull, seldom very faintly glimmering, coarse earthy particles.

It soils strongly.

It is more or less cohering.

It is intermediate between brittle and sectile.

Constituent

Constituent Parts.

Peroxide of Iron, -	83
Water, - - -	12
Silica, - - - -	5
	100

Daubuisson, Ann. de Chim.
September 1810.

Geognostic Situation.

For the geognostic situation, see the observations on the species in general.

Geographic Situation.

It is found at Shotover Hill in Oxfordshire; Kongsberg and Arendal in Norway; Iberg, near Grund in the Hartz; Grosskamsdorf in Saxony; Nassau; Orpes and Kupferberg in Bohemia; Upper Palatinate; Rott in Bavaria; Hüttenberg in Carinthia; Salzburg.

Third Subspecies.

Compact Brown Ironstone.

Dichter Brauneisenstein, *Werner.*

Id. Werner, Pabst. b. i. s. 160. *Id. Wid.* s. 815.—Compact Brown Ironstone, *Kirw.* vol. ii. p. 165.—Dichter Brauneisenstein, *Emm.* b. ii. s. 321.—La Mine de Fer brune compacte, *Broch.* t. ii. p. 259.—Dichter Brauneisenstein, *Reuss,* b. iv. s. 93. *Id. Lud.* b. i. s. 247. *Id. Suck.* 2ter th. s. 247. *Id. Bert.* s. 410. *Id. Mohs,* b. iii. s. 394.—Dichter Brauneisenstein, *Hab.* s. 119. *Id. Leonhard,* Tabel. s. 65.—Fer oxidé brun compacte, *Brong.* t. ii. p. 168.—Gemeiner Brauneisenstein, *Karsten,* Tabel. s. 66.—Dichter Brauneisenstein, *Haus.* s. 108.—Compact Brown Ironstone, *Aikin,* p. 40.

External Characters.

Its colour is clove-brown, of different degrees of intensity. It frequently exhibits a pavonine and bronze-like tarnish.

It occurs most commonly massive and disseminated, frequently also cylindrical, small reniform, and botryoidal, parallel circo-cellular, with pyramidal impressions; and very rarely in supposititious crystals, of which the following are known:

1. Small cube, from common iron-pyrites.
2. Rhomboid, from sparry ironstone.
3. Pentagonal dodecahedron, from common iron-pyrites.
4. Octahedron, from radiated pyrites.

It occurs also in the form of corallites, madreporites, and fungites.

Internally

Internally it is usually semi-metallic glimmering.

The fracture is most commonly even, sometimes also large and flat conchoidal; sometimes fine-grained uneven and fine earthy.

The fragments are indeterminate angular, and more or less blunt-edged.

The streak is yellowish-brown, passing into ochre-yellow.

It is semi-hard, sometimes inclining to soft.

It is not particularly brittle.

It is easily frangible.

Specific gravity, 3.5027, the cubic, } *Brisson.*
3.4771, }
3.551, from Bayreuth, } *Kirwan.*
3.753, from the Tyrol, }
3.073, *Wiedemann.*
3.40, *Daubuisson.*

Chemical Characters.

Before the blowpipe, its colour darkens, and it becomes magnetic: to glass of borax it communicates an olive-green colour.

Constituent Parts.

	Bergzabern.	Vicdessos.	Pyrenees.
Peroxide of Iron,	84	81	81
Water, - -	11	12	11
Oxide of Manganese,	1	..	a trace
Silica, - -	2	4	2
Alumina, - -	..	..	a trace
	98	97	94

Daubauisson, Annal. de Chim. Sept. 1810.

 Geognostic

Geognostic Situation.

It occurs in the same geognostic situation as the following subspecies. It is always accompanied with ochry and fibrous brown ironstone.

Geographic Situation.

Europe.—It is found near Sandlodge in Mainland, the largest of the Zetland Islands; Lauterberg, and Blankenburg in the Hartz; Schmalkalden in Hessia; Saye and Altenkirchen in Westerwald; Schwarzenberg, Schneeberg, Scheibenberg, Grosskamsdorf, Voightsberg in Voightland; Sahlberg, Konitz, and Suhl, in Thuringia; Nassau; Kupferberg, Auspaner mountains near Pressnitz, Wisterschan near Töplitz, Stiahlan near Rakowa in Bohemia; gold mine near Schreiberau, Silesia; Upper Palatinate; Lower Palatinate; Dutchy of Deux-Ponts; Naila in Bayreuth; Suabia; Tyrol; Salzburg; Stiria; Vellach, Hüttenberg, and Eisenaach, in Carinthia; Hungary; Transylvania; France.

Asia.—Beresof and Catharinenburg in Siberia.

America.—United States.

Fonrth

Fourth Subspecies.

Brown Hematite, or Fibrous Brown Ironstone.

Brauner Glasskopf, *Werner*.

Id. Werner, Pabst. b. i. s. 161. *Id. Wid.* s. 817.—Brown Hematite, *Kirw.* vol. ii. p. 163.—Brauner Glaskopf, *Emm.* b. ii. s. 323.—Fer oxidé Hematite brun, *Hauy,* t. iv. p. 105.—L'Hematite brun, *Broch.* t. ii. p. 261.—Fasriger Brauneisenstein, *Reuss,* b. iv. s. 98.—Brauner Glaskopf, *Lud.* b. i. s. 248. *Id. Suck.* 2ter th. s. 273. *Id. Bert.* s. 411. *Id. Mohs,* b. iii. s. 400. —Fasriger Brauneisenstein, *Hab.* s. 120.—Fer oxidé brun fibreux, *Brong.* t. ii. p. 168.—Fasriger Brauneisenstein, *Karsten,* Tabel. s. 66. *Id. Haus.* s. 107.—Brown hematitic Iron-ore, *Kid,* vol. ii. p. 176.—Fer oxidé hematite, & Fer oxidé noire vitreux, *Hauy,* Tabl. p. 98.—Brown Hematite, *Aikin,* p. 40.

External Characters.

The surface of the fresh fracture is clove-brown, which in some varieties passes into steel-grey; in others into blackish-brown, and brownish-black; in others into light yellowish-brown and ochre-yellow. The external surface is tarnished velvet-black and bluish-black; sometimes also steel-grey, pinchbeck-brown, pavonine, and iridescent.

It seldom occurs massive, more frequently stalactitic, coralloidal, reniform, botryoidal, tuberose; sometimes also cylindrical, fructicose, dendritic, large and small cellular, in supposititious six-sided pyramids; and in the following true crystals:

1. Rectangular four-sided prism, which is sometimes so short as to appear like a cube.

 2. Preceding

2. Preceding figure, truncated on all the lateral edges
3. The prism, in which two opposite planes are very broad and two opposite planes very narrow, so that it has a tabular aspect.
4. The flattened prism, or table, in which the smaller lateral planes are bevelled.
5. The preceding figure, in which the angles formed by the meeting of the bevelling planes and the terminal planes are bevelled.
6. Rectangular four-sided prism, flatly acuminated with four planes, which are set on the lateral planes.

The crystals are very small and microcoscopic.

The external surface of the particular external shapes is sometimes smooth, sometimes granulated, but seldom rough or drusy.

Externally it is usually splendent.

Internally it is glistening, which sometimes, however, passes into glimmering, sometimes even into splendent; and the lustre is pearly or silky in the fibrous, and resinous in the conchoidal varieties.

The fracture is fibrous, being either long and delicate, or short and coarse, and straight or curved, and generally stellular or scopiform. The long fibrous sometimes becomes very delicate, and passes into imperfect or perfect conchoidal, and the short fibrous passes into narrow radiated. The conchoidal varieties have the strongest lustre; the delicate fibrous, the darkest colour *.

The

* The following agreements of colour and fracture are deserving of attention:

1. Clove-brown is long and delicate fibrous.
2. The varieties that incline to blue, are short and coarse fibrous.
3. Black, is extremely delicate fibrous, verging on conchoidal.
4 Blackish-brown, is radiated.

The fragments are sometimes splintery; sometimes wedge-shaped; seldom indeterminate angular.

The long and delicate fibrous varieties occur in curved lamellar and longish granular distinct concretions; the lamellar concretions intersect the granular: it occurs also in cuneiform prismatic distinct concretions.

The concretions are not so distinct as those of red hematite.

It is generally opaque; the brownish-black is weakly translucent on the edges.

The streak is ochre-yellow.

It is semi-hard; it is softer than red hematite.

It is brittle.

It is very easily frangible.

Specific gravity, 3.789, *Gellert.* 3.951, *Kirwan.* 4.029, *Wiedemann.*

Chemical Characters.

It becomes black before the blowpipe, and dissolves with some ebullition in glass of borax, to which it communicates an olive-green colour.

Constituent Parts.

Fibrous.

	Bergzabern.	Vicdessos
Peroxide of Iron,	79	82
Water, - -	15	14
Oxide of Manganese,	2	2
Silica, - -	3	1
	99	99

Daubuisson, Ann. de Chim. Sept 1810.

Resinous and Conchoidal.

Oxide of Iron,	80.25
Water, -	15.00
Silica, - -	3.75
	99.00

Vauquelin, Hauy, Tabl. Comp. 274.

Geognostic

Observations.

Brown Ironstone is readily distinguished from *Red Ironstone*, by its yellow streak, and inferior specific gravity; also by the water which it contains, it being a hydrate of iron: further, brown ironstone is generally associated with sparry ironstone, but rarely with red ironstone.

Chemical Properties, Geognostic and Geographic Situations.

These are contained in the following Observations on the Species in General.

A. *Chemical Properties, &c.*

1. This species of ironstone melts easily, and affords usually from 40 to 60 *per cent.* of iron. The cast-iron which it affords is indifferent, and the vessels made of it are not so fine as those manufactured from the cast-iron of red ironstone, and other ores of iron. The wrought iron obtained from it is very malleable, and at the same time hard: hence it is advantageously used in cases where softer iron would not answer. It also affords excellent steel, which is conjectured to be owing to the manganese it contains.

2. When it is intermixed with quartz, it affords a cold-short iron; but if with copper-pyrites, a red-short iron. It would appear, however, to require a greater quantity of sulphur to produce red-short iron from this species, than from most of the other ores of iron, and this is conjectured to be owing to the manganese which it contains.

3. It melts usually without a flux; and when one is necessary, clay-slate is that which is generally used.

4. The brown hematite affords the greatest quantity of iron of any of the subspecies.

B. *Geognostic*

B. *Geognostic Situation.*

It occurs in primitive, transition, and flœtz mountains, but more frequently in the two latter: and when in primitive mountains, in those only which are of newer formation. Its repositories are veins, beds, lying masses (*liegende stöke*), and mountain-masses (*stück gebirge*). When it occurs in veins and lying masses, the compact and ochry subspecies form the principal mass. The brown hematite occurs often in cavities in these veins or beds, but it does not fill them up; it only lines their walls, and is again covered by scaly brown iron-ore; so that here ochry and compact brown ironstone are the oldest, and the scaly brown ironstone the newest formation. It is usually accompanied with sparry ironstone, calcareous-spar, brown-spar, and heavy-spar; less frequently with black ironstone, and rarely with quartz, and red ironstone. Quartz, which occurs so frequently with red ironstone, seldom appears with brown ironstone: on the contrary, it is accompanied with heavy-spar, calcareous-spar, and in some places with fluor-spar.

C. *Geographic Situation.*

Europe.—It occurs in veins in sandstone, along with heavy-spar, at Cumberhead in Lanarkshire; in a similar repository in Mainland, one of the Zetland Islands; and in the Island of Hoy, one of the Orkney group. Small veins of it are met with in the flœtz greenstone of Salisbury Craigs, near Edinburgh. It also occurs at Schneeberg, Scheibenberg and Raschau in the Erzgebirge; and at Kamsdorf, where it (principally the ochry subspecies) occurs in flœtz rocks, in beds, which are sometimes so thick

thick that they nearly form lying masses. A part of this deposition passes into Schwarzburg, as far as Pönitz, and even reaches to Henneberg, where there are very extensive ironworks. Further, it is found in very considerable quantity all around the Fichtelgebirge, and there are ironworks for smelting this ore, both on the Saxon and Bohemian sides, and in that part of it which belongs to Bayreuth. It occurs in beds in the Upper Palatinate, and in Franconia. It is less abundant in the Hartz, where, at Iberg near Grün, the ochry brown ironstone occurs in *putzenwerke* in limestone. Very considerable mines of it are met with in Nassau, Hessia, and Westerwald; and it also occurs in the Tyrol, Carinthia, Stiria, Upper Italy, and in the southern provinces of France.

It may be remarked, that northern countries, such as Sweden and Lapland, which possess so great an abundance of magnetic ironstone and iron-glance, contain but small quantities of this species, which occurs so abundantly in the Hartz, Stiria, Carinthia, Hungary, Saxony, Westphalia, the county of Nassau, and other districts.

10. Umber.

Argile ocreuse brun? *Hauy,* t. iv. p. 446.—Umbra, *Reuss,* b. iv. s. 139.—Umbra, *Leonhard,* Tabel. s. 67. *Id. Karsten,* Tabel. s. 66. *Id. Haus.* Handbuch, b. i. s. 276.

External Characters.

Its colour is clove-brown, which passes into blackish and yellowish brown.

It occurs massive.

Internally

Internally it is dull or glimmering, and resinous.
The fracture is flat conchoidal, passing into fine earthy.
The fragments are blunt-edged.
It is soft, inclining to very soft.
It is rather sectile.
It soils strongly.
It is very easily frangible.
It feels meagre.
It adheres strongly to the tongue.
It readily falls to pieces in water.
Specific gravity, 2.060, *Ullmann*.

Constituent Parts.

	From Cyprus.
Oxide of Iron, - -	48
Oxide of Manganese, -	20
Silica, - - - - -	13
Alumina, - - -	5
Water, - - - - -	14
	100

Klaproth, Beit. b. iii. s. 140.

Geognostic and Geographic Situations.

It occurs in beds in the Island of Cyprus.

Use.

It is used as a pigment.

Observation.

Other minerals are known under the name Umber, particularly Earth-coal, the Humus umbra of Wallerius.

11. Black

11. Black Ironstone.

Schwarzeisenstein, *Werner.*

This species is divided into two subspecies, viz. Compact Black Ironstone, and Black Hematite.

First Subspecies.

Compact Black Ironstone.

Dichter Schwarzeisenstein, *Werner.*

Black Ironstone, *Kirw.* vol. ii. p. 167.—Mine de Fer noire compacte, *Broch.* t. ii. p. 268.—Dichter Schwarzeisenstein, *Reuss,* b. iv. s. 103. *Id. Mohs,* b. iii. s. 414. *Id. Leonhard,* Tabel. s. 66. *Id. Karsten,* Tabel. s. 66.—Dichter Manganschwärze, *Haus.* s. 109.—Black Hematitic Iron-ore, *Kid,* vol. ii. p. 176. —Black Iron-ore, *Aikin,* p. 41.

External Characters.

Its colour is intermediate between bluish-black and dark steel-grey, but more inclining to the first.

It occurs massive, tuberose, small reniform, botryoidal, fructicose, and claviform.

The external shapes have a rough glimmering, or faintly glistening surface.

Internally it is glimmering, passing into glistening, and the lustre is semi-metallic.

The fracture is usually conchoidal, but sometimes passes into fine and small grained uneven.

The

The fragments are indeterminate angular, and more or less sharp-edged.

It sometimes occurs in distinct concretions, which are thin and concentric curved lamellar, and bent in the direction of the external surface.

The streak is shining, but its colour remains unchanged.

It is semi-hard.

It is brittle.

It is easily frangible.

It is heavy.

Specific gravity, 4.750, *Ullmann*.

Second Subspecies.

Black Hematite.

Schwarzer Glaskopf, *Werner*.

Black Ironstone, *Kirw.* vol. ii. p. 167.—Mine de Fer noire compact, *Broch.* t. ii. p. 268.—Fasriger Schwarzeisenstein, *Reuss*, b. iv. s. 105. *Id. Mohs*, b. iii. s. 415. *Id. Leonhard*, Tabel. s. 66. *Id. Karsten*, Tabel. s. 66.—Fasriger Manganschwärze, *Haus.* s. 109.—Black Iron-ore, *Aikin*, p. 41.

External Characters.

Its colour inclines more to steel-grey than the preceding subspecies.

It occurs massive and reniform.

Internally it is glimmering, often even glistening, and the lustre is semi-metallic.

The fracture is extremely delicate fibrous, which passes into

into even ; the fibrous is either curved or straight, and scopiform or stellular diverging.

The fragments are cuneiform.

It occurs in large and coarse granular distinct concretions.

The streak is shining.

In other characters it agrees with the preceding subspecies.

Chemical Characters.

When melted before the blowpipe with borax, it yields a violet-blue coloured glass.

Constituent Parts.

Its composition is not known. It is supposed to contain more Manganese than brown ironstone, and alse Alumina and Calcareous Earth.

Geognostic Situation.

It occurs in veins, in primitive, transition, and flœtz mountains, and is usually accompanied with brown ironstone and quartz.

Geographic Situation.

It occurs in several places in the Saxon Erzgebirge, and the Hartz ; but more frequently in Thuringia and Westphalia.

Uses.

It is very easily fusible, and yields a good iron ; but it has the inconvenience of acting very powerfully on the sides of the furnace.

Observations.

Observations.

1. It does not occur any where in considerable quantity; and of the two subspecies, the hematitic is the rarest.

2. Daubuisson analysed an ore sent from Freyberg as black ironstone, but which contained no iron. Its specific gravity was only 3.6, a proof of its being a distinct substance from that now described.

12. Sparry Ironstone.

Spatheisenstein, *Werner.*

Minera Ferri alba, *Wall.* t. ii. p. 251.—*P. J. Hjelm,* Chemisk och Mineralogisk Afhandling om huita Järnmalmer, Upsala, 1774.—Mine de Fer spathique, *Romé de Lisle,* t. iii. p. 281.—Spathiger Eisenstein, *Werner,* Pabst. b. i. s. 164. *Id. Wid.* s. 820.—Calcareous or Sparry Iron-ore, *Kirw.* vol. ii. p. 190.—Fer spathique, ou Mine de Fer blanche, *De Born,* t. ii. p. 290.—Fer spathique, *Lam.* t. i. p. 263.—Chaux carbonatée ferrifere avec Manganese, *Hauy,* t. iv. p. 117, 118.—La Mine de Fer spathique, ou le Fer spathique, *Broch.* t. ii. p. 264.—Spatheisenstein, *Reuss,* b. iv. s. 107. *Id. Lud.* b. i. s. 249. *Id. Suck.* 2ter th. s. 278. *Id. Bert.* s. 428. *Id. Mohs,* b. iii. s. 407. *Id. Hab.* s. 124. *Id. Leonhard,* Tabel. s. 66.—Fer spathique, *Brong.* t. ii. p. 175.—Spath Eisenstein, *Karsten,* Tabel. s. 66.—Eisenspath, *Haus.* s. 129.—Sparry Iron-ore, *Kid,* vol. ii. p. 188.—Fer oxidé carbonatée, *Hauy,* Tabl. p. 99.—Sparry Iron-ore, *Aikin,* p. 41.

External Characters.

Its colour is pale yellowish-grey, bordering on greyish-white, and sometimes greenish-grey. On exposure to the

the weather, the colour changes more or less rapidly, from the surface to the interior, becoming first cream or ochre yellow, then blackish-brown, and lastly, brownish-black. The surface is sometimes tarnished gold-yellow, columbine, and pavonine.

It occurs massive, disseminated, with pyramidal impressions, in plates, and crystallised.

The crystallisations are the following:

1. Rhomboid, which is either perfect, or truncated on the two diagonally opposite obtuse angles, and the faces are generally curved, and either convex or concave.
2. Lens, either spherical or saddle-shaped.
3. Octahedron, in which the faces are either straight or convex, and sometimes truncated on the angles.
4. Garnet or rhomboidal dodecahedron, which is formed by an aggregation of rhombs.

The crystals are seldom large, commonly middle-sized and small, sometimes even very small.

The surface of the octahedron is smooth, specular, and splendent; that of the other crystallisations is rough, or drusy and glistening.

Internally it varies from splendent to glistening and glimmering, and the lustre is pearly.

The fracture is more or less perfect foliated, sometimes straight, more frequently curved foliated. The cleavage is threefold, and the folia intersect each other obliquely *. Some varieties are fibrous †.

The

* According to Dr Wollaston, the obtuse angles of the rhomboid measure 107°. It may, however, be remarked, that the curvilinear direction of the faces must affect the accuracy of goniometrical measurements.

† Leonhard, in Annal. d. Wetterauische Gesellsch. b. iii. s. 13.

The imperfect foliated fracture is sometimes conjoined with the splintery, and this occurs principally in the greenish-grey varieties.

The fragments are very oblique rhomboids, like those of calcareous-spar.

It is generally translucent on the edges, also translucent; but the black varieties are opaque.

It occurs in granular distinct concretions, which are of all degrees of magnitude, from large to fine.

The streak is yellowish-brown.

It is semi-hard, (harder than calcareous-spar), which in the darker varieties inclines to soft.

It is not particularly brittle.

It is easily frangible.

It is heavy.

Specific gravity, 3.784, *Gellert.* 3.640,–3.810, *Kirwan.* 3.672, *Brisson.* 3.300 to 3.600, the decomposed, *Kirwan.* 3.693, *Guyton.* 3.600—3.900, *Collet-Descotils.*

Chemical Characters.

It blackens, and becomes magnetic before the blowpipe, but does not melt: it effervesces with muriatic acid. It dissolves with ebullition in glass of borax, and communicates to it an olive-green colour.

Constituent Parts.

	Dankerode.	Baireuth.	Baireuth.		Fibrous from Steinheim.
Oxide of Iron,	57:50	58.00	59.50		63.75
Carbonic Acid,	36.00	35.00	36.00		34.00
Oxide of Manganese,	3.50	4.25	a trace		0.75
Lime, - -	1.25	0.50	2.50		...
Magnesia, -	-	0.75			0.25
Water, - -	-	-	2.00	Loss,	1.25
	98.25	98.50	99.00		100.00
	Klaproth, Beit. b. iv. s. 115.	Ibid. s. 115.	*Bucholz.*		*Klaproth,* in Magaz. Nat. Fr. b. v. s. 335.

Geognostic Situation.

It occurs in veins and beds, and mountain-masses, in primitive, transition, and flœtz rocks; and in all these situations, it is frequently associated with brown, red, and black ironstone, calcareous-spar, and quartz. In other venigenous formations which traverse gneiss, mica-slate, clay-slate, and grey-wacke, it is associated with ores of silver, lead, cobalt, copper, seldomer with ores of nickel, and bismuth; more frequently with galena or lead-glance, grey copper-ore, iron-pyrites, and grey antimony-ore. It also occurs filling up amygdaloidal cavities in trap rocks,

Geographic Situation.

Europe.—It is found in small quantities in different places in England, Scotland, and Ireland; also in Saxony, Bohemia, Bayreuth, Upper Palatinate, Silesia, Koburg, Savoy, Switzerland, Sweden, and Norway: but it is only in the following countries where it is found in such quantity as to be employed as an ore of iron:—In the Fichtelgebirge; the black variety occurs in great quantity at Schmalkalden in Hessia, where it has been mined and smelted for many centuries; in the Hartz, as at Clausthal, Iberg, Blankenburg, and Stollberg, it occurs less abundant; in Westphalia, the light-coloured is mined in great quantity; Eisenerz and Schladinrig in Stiria, affords it in considerable quantity; Hüttenberg in Carinthia, Schwatz in the Tyrol, and Jauberling in Carniola, are well known for mines of sparry ironstone; in many places in Salzburg, in Hungary, as Schemnitz, Schmolnitz, Dopshau, and Siowinka, it occurs in small quantity; mines of it also exist in Piedmont, and France; and at Somororstro, in the province of Biscay in Spain, there

there is a whole hill composed of this species of ironstone, which has been worked for several hundred years. It is there accompanied with red ironstone, which renders the smelting very advantageous.

Asia.—In the mines of Catharinenburg.

America.—West Greenland; and Mexico.

Uses.

It affords an iron which is excellently suited for steel making. The black variety is said to afford the best kind of iron.

Observations.

1. It is nearly allied to brown ironstone, and brown-spar, and there is a transition from calcareous-spar through brown-spar, sparry ironstone to brown ironstone.

2. Cast-iron obtained from this species, or from brown ironstone, presents a whitish colour and radiated fracture; whereas that obtained from red ironstone, and several other ores of iron, has a dark grey colour, and a granular fracture. Further, the cast-iron obtained from this species can be converted into steel; but a great portion of that obtained from red ironstone, &c. passes to the state of malleable iron, long before the mass in the furnace has become steel. The steel obtained from this ore is said to contain a small portion of manganese, which is supposed to be the cause of its durability in the fire, and what renders it less liable to become soft and irony.

3. It generally occurs more or less weathered. By exposure to the air, it experiences a gradual decomposition, which has a great effect on its external aspect. This decomposition at first affects only the external colour, ex-

ternal lustre, and the transparency; but as it advances, it also changes even the structure, hardness, solidity, and weight of the mineral. The oxidation of the iron and manganese destroys the weak combination of these metals with the carbonic acid, and there is formed a hydrate of iron, sometimes also an oxide of iron, and hydrate of manganese. The whole mass is disintegrated by the escape of the carbonic acid, this acid combining with percolating water dissolves the small portion of lime in the ore, and also portions of the still undecomposed carbonate of iron and oxide of manganese. A knowledge of these changes enables us to understand the very different results obtained in the analysis of specimens more or less weathered or decomposed, and also throws some light on the different results obtained in the smelting of sparry ironstone more or less decomposed.

4. The analysis of Hielm, published under the sanction of Bergmann, is the earliest we possess of this ironstone: it gives as the constituent parts, 22.38 Oxide of Iron: 24.28 Oxide of Manganese: 29.43 Carbonate of Lime; and 6.9 Water. The errors of this analysis have been pointed out and corrected by the labours of Drappier, Descotils, Berthier, Klaproth, and Bucholz.

13. Clay Ironstone.

Thoneisenstein, *Werner*.

This species is divided into the following subspecies, viz. Reddle, Columnar Clay-Ironstone, Lenticular Clay-Ironstone, Jaspery Clay-Ironstone, Common Clay-Ironstone, Reniform or Kidney-shaped Clay-Ironstone, and Pea-ore or Pisiform Clay-Ironstone.

First

First Subspecies.

Reddle.

Roethel, *Werner*.

Ochra Ferri rubra, cretacea solida, rubrica, *Wall.* t. ii. p. 260. —Rother Eisenokker, *Wid.* s. 813.—Argile martiale rouge, Sanguine ou Crayon rouge, *De Born,* t. ii. p. 230.—Röthel, *Emm.* b. ii. s. 350.—Argile ocreuse rouge graphique, *Hauy,* t. iv. p. 445, 446.—Le Crayon rouge, *Broch.* t. ii. p. 271.— Röthel, *Reuss,* b. iv. s. 124. *Id. Lud.* b. i. s. 251.—Rother Thoneisenstein, *Suck.* 2ter th. s. 289.—Röthel, *Bert.* s. 425. *Id. Mohs,* b. iii. s. 418. *Id. Leonhard,* Tabel. s. 66.— Ochriger Thoneisenstein, *Karsten,* Tabel. s. 66.

External Characters.

Its colour is light brownish-red, which sometimes inclines to steel-grey; seldomer reddish-brown, and intermediate between brick-red and blood-red.

It occurs massive.

The principal fracture is glimmering; the cross fracture is dull.

The principal fracture is fine slaty; the cross fracture is earthy.

The fragments are sometimes tabular, and sometimes splintery.

The streak is lighter and more shining than that of the fracture surface.

It is soft, and very soft.

It soils strongly, and writes.

It is sectile.

It

It is easily frangible.

It adheres strongly to the tongue.

It feels meagre.

Specific gravity, 3.391, *Blumenbach*. 3.1391, *Brisson*. 3.805, *Ullmann*.

Chemical Characters.

Exposed to a red heat, it decrepitates and becomes black; at the temperature of 159° it melts into a greenish-grey spumous enamel.

Geognostic Situation.

It occurs in beds in newer clay-slate, and in nests and kidneys in those varieties of slate-clay that incline to clay-slate; also in sandstone, and limestone.

Geographic Situation.

Europe.—It occurs in Hessia, Thuringia, Upper Lusatia, Silesia, and Salzburg.

Asia.—Jelschansk in Siberia.

Uses.

It is principally used for drawing. The coarser varieties are used by the carpenter, the finer by the painter. It is either used in its natural state, or it is pounded, washed, and mixed with gum, and cast into moulds. The crayons thus formed, when intended for coarse drawings, are mixed with but a small portion of gum; but those which are to be used for small and delicate drawings, with a much greater proportion, in order to give them sufficient hardness.

Observations.

Observations.

1. It is usually called *Red Chalk.*
2. It is never smelted as an ore of iron.

Second Subspecies.

Columnar Clay-Ironstone.

Stänglicher Thoneisenstein, *Werner.*

Id. Werner, Pabst. b. i. s. 167.—Var. of Gemeiner Thoneisenstein, *Wid.* s. 825.—Columnar or Scapiform Iron-ore, *Kirw.* vol. ii. p. 176.—Fer oxidé rouge bacillaire, *Hauy,* t. iv. p. 107.—Le Fer argilleux scapiforme, *Broch.* t. ii. p. 273.—Stänglicher Thoneisenstein, *Reuss,* b. iv. s. 115. *Id. Lud.* b. i. s. 251. *Id. Suck.* 2ter th. s. 283. *Id. Bert.* s. 422. *Id. Mohs,* b. iii. s. 419. *Id. Leonhard,* Tabel. s. 66.—Fer terreux argilleux bacillaire, *Brong.* t. ii. p. 173.—Stänglicher Blutstein, *Haus.* s. 106.—Stänglicher Thoneisenstein, *Karsten,* Tabel. s. 66.—Columnar Clay-Ironstone, *Aikin,* p. 42.

External Characters.

Its colour is brownish-red, which passes on the one side into cherry-red, and on the other into clove-brown.

It occurs massive, and in angular pieces.

Internally it is dull.

The fracture is fine earthy.

It occurs almost always in columnar distinct concretions, which are straight or curved, and thick or thin; usually parallel; sometimes scopiform diverging; and also jointed.

The surface of the concretions is rough and dull.

The streak is blood-red.
It is soft.
It is brittle.
It is uncommonly easily frangible.
It adheres slightly to the tongue.
In single pieces, it gives a ringing sound?
Specific gravity, 4.026, *Ullmann.*

Chemical Characters.

It becomes black before the blowpipe, bubbles up with borax, and communicates to it an olive green and blackish colour.

Constituent Parts.

Oxide of Iron, - -	50.00
Water, - - -	13.00
Silica, - -	32.50
Alumina, - - -	7.00

Brocchi, Trattato, &c. vol. ii. p. 119.

Geognostic and Geographic Situations.

It is a rare mineral, and is probably in some cases a pseudo-volcanic product; for it is found along with earthy-slag, porcelain-jasper, and burnt clay, in the neighbourhood of pseudo-volcanoes. It is also found in other countries, where there are no volcanoes, as in the Island of Arran, in the Frith of Clyde.

Besides the Island of Arran, already mentioned, it is also found at Hoschnitz and Delau, in the Saatzer circle, Straska and Schwintschitz in the circle of Leutmeritz in Bohemia; Amberg in the Upper Palatinate; Dutweiler in Saarbruck.

Third

Third Subspecies.

Lenticular Clay-Ironstone.

Linsenförmiger Thoneisenstein, *Werner.*

Id. Werner, Pabst. b. i. s. 167. *Id. Wid.* s. 826.—Acinose Iron-ore, *Kirw.* vol. ii. p. 177.—Körniger Thoneisenstein, *Emm.* b. ii. s. 342.—Le Fer argileux grenu ou lenticulaire, *Broch.* t. ii. p. 274.—Körniger Thoneisenstein, *Reuss,* b. iv. s. 120. —Linsenförmiger Thoneisenstein, *Lud.* b. i. s. 252. *Id. Suck.* 2ter th. s. 285. *Id. Bert.* s. 423. *Id. Mohs,* b. iii. s. 420. *Id. Leonhard,* Tabel. s. 66.—Fer oxide brun granuleux, *Brong.* t. ii. s. 170.—Körniger Thoneisenstein, *Karsten,* Tabel. s. 66. —Körniger Blutstein, *Haus.* s. 106.——Lenticular Clay Ironstone, *Aikin,* p. 42.

External Characters.

Its colour is brownish-red, which passes on the one side into steel-grey, on the other into reddish-brown, yellowish-brown, blackish-brown, and greyish-black.

It occurs massive.

Internally it is always strongly semi-metallic glimmering, which passes into glistening.

On account of the smallness of the concretions, it is difficult to ascertain the kind of fracture, yet it appears to be sometimes fine earthy, and sometimes (particularly the red varieties) slaty.

The fragments are indeterminate angular, and blunt-edged.

It occurs in distinct concretions, which are sometimes small, sometimes fine and round granular, but more frequently lenticular.

The

The red yields a light-red coloured streak; the yellowish a light yellowish-brown streak; and the black a greyish-black streak.

It is soft; some varieties pass into very soft; others into semi-hard.

It is brittle; but inclining to sectile in the black variety.

It is very easily frangible.

It soils.

Specific gravity, 2.673? *Kirwan.* Red variety, 3.770–3.810, *Ullmann.* Brown variety, 3.018, *Ullmann.*

Physical Character.

The black variety is slightly affected by the magnet.

Constituent Parts *.

From Radnitz in Bohemia.	
Oxide of Iron,	64.0
Alumina, -	23.0
Silica, - -	7.5
Water, -	5.0
	99.5

Lampadius.

From Doubs.	
Peroxide of Iron,	73
Water, - -	14
Silica, - -	9
Peroxide of Manganese,	1
Loss, - -	3
	100

Daubuisson, Journ. des Mines, 1810.

Geognostic Situation.

The red, the brown, and the black varieties, appear to occur in different geognostic situations.

The red variety, which is usually in lenticular distinct concretions, occurs, according to the observations of Werner,

* The red Bohemian ore affords in the furnace 60 *per cent.* of iron; the brown from 30 to 36 *per cent.*

ner, in mountain-masses in transition mountains. The brown variety, which has usually round granular distinct concretions, occurs in thick beds, in shell limestone, and is sometimes so abundantly intermixed with the limestone, as to form the greater portion of the bed: it also occurs in beds in sandstone, and these beds, like the sandstone, contain petrified shells; thus occupying the place of the second flœtz gypsum formation. The black variety appears to occur in the same geognostic situation as the brown.

Geographic Situation.

The red occurs particularly abundant in Bohemia; the brown is found in Franconia, Bavaria, Salzburg, and extends into Switzerland, France, and even to the Netherlands. The black has been hitherto found only in the canton of Berne.

Uses.

The red variety melts excellently, and affords a malleable iron nearly as good as that obtained from the best kinds of red ironstone. It also affords excellent cast-iron.

The brown variety melts excellently, and affords both good cast and malleable iron.

The black variety, which is said to afford 90 *per cent.* of iron, melts badly, and affords an indifferent iron.

Observations.

1. The black variety appears to be the link which connects brown ironstone with magnetic ironstone. It is also magnetic.

2. The

2. The brown and red varieties appear to be intermediate between red and brown ironstone, and to form, as it were, the links that connect them.

Fourth Subspecies.

Jaspery Clay-Ironstone.

Jaspisartiger Thoneisenstein, *Werner.*

Jaspisartiger Thoneisenstein, *Reuss,* b. iv. s. 126. *Id. Lud.* b. i. s. 252. *Id. Suck.* 2ter th. s. 290. *Id. Mohs,* b. iii. s. 422. *Id. Leonhard,* Tabel. s. 66. *Id. Karsten,* Tabel. s. 66.—Jaspisartiger Gelbeisenstein, *Haus.* s. 107.

External Characters.

Its colour is brownish-red.

It occurs massive.

Internally it is feebly glimmering.

The fracture is flat conchoidal, which sometimes passes into even.

The fragments are rhomboidal, and sometimes cubical and trapezoidal.

In the streak it becomes somewhat lighter.

It is soft, passing into semi-hard.

It is brittle.

It is rather easily frangible.

It is heavy.

Geognostic and Geographic Situations.

It is found near Wienerisch-Neustadt, in beds that rest on transition limestone, and covered with sandstone belonging to a coal formation.

Observations.

Observations.

1. Its hardness, and the shape of its fragments, distinguish it from the other subspecies of Clay-Ironstone.

2. It is termed *Jaspery*, by reason of its resemblance in external aspect to jasper.

Fifth Subspecies.

Common Clay-Ironstone.

Gemeiner Thoneisenstein, *Werner.*

Id. Werner, Pabst. b. i. s. 165. *Id. Wid.* s. 823.—Common argillaceous Iron-orc, *Kirw.* vol. ii. p. 173.—Gemeiner Thoneisenstein, *Emm.* b. ii. s. 337.—Le Fer argileux commun, *Broch.* t. ii. p. 276.—Gemeiner Thoneisenstein, *Reuss,* b. iv. s. 127. *Id. Lud.* b. i. s. 259. *Id. Suck.* 2ter th. s. 281. *Id. Bert.* s. 421. *Id. Mohs,* b. iii. s. 422. *Id. Leonhard,* Tabel. s. 67.—Fer terreux argileux commun, *Brong.* t. ii. p. 172.—Gemeiner Thoneisenstein, *Karsten,* Tabel. s. 66.—Gemeiner Gelbeseinstein, *Haus.* s. 107.—Fer oxidé massif, *Hauy,* Tabl. p. 98.—Clay-Ironstone, *Aikin,* p. 41.

External Characters.

Its colour is light yellowish-grey, which inclines to light ash-grey, and passes on the one side into bluish-grey, on the other into yellowish, reddish, and clove brown, and into brownish-red.

The lightest coloured varieties change their colour on exposure to the air; they become first yellowish, then brownish, dark-brown, and lastly black. Some varieties (from Poland) change, on exposure, to bluish-grey and pearl-

pearl-grey. This change of colour is not confined to the surface, but extends to the centre of the mass.

It occurs massive, globular *, reniform, ovoidal, and frequently in extraneous shapes, as in that of bivalves, multivalves, with vegetable impressions, &c.

Internally it is dull or glimmering.

The fracture is generally earthy; sometimes, however, it is flat conchoidal and even; and other varieties shew a tendency to the slaty fracture.

The fragments are indeterminate angular and blunt-edged.

It is soft, and often very soft, and sometimes inclines to semi-hard.

It is rather brittle.

It is more or less easily frangible.

It adheres slightly to the tongue.

It feels meagre.

Specific gravity, 2.936, from Cathma at Raschau, *Kirwan*. 3.471, county of Roscommon in Ireland; and 3.205 to 3.357, Carron in Scotland, *Dr Rotheram*. 2.786—3.367, *Ullmann*.

Chemical Characters.

It becomes black and magnetic before the blowpipe, and gives with glass of borax, after a little ebullition, a dark red or blackish-green glass.

Constituent

---

* The globular and ovoidal masses, when split, frequently exhibit in their interior a miniature representation of basaltic columns, and the spaces between the little columns or prisms, are sometimes filled with different minerals, as calcareous-spar, sparry ironstone, galena or lead-glance, iron-pyrites, copper-pyrites, blende, mineral pitch, &c.

Constituent Parts.

	Blancheland.	Geis-lautern.	Coalbrooke-dale.
Oxide of Iron,	54.00	38.6	50.0
Oxide of Manganese,	2.40	1.8	2.6
Silica, -	13.00	32.0	10.6
Alumina, -	1.00	4.0	2.0
Lime, -	4.20	1.8	1.6
Magnesia, -	2.00	4.3	2.4
Carbonic Acid, and Water,	24.60	20.0	32.0
	101.2	102.5	101.2

Descotils, Ann. de Chim. for 1812, N. 251. p. 188.

	Brandau.	Brandau.
Oxide of Iron,	35	39
Silica, -	11	5
Alumina,	39	40
Magnesia,	2	6
Sulphur, -	3	1
Water, -	10	9
	100	100

Lampadius.

In the three first analyses, the iron is in the state of carbonate, in the two last in the state of oxide. It would appear, that the carbonated ironstones, by the process of decomposition, lose their carbonic acid, and are thus converted into oxidated varieties. In those common clay ironstones which have a yellow or brown streak, the iron is in the state of hydrate: in those having a red streak, in the oxidised state; and in most of the varieties with a grey streak, the iron is carbonated. When the carbonated varieties begin to decay, they assume a liver or reddish-brown colour, and become soft.

Geognostic Situation.

Common clay ironstone occurs in beds, in a variety of clay-slate belonging to the transition class of rocks, but most frequently and abundantly as a member of the flœtz class. The great coal-fields met with in this island, contain vast quantities of it, disposed in variously-shaped masses, in beds, or in veins, imbedded either in bituminous-shale, slate-clay, limestone, sandstone, or even coal, or or the trap rocks subordinate to the general coal-formation. It also occurs in beds, or imbedded in some of the flœtz limestone formations.

Geographic Situation.

Europe.—It occurs in vast quantity in different parts of Scotland and England, as in the coal-fields of the Forth and Clyde; in those of Shropshire, Staffordshire, Yorkshire, and South Wales; in the Faroe Islands it occurs in beds in flœtz-trap rocks; Westphalia; at Wehrau in Saxony, in a new flœtz formation; Silesia, in flœtz limestone; Bohemia; Franconia; Upper Palatinate; Salzburg; Poland; and Russia.

Asia.—In Siberia.

America.—In extensive beds near Baltimore in the United States.

Uses.

It is smelted as an ore of iron, and affords from 30 to 40 *per cent.* of good iron.

Observations.

1. The brown varieties sometimes pass into Compact Brown Ironstone, and the red varieties into Red Ironstone.

2. It bears considerable resemblance to Compact Limestone, and Indurated Clay, but is distinguished from them by its greater specific gravity, and complete opacity.

3. When a nodule of ironstone is hollow, and contains a loose nucleus or kernel, it is named *Ætites*, or *Eaglestone*, from an opinion that it was found in eagles nests The ancients held it in great veneration, for its fancied medicinal properties.

Sixth

Sixth Subspecies.

Reniform or Kidney shaped Clay-Ironstone.

Eisenniere, *Werner.*

Ætites, *Wall.* t. ii. p. 614.—Pierre d'Aigle, *Romé de Lisle,* t. iii. p. 300.—Eisenniere, *Werner,* Pabst. b. i. s. 167.—Var. of Bohnerz, *Wid.* s. 827.—Nodular Ironstone, *Kirw.* vol. ii. p. 178.—Fer limoneux spheroidal, *De Born,* t. ii. p. 283.—Eisenniere, *Emm.* b. ii. s. 344.—Pierre d'Aigle, *Lam.* t. i. p. 245.—Fer oxydé rubigineux geodique, *Hauy,* t. iv. p. 107, &c.—La Fer reniforme, *Broch.* t. ii. p. 278.—Eisenniere, *Reuss,* b. iv. s. 132. *Id. Lud.* b. i. s. 253. *Id. Suck.* 2ter th. s. 286. *Id. Bert.* s. 423. *Id. Mohs,* b. iii. s. 425. *Id. Leonhard,* Tabel. s. 67.—Fer oxidé brun ætite, *Brong.* t. ii. p. 169.—Schaaliger Thoneisenstein, *Karsten,* Tabel. s. 66.—Schaaliger Gelbeisenstein, *Haus.* s. 107.—Ætites or Eaglestone, *Kid,* vol. ii. p. 181.—Fer oxydé geodique, *Hauy,* Tabl. p. 98.

External Characters.

Its colour is yellowish-brown, but it shews various degrees of intensity, even in the same specimen: externally it is darker, approaching to blackish-brown; internally the colour is very light, and sometimes it includes an ochre-yellow kernel.

It occurs in masses, which are blunt-angular, roundish, tuberose, or more or less inclining to reniform.

The lustre of the external layers is glimmering and semi-metallic; that of the internal layers is dull.

The fracture towards the interior is fine earthy; towards the exterior, even; in the dark yellowish-brown varieties, nearly conchoidal; that of the ochre-yellow, even.

The fragments are indeterminate angular.

The individual reniform and other shaped masses are disposed in thick concentric lamellar concretions, which often include a loose nodule; but the different masses themselves joined together, form whole beds, which are thus composed of large granular concretions.

The surface of the concretions is rough and glimmering.

The external layers are soft, sometimes inclining to semi-hard; the internal very soft, sometimes inclining to friable.

It is brittle.

The streak is pale yellowish-brown, bordering on ochre-yellow, and is glistening.

It is easily frangible.

It feels meagre.

It adheres to the tongue.

Specific gravity, 2 574, *Wiedemann*.

Constituent Parts.

Peroxide of Iron, -	76	78
Water, - - -	14	13
Silica, - - - - -	5	7
Oxide of Manganese, -	2	trace
Alumina, - - -		1
Lime, - - - -		trace
	97	99

Daubuisson, Ann. d. Chimie for 1810.

Geognostic Situation.

It occurs imbedded in ironshot clay, in flœtz rocks of the newest formation, and also in loam and clay beds that lie over black coal.

Geographic

Geographic Situation.

Europe.—It is found in different places in the counties of Mid Lothian and East Lothian; at Colebrookedale in England; Norway; Denmark; at Wehrau in Upper Lausitz; Bohemia; Upper Palatinate; Oppeln, Beuthen, Tarnowitz, in Silesia; Mountains of Cracau in Poland; Transylvania; and France.

Asia.—Siberia.

Uses.

It is one of the best kinds of ironstone, yields an excellent iron, and is smelted in many places.

Seventh Subspecies.

Pea-Ore, or Pisiform Clay-Ironstone.

Bohnerz, *Werner.*

Minera Ferri subaquosa globosa, *Wall.* t. ii. p. 257.—Mine de Fer en grains, *Rome de Lisle,* t. iii. p. 300.—Bohnerz, *Werner,* Pabst. b. i. s. 168. *Id. Wid.* s. 827.—Pisiform or granunular Ironstone, *Kirw.* vol. ii. p. 178.—Bohnerz, *Emm.* b. ii. s. 347.—Fer oxyde rubigineux globuliforme, *Hauy,* t. iv. p. 111.—Le Fer pisiforme, *Broch.* t. ii. p. 280.—Kuglicher Thoneisenstein, *Reuss,* b. iv. s. 135. *Id. Lud.* b. i. s. 254. *Id. Suck.* 2ter th. b. ii. s. 288. *Id. Bert.* s. 424. *Id. Mohs,* b. iii. s. 426.—Hagelförmig, körniger, thoniger Brauneisenstein; Braunes Bohnerz, *Hab.* s. 122.—Bohnerz, *Leonhard,* Tabel. s. 67.—Fer oxide brun granuleux, *Brong.* t. ii. p. 170. —Kuglicher Thoneisenstein, *Karsten,* Tabel. s. 66.—Kuglicher Gelbeisenstein, *Haus.* s. 107.—Pea-ore, *Kid,* vol. ii. p. 181.—Fer oxide globuliforme, *Hauy,* Tabl. p. 98.—Pisiform Clay-Ironstone, *Aikin,* p. 42.

External Characters.

Internally its colour is yellowish-brown, of different

 degrees

degrees of intensity, which sometimes passes into blackish-brown. Externally it is reddish, yellowish, and liver brown, and sometimes yellowish-grey, which are, however, accidental, as they depend on the kind of clay in which it is imbedded.

It occurs in small round and spherical grains.

Internally it passes from dull to glistening, in such a manner that the centre of the grain is dull, and the lustre increases in strength towards the surface; the lustre is resinous.

The fracture is fine earthy in the centre of the grain, but towards the surface even.

The fragments are indeterminate angular, and not particularly sharp-edged.

It occurs in thin concentric lamellar distinct concretions, in which the surface is usually smooth and glistening.

The streak is yellowish-brown.

It is soft.

It is not very brittle.

It is easily frangible.

Specific gravity, 5.207, *Mollinghof.* 4.423, *Ullmann.*

Constituent Parts.

	Pene, in the district of Gaillac.	Mardorf.
Oxide of Iron,	48	60
Alumina,	31	13
Silica, -	15	12
Water, -	6	15
	100	100
	Vauquelin, Journ. des Mines, xii.	*Mollinghof*, Crell's Analen 1802, s. 110.

	Hogau.	Berri.
Oxide of Iron,	53.00	70
Oxide of Manganese,	1.00	trace
Alumina, -	23.00	7
Silica, - -	6.50	6
Water, -	14.50	15
	98.00	98
	Klaproth, Beit. b. iv. s. 131.	*Daubuisson*, Annal. de Chim. for 1810.

Geognostic

Geognostic Situation.

It is said to occur in floetz limestone, and also to form a kind of amygdaloid, in which the globular concretions of the ironstone are imbedded in a calcareous clayey basis.

Geographic Situation.

It is found in upper district of Ayrshire. On the Continent, it occurs at Eichstadt in Franconia; Mardorf near Homburg in Hessia; Nardern Duttlengen, Heerbrechtlingen in Suabia; Basle, Aarau near Bern, and in the Jura mountains, where it occurs in an extensive bed, which rests on the Jura limestone, in Switzerland; Salzburg; Alsace, Burgundy, Languedoc, &c. in France; Dalmatia.

Uses.

It yields from 30 to 40 *per cent.* of iron; and at Aarau it supplies very considerable ironworks. In Dalmatia, it is said to be used by the inhabitants in place of shot.

14 Bog Iron-Ore.

Raseneisenstein, *Werner.*

This species is divided into three subspecies, viz. Morass-ore, Swamp-ore, and Meadow-ore.

First Subspecies.

Morass-ore, or Friable Bog Iron-ore.

Morasterz, *Werner.*

Id. Werner, Pabst. b. i. s. 168. *Id. Wid.* s. 830.—Morassy Iron-ore, *Kirw.* vol. ii. p. 183.—Morasterz, *Emm.* b. ii. s. 352. —Fer oxydé rubigineux massif, *Hauy,* t. iv. p. 138.—La Mine des Marais, ou le Morasterz, *Broch.* t. ii. p. 283.—Morasterz, *Reuss,* b. iv. s. 138. *Id. Lud.* b. i. s. 254. *Id. Mohs,* b. iii. s. 431. *Id. Leonhard,* Tabel. s. 67.—Fer oxidé limoneux, le Mine des marais, *Brong.* t. ii. p. 174.—Zerreiblicher Raseneisenstein, *Karsten,* Tabel. s. 66.—Lowland Iron-ore, *Kid,* vol. ii. p. 182.—Earthy Bog Iron-ore, *Aikin,* p. 43.

External Characters.

Its colour is yellowish-brown.

It is sometimes friable, sometimes nearly coherent.

The coherent varieties occur massive, corroded, in grains, and sometimes tuberose. The friable is composed of dull dusty particles.

The coherent varieties are externally and internally dull.

The fracture is earthy.

It soils pretty strongly.

It feels meagre, but fine.

It is light, and extending to not particularly heavy.

Second

Second Subspecies.

Swamp-ore, or Indurated Bog Iron-ore.

Sumpferz, *Werner.*

Id. Werner, Pabst. b. i. s. 168. *Id. Wid.* s. 831.—Swampy Iron-ore, *Kirw.* vol. ii. p. 183.—Sumpferz, *Emm.* b. ii. s. 353.—La Mine des Lieux bourbeux, ou le Sumpferz, *Broch.* t. ii. p. 283.—Sumpferz, *Reuss,* b. iv. s. 140. *Id. Lud.* b. i. s. 254. *Id. Mohs,* b. iii. s. 43. *Id. Leonhard,* Tabel. s. 67.—Fer oxidé limoneux, la Mine des lieux bourbeux, *Brong.* t. ii. p. 174.—Verhärteter Raseneisenstein, *Karsten,* Tabel. s. 66.

External Characters.

Its colour is dark yellowish-brown, sometimes passing into dark yellowish-grey.

It occurs corroded and vesicular, also amorphous.

Internally it is commonly dull, but the darker varieties are glimmering, and sometimes even glistening.

The fracture is earthy, sometimes passing into fine-grained uneven.

The fragments are indeterminate angular, and blunt-edged.

The streak is yellowish-brown.

It is very soft.

It is sectile.

It is easily frangible.

Specific gravity, 2.944, from Sprottau, *Kirwan.*

Third Subspecies.

Meadow-Ore, or Conchoidal Bog Iron-Ore.

Weisenerz, *Werner*.

Id. Werner, Pabst. b. i. s. 168. *Id. Wid.* s. 832.—Meadow Iron-ore, *Kirw*. vol. ii. p. 182.—Wiesenerz, *Emm*. b. ii. s. 354.—La Mine des Prairies, ou le Wiesenerz, *Broch*. t. ii. p. 284.—Wiesenerz, *Reuss,* b. iv. s. 142. *Id. Lud.* b. i. s. 256. *Id. Mohs,* b. iii. s. 432. *Id. Leonhard,* Tabel. s. 67.—Fer terreux limoneux, la Mine des Prairies, *Brong*. t. ii. p. 174.—Muschlicher Raseneisenstein, *Karsten,* Tabel. s. 66.—Limonite, *Haus*. s. 107.

External Characters.

On the fresh fracture it is blackish-brown, which sometimes passes into brownish-black. Externally it has different colours, according to the earth in which it is found.

It occurs massive, in roundish grains, perforated, tuberose, and amorphous.

Internally it extends from shining to glistening, and the lustre is resinous.

The fracture is usually imperfect and small conchoidal, from which it sometimes passes into small-grained uneven; the uneven sometimes inclines to earthy.

The fragments are indeterminate angular, and blunt-edged.

It yields a light yellowish-grey streak.

It is soft.

It is rather brittle.

It is very easily frangible

Constituent

Constituent Parts.

Oxide of Iron,	66.00	Oxide of Iron,	61.0
Oxide of Manganese,	1.50	Oxide of Manganese,	7.0
Phosphoric Acid,	8.00	Phosphoric Acid, with a trace of Sulphur,	2.5
Water, -	23.00	Water, -	19.0
	98.50	Silica, - -	6.0
Klaproth, Beit. b. iv. s. 127.		Alumina, -	2.0
			97.5
		Daubuisson, Annal. de Chim. 1800.	

It would appear from the experiments of Vauquelin, that this ore also contains Chrome, Magnesia, Silica, Alumina, and Lime; and the late experiments of Lescherin shew, that Zinc and Lead also occasionally occur in it. These last mentioned ingredients would seem to be accidental.—Vid. Annal. du Mus. t. viii. p. 435,–460.; also Journ. des Mines, t. 31. p. 45. to 54.

Geognostic Situation.

This species belongs to a very new formation. According to Werner, it is formed in the following manner:—The water which flows into marshy places is impregnated with a vegetable acid, formed from decaying vegetables, which enables it to dissolve the iron in the rocks over which it flows, or over which it stands. This water having reached the lower points of the country, or being poured into hollows, becomes stagnant, and by degrees evaporates; the dissolved iron being accumulated in quantity by fresh additions of water, there follow successive depositions, which at first are yellowish, earthy, and of of little consistence, and this is *Morass-ore;* but in course of time they become harder, their colour passes to brown, and thus *Swamp-ore* is formed. After the water has completely evaporated, and the swamp is dried up, the swamp-ore becomes much harder, and at length passes into *Meadow-ore,* which is already covered with soil and grass *.

From the preceding observations, it is evident that there is a complete transition of the different subspecies of bog iron-ore into each other, and that masses may be found in which we can observe the different degrees of induration.

Geographic Situation.

It is found in various places in the Highlands of Scotland, in the Hebrides, and Orkney and Zetland Islands. In Saxony it occurs at Torgau; in Upper and Lower Lusatia; in a part of the Mark Brandenburg; in Mecklenburg; Pomerania; and in the kingdom of Hanover. It also extends through Prussia, Poland, Courland, Liefland, into Russia, and the southern parts of Sweden, particularly in Smoland, where it is found in very considerable quantity. It is also found in small quantity in the northern parts of Westphalia; in Silesia; in the island of Seeland in the Baltic; in the Upper Palatinate; and Hungary.

It occurs in general more abundantly in the northern than in the western and southern European countries.

Uses.

* In some of the Swedish lakes, this ore is deposited so abundantly, that it is dredged up every twenty or thirty years.—Vid. Swedenborg's Regnum Subterraneum.

Uses.

The three subspecies of ore appear different in working. The Morass-ore is the most easily fusible, and also affords the best iron. The Meadow-ore is more difficultly fusible. When melted with other ores of iron, red and brown ironstone are to be preferred. Of these he ochry subspecies smelt the most advantageously; but where these cannot be obtained, and we are obliged to use the compact and hematitical subspecies, we must be careful that they be previously well roasted. Ev n in the first melting, bog-ore affords an iron for the finest kinds of cast-ware. Owing, however, to the phosphoric acid it contains, it is not so tenacious as that obtained from some other ores. The malleable iron prepared from this ore has always a tendency to be cold-short, and can scarcely be used for plate-iron, and never for iron-wire. It is however well fitted for nails, because it takes a good point, and welds well. The usual flux is limestone.

15. Pitchy Iron-Ore.

Eisenpecherz, *Werner.*

Eisenpecherz, *Karsten,* in Magaz. d. Ges. Natf. Fr. Berlin, 1808, s. 111. *Id. Karsten,* Tabel. s. 66. *Id. Klaproth,* Beit. b. iv. s. 217.—Pittizit, *Haus.* s. 107.—Fer oxydé resinit, *Hauy,* Tabl. p. 98.—Pitchy Bog Iron-ore, *Aikin,* 2d edit. p. 105.

External Characters.

Its colour is greyish-black, which passes into dark liver-brown; sometimes it is yellowish-brown.

It

It appears to occur in crusts.

Internally it is splendent or shining, and the lustre is resinous.

The fracture is imperfect conchoidal.

The fragments are indeterminate angular, and sharp-edged.

It occurs in small granular distinct concretions.

It is translucent on the edges.

It is soft.

The streak is lemon-yellow.

Specific gravity, 2.407.

When placed in water, it becomes red, semitransparent, and vitreous.

Constituent Parts.

Oxide of Iron, - -	67
Sulphuric Acid, - -	8
Water, - - - -	20
	100

Klaproth, Beit. b. v. s. 221.

Geographic Situation.

It was formerly met with in the mine named Christbescherung near Freyberg, and has been lately found in the district of Pless in Upper Silesia *.

16. Blue

* Leonhard's Taschenbuch, p. 599. for 1815.

16. Blue Iron-Ore.

Eisenblau, *Hausmann.*

This species is divided into three subspecies, viz. Foliated Blue Iron-ore, Fibrous Blue Iron-ore, and Earthy Blue Iron-ore.

First Subspecies.

Foliated Blue Iron-Ore.

Blättriehes Eisenblau, *Hausmann.*

Blättriches Eisenblau, *Uttinger,* Moll's Eph. b. iv. s. 71.—Fer phosphaté crystallisé, *Hauy,* Tabl. p. 99.—Fer phosphaté crystallisé ou laminaire, *Lucas,* t. ii. p. 413.

External Characters.

Its colour is dark indigo-blue, and sometimes bluish-grey.

It occurs crystallised in the following figures:

1. Oblique four-sided prism.
2. Eight-sided prism, acuminated with four planes.

The crystals are small, and irregularly aggregated.

Internally it is shining, and the lustre is vitreous.

The fracture is foliated, with a single cleavage.

It is translucent, but transparent in thin pieces.

It is soft.

The streak is pale smalt-blue.

It is intermediate between brittle and sectile.

It is not particularly heavy.

Constituent

Constituent Parts.

	From the Isle of France.
Oxide of Iron, -	41.25
Phosphoric Acid, -	19.25
Water, - - -	31.25
Ironshot Silica, - -	1.25
Alumina, - - -	5.00
	98

Fourcroy and *Laugier*, in Ann. du Mus. t. iii. p. 405.

Geognostic and Geographic Situations.

Europe.—It occurs along with iron-pyrites, and magnetic pyrites, in gneiss, in the Silberberg at Bodenmais, in Bavaria; and in the department of Allier in France.

Africa.—In the Isle of France.

America.—Brazil.

Observations.

This mineral is described by Reuss as Kyanite *; and by Brunner as Foliated Gypsum †. Its true nature was first ascertained by Uttinger of Sonthofen, in a paper in Von Moll's Ephemeriden, already quoted.

Second

* Reuss, Lehrbuch der Mineralogie.

† Annalen der Berg et Hüttenkunde, b. iii. lif. 2. s. 296.

Second Subspecies.

Fibrous Blue Iron-Ore.

Fasriges Eisenblau, *Hausmann*.

Fasriges Eisenblau, *Haus.* s. 138.

External Characters.

Its colour is indigo-blue.

It occurs massive, and sometimes intimately connected with hornblende, and in roundish blunt angular pieces.

Internally it is glimmering and silky.

The fracture is fibrous, and is delicate scopiform or promiscuous.

It is opaque.

It is soft.

Geognostic and Geographic Situations.

Europe.—It occurs in transition syenite at Stavern in Norway *.

America.—In West Greenland †.

Third

* Hausmann's Reise durch Scandinavien, b. ii. s. 109.

† Schumacher, Verz. s. 139.

Third Subspecies.

Earthy Blue Iron Ore, or Blue Iron-Earth.

Blaue Eisenerde, *Werner.*

Erdiges Eisenblau, *Hausmann.*

Cœruleum berolinense naturale, *Wall.* t. ii. p. 260.—Ocre martiale bleu; Bleu de Prusse natif, *Romé de Lisle,* t. iii. p. 295. —Prussiate de Fer natif, *De Born,* t. ii. p. 275.—Blaue Eisenerde, *Werner,* Pabst. b. i. s. 169. *Id. Wid.* s. 835.—Blue Martial Earth, *Kirw.* vol. ii. p. 185.—Blaue Eisenerde, *Emm.* b. ii. s. 359.—Prussiate de Fer natif, *Lam.* t. i. p. 247.—Fer azure, *Hauy,* t. iv. p. 119,–122.—Le Fer terreux bleu, *Broch.* t. ii. p. 288.—Blaue Eisenerde, *Reuss,* b. iv. s. 146. *Id. Lud.* b. i. s. 257. *Id. Mohs,* b. iii. s. 433. *Id. Leonhard,* Tabel. s. 68.—Fer phosphaté azure, *Brong.* t. ii. p. 179.—Blaue Eisenerde, *Karsten,* Tabel. s. 66.—Erdiges Eisenblau, *Haus.* s. 138.—Phosphate of Iron; Native Prussian Blue, *Kid,* vol. ii. p. 189.—Fer phosphaté terreux, *Hauy,* Tabl. p. 99.—Blue Iron-ore, *Aikin,* p. 43.

External Characters.

In its original repository it is said to be white, but afterwards becomes indigo-blue, of different degrees of intensity, which sometimes passes into smalt-blue.

It is usually friable, sometimes loose, and sometimes cohering.

It occurs massive, disseminated, and thinly coating.

Its particles are dull and dusty.

It soils slightly.

It feels fine and meagre.

It is rather light.

Chemical

Chemical Characters.

Before the blowpipe, it immediately loses its blue colour, and becomes reddish-brown, and, lastly, melts into a brownish-black coloured slag, attractable by the magnet.

It communicates to glass of borax a brown colour, which at length becomes dark yellow. It dissolves rapidly in acids.

Constituent Parts.

	From Eckartsberg.
Oxide of Iron, - -	47.50
Phosphoric Acid, -	32.00
Water, - - -	20.00
	99.50

Klaproth, Beit. b. iv. s. 122.

Geognostic Situation.

It occurs in nests and beds in clay-beds, also disseminated in bog iron-ore, or incrusting turf and peat.

Geographic Situation.

Europe.—On the surface of peat-mosses in several of the Zetland Islands; and in river-mud at Toxteth, near Liverpool; Iceland; Helsingor on the Island of Seeland; Schonen in Sweden; Russia; Maschen in Hanover; Steinbach, Oberlichtenau, and Weissig in Upper Lusatia; Silesia; Suabia; Upper Palatinate; Bavaria; Carniola *; France.

 Asia.

* Dr Clarke, upon the subject of this mineral, in a letter to Dr Bruce, remarks, " That it occurs in the mouth of the Cimmerian Bosphorus, now called

Asia.—Borders of the Lake Baikal in Siberia.

America.—Along with bog iron-ore in alluvial soil in New Jersey *.

Uses.

It is sometimes used as a pigment. It is principally employed in water-colours, because, when mixed with oil, the colour is said to change into black †. Beautiful green and olive colours have been formed, by mixing it with other colours. It would appear that this mineral was known to the ancients, for a substance answering to blue iron-earth is mentioned by Pliny, as being collected in the marshes of Egypt, and ground and washed, and used as a pigment.

17. Chromate

---

called the Straits of Taman, between the Sea of Azoph and the Black Sea It lies there associated with extraneous fossil remains of animals, whose decomposition, it is conjectured, afforded phosphoric acid to the metal."—Bruce's Journal, p. 123.

* Cutbush, in Bruce's American Mineralogical Journal, p. 86.

† Mr Cutbush was informed, that a piece of this mineral, by grinding with oil, afforded a beautiful blue colour, which shews, that the American variety is different from that used by painters in Europe.—Vid. Cutbush, in Bruce's Journal, p. 87, 88.

17. Chromate of Iron.

Chromeisenstein, *Hausmann.*

Eisenchrom, *Reuss,* Min. b. iv. s. 625.—Fer chromaté, *Brong.* t. ii. p. 181. *Id. Brard,* p. 33.—Eisenchrom, *Karsten,* Tabel. s. 74.—Fer chromaté, *Hauy,* Tabl. p. 99.—Chromate of Iron, *Aikin,* p. 43.—Chromeisenstein, *Haus.* Hand. b. i. s. 252.

External Characters.

Its colour is pitch-black, or of a colour intermediate between steel-grey and iron-black.

It occurs massive, disseminated, and crystallised in ectahedrons.

Internally it is shining, and the lustre is intermediate between resinous and metallic.

The fracture is uneven, or imperfect small conchoidal, and sometimes imperfect foliated.

It sometimes occurs in granular distinct concretions.

It is opaque.

It scratches glass.

The colour of the streak is ash-grey or brownish.

Specific gravity, 4.03.

Physical Characters.

It is sometimes magnetic, sometimes infusible; and tinges borax of a very beautiful green colour, very different from the dark green colour which it receives when melted with magnetic ironstone.

Constituent Parts.

	France.	Siberia.
Oxide of Iron,	34.7	34
Oxide of Chrome,	43.0	53
Alumina, -	20.3	11
Silica, - -	2.0	1
Oxide of Manganese, -		1
	100.00	100
	Hauy, Traité, t. iv. p. 130.	*Laugier*, Ann. du Mus. t. iv. p. 325.

	Stiria.
Oxide of Iron,	33.00
Oxide of Chrome,	55.50
Alumina, -	6.00
Silica, - -	2.00
Loss by heating,	2.00
	98.50
	Klaproth, Beit. b. iv. s. 132.

Geognostic Situation.

It occurs in beds, veins, or imbedded in primitive serpentine; in talc-slate, to which it has communicated a beautiful colour, intermediate between cochineal-red and peach-blossom-red; also in beds between clay-porphyry and wacke.

Geographic Situation.

Europe.—It occurs in serpentine in the islands of Unst and Fetlar in Zetland; and also in the serpentine of Portsoy in Banffshire. On the Continent, it occurs in serpentine near to Gassin, in the department of Var, and in serpentine in the vicinity of Nantes; and at Krieglach in Stiria, it is imbedded in talc-slate, to which it has communicated a beautiful red colour.

Asia.—It is said to occur in beds between clay-porphyry and wacke in the Uralian Mountains.

America.—It occurs in considerable quantity in serpentine in the Bare Hills, near Baltimore; and it is said also to have been discovered near Pennsylvania *.

Uses.

* Hayden, in Bruce's American Min. Journal, p. 243,—248.

Uses.

When the chromic acid is combined with lead, it forms an uncommonly beautiful yellow pigment. In America, where the chromate of iron occurs in considerable quantity, the chromic acid is obtained from it, and combined with lead, forms the Chromic-yellow, which is now becoming an article of trade.

18. Cube-Ore, or Arseniate of Iron.

Wurfelerz, *Werner.*

Wurfelerz, *Reuss,* b. iv. s. 163. *Id. Lud.* b. i. s. 183.—Arseniksaures Eisen, *Suck.* 2ter th. s. 297.—Wurfelerz, *Bert.* s. 420. *Id. Mohs,* b. iii. s. 437.—Fer arseniaté, *Lucas,* p. 148. —Wurfelerz, *Leonhard,* Tabel. s. 68.—Fer arseniaté, *Brong.* t. ii. p. 182. *Id. Brard,* p. 332.—Wurfelerz, *Karsten,* Tabel. s. 66.—Pharmakosiderit, *Haus.* s. 138.—Arseniate of Iron, *Kid,* vol. ii. p. 101.—Fer arseniaté, *Hauy,* Tabl. p. 100.—Arseniate of Iron, *Aikin,* p. 44.

External Characters.

Its colour is olive-green, of different degrees of intensity, which sometimes passes on the one side into emerald and grass green, on the other into yellowish-brown, and blackish-brown.

It occurs massive; and crystallised in the following figures:

1. Perfect cube.
2. Cube, in which four opposite angles are truncated.
3. Preceding figure, in which the angles formed by

the truncating planes and the neighbouring planes are truncated.

4. Cube truncated on all the edges.
5. Cube truncated on all the edges and angles.

The planes of the crystals are smooth, shining, adamantine, and generally diagonally streaked.

Internally it is glistening, and the lustre is intermediate between vitreous and resinous.

The fracture is imperfect foliated, and sometimes uneven, or imperfect conchoidal.

The fragments are indeterminate angular.

It occurs in distinct concretions, which are small granular.

It is translucent.

The streak is straw-yellow.

It is soft, inclining to very soft.

It is brittle.

Specific gravity, 3.000, *Bournon*.

Chemical Characters.

Before the blowpipe it melts, and gives out arsenical vapours.

Constituent Parts.

Iron, - -	48	Arsenic Acid,	31.0
Arsenic Acid,	18	Oxide of Iron,	45.5
Water of crystallisation, -	32	Oxide of Copper,	9.0
		Silica, - -	4.0
Carbonate of Lime,	2 to 3	Water, -	10.5
	100		100
Vauquelin, in Brong. Min. t. ii. p. 183.		*Chenevix*, in Phil. Trans. for 1801.	

Geognostic

Geognostic Situation.

It is found in veins, accompanied with ironshot quartz, copper-glance or vitreous copper-ore, copper-pyrites, and brown ironstone.

Geographic Situation.

It occurs in Tincroft, Carrarach, Muttrel, Huel-Gorland, and Gwenap mines in Cornwall; and at St Leonard, in the department of Haut-Vienne in France.

Observations.

Proust discovered a white arseniate of iron in Spain; and it is said that more of the same description occurs in Chili.

19. Pyrosmalite, or Native Muriate of Iron.

Pyrodmalite, *V. Moll's* Eph. iv. s. 390.—*Hisinger,* Samling till en Mineralogisk Geograffi öfver Swerige, 175.—Pyrosmalith, *Karsten,* Tabel. s. 103. *Id. Haus.* Handb. s. 1068.—Fer muriaté, *Hauy,* in Lucas, t. ii. p. 418.

External Characters.

Its colour is liver-brown, inclining to pistachio-green.

It occurs crystallised in the following figures:

1. Regular six-sided prism, which is sometimes so short as to form a six-sided table.
2. The six-sided prism, truncated on the terminal edges.

The terminal planes of the crystals are shining and pearly;

pearly; the lateral planes, when not covered with a rough dull crust, are shining and vitreous.

The principal fracture is foliated, with a fourfold cleavage: the most distinct cleavage is that parallel with the terminal planes; the other three, which are parallel with the lateral planes, are less distinct: the cross fracture is uneven, passing into fine splintery.

Internally the lustre of the principal fracture is shining and pearly; the cross fracture is glimmering.

It occurs in straight lamellar distinct concretions.

It is translucent on the edges.

It is semi-hard.

Its streak is brownish-white.

It is brittle.

Specific gravity, 3.081.

Chemical Characters.

It is insoluble in water. It is soluble in muriatic acid, with exception of a small siliceous residuum. Before the blowpipe, it gives out vapours of oxygenated muriatic acid *, and is converted into a magnetic oxide of iron †.

Constituent Parts.

It is Muriate of Iron, combined with a small portion of Silica.

Geognostic

---

* A small piece of this mineral will fill a whole room with the smell of oxygenated muriatic acid. Its name is borrowed from this property.

† By heating, the iron parts with a part of its oxygen to the muriatic acid, and converts it into oxygenated acid.

Geognostic and Geographic Situations.

It occurs in a bed of magnetic ironstone, along with calcareous-spar and hornblende, in Bjelke's mine in Nordmark, near Philipstadt in Wermeland.

Observations.

It was discovered by Messrs Henry, Gahn and Clason, during a mineralogical journey through Wermeland.

IRON-MINES.

The most considerable Iron-mines in the world, are those in Great Britain, and France: the next are the Russian, Swedish, and Austrian ; the others in the order of their relative magnitude, are in the United States of America, Prussia, Kingdom of Westphalia, Spain, the Danish States, Bavaria, including the Tyrol, and the Kingdom of Saxony.

TABLE

TABLE of the ANNUAL QUANTITY of IRON raised from the Earth in different countries.

		Quintals.
1.	Great Britain, - - - - -	5,000,000
2.	France, - - - - - -	4,500,000
3.	Russia, - - - - - -	1,675,679
4.	Sweden, - - - - - -	1,500,000
5.	Austria, - - - - - -	1,010,400
6.	United States, without including Louisiana and the Indian territory *, -	480,000
7.	Prussia, after the treaty of Tilsit, -	322,053
8.	Kingdom of Westphalia in 1808, -	187,411
9.	Spain, - - - - - - -	180,000
10.	Danish States, - - - - -	135,000
11.	Bavaria, and the Tyrol, - - -	110,000
12.	Kingdom of Saxony, - - - -	80,000
	Total Annual Amount,	15,180,543

ORDER VII.

* In Dr Bruce's Journal, we are informed, that the value of all the iron, and all the manufactures of iron, annually made in the United States, is from Twelve to Fifteen millions of dollars. The annual importation, including bar-iron, and every description of manufactures of iron and steel, are estimated at near Four millions of dollars.

ORDER VII.—MANGANESE.

This Order contains six species, viz. Grey Manganese-ore, Black Manganese-ore, Wad, Sulphuret of Manganese, Phosphate of Manganese, and Red Manganese-ore.

1. Grey Manganese-Ore.

Grau Braunsteinerz, *Werner*.

This species is divided into five subspecies, viz. Fibrous Grey Manganese-ore, Radiated Grey Manganese-ore, Foliated Grey Manganese-ore, Compact Grey Manganese-ore, and Earthy Grey Manganese-ore.

First Subspecies.

Fibrous Grey Manganese-Ore.

Faseriges Grau-Braunsteinerz, *Ullmann*.

Haarförmiges Grau-Braunsteinerz, *Mohs*, b. iii. s. 449. *Id. Haus.* Handb. b. i. s. 290.—Faseriges Grau-Braunsteinerz, *Ullmann*, System. Tabell. Ubers. s. 402.

External Characters.

Its colour is dark steel-grey, passing into iron-black.

It occurs massive, disseminated, in crusts, reniform, botryoidal, stalactitic, and crystallised in very delicate capillary

capillary and acicular crystals; in very thin and long rectangular four-sided tables, in which the longer terminal planes are set on obliquely.

The crystals are small and very small, and scopiformly or promiscuously aggregated.

The lateral planes of the crystals are generally longitudinally streaked; the surface of the particular external shapes is very delicately drusy.

Externally it is glistening, passing into glimmering the crystals are shining and splendent.

Internally it is glistening, shining, and the lustre is metallic.

The fracture is delicate fibrous, sometimes scopiform, sometimes stellular, or promiscuous.

The fragments are indeterminate angular, and blunt-edged, or wedge-shaped.

It occurs in distinct concretions, which are coarse, or small granular, or wedge-shaped.

The streak is dull and black.

It soils strongly.

It is soft.

It is brittle.

It is easily frangible.

It is heavy.

Geognostic and Geographic Situations.

It occurs in several veins of brown ironston in the Westerwald; also at Stahlberg, near Schmalkalden in Saxony; and Christiansand in Norway.

Second

Second Subspecies.

Radiated Grey Manganese-Ore.

Strahliges Grau Braunsteinerz, *Werner.*

Magnesia fuliginosa striata, *Wall.* Syst. Min. t. i. p. 329.—Strahliges Grau Braunsteinerz, *Werner,* Pabst. b. i. s. 216.—*Id. Wid.* s. 948.—Striated Grey Ore of Manganese, *Kirw.* vol. ii. p. 291.—Strahliges grau Braunsteinerz, *Emm* b. ii. s. 522.—Le Manganese gris rayonné, *Broch.* t. ii. p. 414.—Manganese oxidé metalloide, *Hauy,* t. iv. p. 246.—Strahliges Graubraunsteinerz, *Reuss,* b. i. s. 448. *Id. Lud.* b. i. s. 291. *Id, Mohs,* b. iii. s. 442. *Id. Leonhard,* Tabel s. 69.—Manganese metalloide chalybin, *Brong.* t. ii. p. 107 —Strahliges Grau Manganerz, *Karsten,* Tabel. s. 72.—Strahliger Braunstein, *Haus.* s. 108.—Manganese oxdé metalloide gris, *Hauy,* Tabl. p. 110.—Grey Manganese, *Aikin,* p. 61.

External Characters.

Its colour is dark steel-grey, which inclines more or less to iron-black. It is sometimes tarnished with pitch-black, velvet-black, or tempered-steel colours.

It occurs massive, disseminated, stalactitic; and crystallised in the following figures:

1. Long oblique four-sided prism, rectangular four-sided prism, six-sided prism, or eight-sided prism, which are either bevelled on the extremities, or acuminated with four or eight planes *.

2. Sometimes

---

* The lateral planes of the prism, according to Hauy, meet under angles of 100° and 80°: According to Hausmann, under angles of 115° and 65°.

2. Sometimes the prisms are acicular, and when the obtuse lateral edges are rounded off, they have a reed-like shape.

The surface of the crystals is longitudinally streaked, and shining, passing to splendent.

Internally it is glistening and shining, and the lustre is metallic.

The fracture is narrow, straight, scopiform and stellular radiated, which sometimes inclines to fibrous, sometimes to foliated.

The fragments are wedge-shaped and splintery, but in the great, indeterminate angular and blunt-edged.

It occurs in distinct concretions, which are coarse, large, and small granular, and sometimes inclining to wedge-shaped.

The colour is not changed in the streak.

When rubbed it soils strongly.

It is soft, inclining to very soft.

It is brittle.

It is rather difficultly frangible.

Specific gravity, 3.530 to 4.325, *Muschenbröck.* 4.143, *Hagen.* 4.2491 to 4.7563, *Brisson.* 4.181, *Rinmann.*

Constituent Parts.

	Ilefeld.	Moravia.
Black Oxide of Manganese,	90.50	89.00
Oxygen, - - - - -	2.25	10.25
Water, - - - - -	7.00	0.50
	99.75	99.75

Klaproth, Beit. b. iii. s. 308. & 310.

Geognostic

Geographic Situation.

Europe.—It occurs in the vicinity of Aberdeen; also in Cornwall, Devonshire, Somersetshire and Derbyshire; Christiansand in Norway; Nassau; Ilefeld in the Hartz; Ilmenau and Saalfeld in Thuringia; Konradswaldau, Kupferberg, &c. in Silesia; Miess in Bohemia; Hüttenberg in Carinthia; St Gothard in Switzerland; Piedmont; and Ischio near Vicenza in Italy.

Asia.—Kolyvan in Siberia.

Third Subspecies.

Foliated Grey Manganese-Ore.

Blättriges Grau Braunsteinerz, *Werner.*

Id. Werner, Pabst. b. i. s. 218. *Id. Emm.* b. ii. s. 525.—Le Manganese gris lamelleux, *Broch.* t. ii. p. 417.—Blättriges Grau Braunsteinerz, *Reuss,* b. iv. s. 453. *Id. Lud.* b. i. s. 292. *Id. Mohs,* b. iii. s. 447. *Id. Leonhard,* Tabel. s. 69.—Manganese metalloide chalybin, texture lamelleuse, *Brong.* t. ii. p. 108. —Blättriges Grau Manganerz, *Karsten,* Tabel. s. 72.—Blättricher Braunstein, *Haus.* s. 108.

External Characters.

Its colour is steel-grey, inclining to iron-black.

It occurs massive, disseminated; and crystallised in the following figures:

1. Oblique four-sided prism.
2. Rectangular four-sided prism.

3. Cube.

3. Cube.

4. Lenticular crystals.

Internally it alternates from shining to splendent, and the lustre is metallic.

The fracture is foliated and delicately streaked; sometimes a threefold cleavage is to be observed, the folia in the direction of the planes of the oblique four-sided prism.

The fragments are indeterminate angular, and blunt-edged.

It occurs in distinct concretions, which are small, and sometimes fine granular.

It yields a dull black streak.

It soils.

It is soft.

It is brittle.

It is more easily frangible than the preceding subspecies.

Specific gravity, 3.742, *Hagen*.

Geographic Situation.

It is found in Devonshire; Ilefeld in the Hartz; Johanngeorgenstadt in the kingdom of Saxony; Bohemia; Salzburg; and Transylvania.

Fourth

Fourth Subspecies.

Compact Grey Manganese-Ore.

Dichtes Grau Bransteinerz, *Werner.*

Id. Werner, Pabst. b. i. s. 219.—Indurated Grey Ore of Manganese, *Kirw.* vol. ii. p. 249.—Le Manganese gris compacte, *Broch.* t. ii. p. 418.—Dichtes Graubraunsteinerz, *Reuss,* b. iv. s. 454. *Id. Lud.* b. i. s. 293. *Id. Mohs,* b. iii. s. 447. *Id. Leonhard,* Tabel. s. 69.—Manganese terne compact, *Brong.* t. ii. p. 109.—Dichtes Graumanganerz, *Karsten,* Tabel. s. 72. —Dichter Braunstein, *Haus.* s. 109.—Manganese oxydé gris compact, *Hauy,* Tabl. p. 110.—Compact Grey Manganese, *Aikin,* p. 61.

External Characters.

Its colour is steel-grey, inclining to bluish-black.

It occurs massive, disseminated, and small botryoidal.

Internally it is glistening, passing into glimmering, and the lustre is metallic.

The fracture is even, sometimes inclining to flat conchoidal, and uneven.

The fragments are indeterminate angular, and rather sharp-edged.

It sometimes occurs in distinct concretions, which are thick and curved lamellar.

It becomes darker, and dull in the streak.

It soils.

It is soft.

It is brittle.

It is easily frangible.

Specific gravity, 4.407, *Karsten.* 4.073, *Vauquelin.*

 Constituent

Constituent Parts.

The four following analyses made by Cordier, Beaunier, Vauquelin and Dolomieu, are said by Brochant to be of this subspecies.

	St Micaud.	Perigueux.	Romaneche.	Laveline.
Yellow Oxide of Manganese,	35	50.0	50.0	65
Oxygen,	33	17.0	33.7	17
Red Oxide of Iron,	18	13.5		
Charcoal,			0.4	
Lime, with Magnesia, Iron and Manganese,	7	6.0		
Carbonate of Lime,	-			7
Barytes,	4	5.0	14.7	9
Silica,	3	7.0	1.2	6
Loss,		1.5		6
	100	100	100	100

Journal des Mines, N. 58. p. 778.

Geographic Situation.

It occurs at Upton Pyne in Devonshire; Wurzelberg in the Hartz; Nassau; and at Christiansand in Norway.

Fifth

Fifth Subspecies.

Earthy Grey Manganese-Ore.

Erdiches Grau Braunsteinerz, *Werner*.

Erdiger Braunstein, *Wid.* s. 953.—Ochre of Manganese, *Kirw.* vol. ii. p. 293.—Erdiches Graubraunsteinerz, *Emm.* b. ii. s. 529.—Le Manganese gris terreux, *Broch.* t. ii. p. 420.—Erdiches Graubraunsteinerz, *Lud.* b. i. s. 293. *Id. Suck.* 2ter th. s. 419. *Id. Bert.* s. 492. *Id. Mohs,* b. iii. s. 450.—Manganese terne terreux, *Brong.* t. ii. p. 110.—Zerreibliches Graubraunsteinerz, *Karsten,* Tabel. s. 72.—Ochriger Braunstein, *Haus.* s. 108.

External Characters.

Its colour is blackish grey.

It is composed of dull, generally loose, dusty particles.

It soils strongly.

It feels fine, but meagre.

It is not particularly heavy.

Geographic Situation.

It occurs in the mine Johannis, near Langeberg in the Saxon Erzgebirge.

Chemical Characters of the Species.

It is infusible without addition before the blowpipe. It tinges borax purple: it effervesces with muriatic acid, giving out oxymuriatic acid.

 Geognostic

Geognostic Situation of the Species.

This mineral occurs in granite, gneiss, mica-slate, porphyry, and sandstone, either in veins, or in large imbedded cotemporaneous masses. Several different formations are enumerated and described by mineralogists: in one formation, which is in porphyry, the ores, which are principally the radiated and foliated subspecies, occur in veins, along with heavy-spar; and in another, the ores, principally the compact and earthy subspecies, are in veins, along with red and brown ironstone.

Uses.

It is added to glass, in small quantity, when we wish to destroy the brown colour which that material receives from intermixed inflammable substances, or in larger quantity when we wish to give to it a violet-blue colour. It affords a fine brown colour, which is used for painting on porcelain. It is employed in the laboratory, as the cheapest and most convenient material from which to procure oxygen gas. All the oxymuriatic acid used in bleacheries, and for the purpose of destroying contagious matter, is prepared from manganese, and the usual materials of muriatic acid.

2. Black Manganese-Ore.

Schwarz Braunsteinerz, *Werner.*

This species is divided into three subspecies, viz. Friable Black Manganese-ore, Foliated Black Manganese-ore, and Dendritic Black Manganese-ore.

First

First Subspecies.

Friable Black Manganese-Ore.

Zerreibliches Schwarz Braunsteinerz, *Karsten.*

Zerreibliches Schwarzbraunsteinerz, *Reuss,* b. iv. s. 459. *Id. Karsten,* Tabel. s. 72.—Zerreibliches Schwarzmanganerz, *Haus.* s. 109.—Manganese oxydé noire brunâtre pulverulent, *Hauy,* Tabl. p. 110. (in part).—Ochriger Schwarz Braunsteinerz, *Haus.* Handb. b. i. s. 294.

External Characters.

Its colour is brownish-black.

It occurs fruticose, in crusts, and in imbedded roundish kidneys.

It is dull, or glimmering and resinous.

The fracture is earthy.

It occurs in curved lamellar concretions.

It is very soft, passing into friable.

It soils slightly.

It is light.

Geognostic and Geographic Situations.

It occurs along with grey manganese-ore, compact black ironstone, and lithomarge. The fruticose variety is found at Schmalkalden; and the imbedded roundish kidneys in snow-white lithomarge at Johanngeorgenstadt. It is also met with at Ilefeld.

Second Subspecies.

Foliated Black Manganese-Ore.

Blättricher Schwarzbraunsteinerz, *Hausmann.*

Verhärtetes Schwarzbraunsteinerz, *Reuss,* b. iv. s. 463. *Id. Karsten,* Tabel. s. 72.—Blättriche Manganschwarze, *Haus.* s. 109.—Blättricher Schwarz Braunsteinerz, *Haus.* Handb. s. 293.

External Characters.

Its colour is intermediate between brownish-black and iron-black.

It occurs massive, disseminated, and crystallised in elongated acute double four-sided pyramids, which are small and very small, and aggregated in rows.

The surface of the crystals is seldom smooth, and shining or glistening, generally drusy, or rough and dull.

Internally it is intermediate between shining and glistening, and the lustre is adamantine.

The fracture is generally imperfect foliated, with a single cleavage, but sometimes inclining to diverging radiated, sometimes to uneven.

The fragments are indeterminate angular, and blunt-edged.

It occurs in small and fine granular concretions.

It is opaque.

It affords a reddish-brown dull streak.

It is semi-hard.

It is brittle.

It is easily frangible.

It is heavy.

Geognostic

Geognostic and Geographic Situations.

It is a rare mineral, and generally occurs along with grey manganese-ore ; at Ehrenstock near Ilmenau, incrusting radiated grey antimony-ore.

Third Subspecies.

Dendritic Black Manganese-Ore.

Dendritischer Schwarzbraunstein, *Hausmann.*

Manganese oxidé noire brunâtre ramuleux, *Hauy,* Tab. p. 111. (in part).—Dendritischer Schwartz Braunstein, *Haus.* Handb. s. 295.

External Characters.

Its colour is bluish-black, which passes into coal-black or iron-black.

It occurs in dendritical external forms, which are disposed either in the surface of rents, or distributed through the mass of minerals of different kinds.

It is dull.

Geognostic and Geographic Situations.

It occurs upon the surface or in the interior of many different minerals, as quartz, and marl-slate. It is found in the Iberg near Grund ; and at Maria-Sprung near Göttingen.

3. Wad.

This species is divided into three subspecies, viz. Fibrous Wad, Ochry Wad, and Dendritic Wad.

First Subspecies.

Fibrous Wad.

Fasriges Wad, *Hausmann.*

Fasriges Wad, *Haus.* Handb. b. i. s. 296.

External Characters.

Its colour is nut-brown.

It occurs massive.

Internally it is glimmering and metallic.

The fracture is diverging fibrous.

It occurs in curved lamellar concretions, which intersect the fibrous fracture.

It is opaque.

It soils.

It is very light.

Geognostic and Geographic Situations.

It occurs along with the ochry wad, at Iberg, near Grund in the Hartz.

Second Subspecies.

Ochry Wad.

Ochriges Wad, *Hausmann.*

Manganese oxidé brun, *Hauy*, t. iv. p. 245.—Wad, *Leonhard*, Tabel. s. 70. (in part). *Id. Karsten*, Tabel. s. 72.—Ochriges Wad, *Haus.* s. 109. *Id. Haus.* Handb. b. i. s. 297.

This

This subspecies is divided into two kinds, viz. Indurated, and Pulverulent.

First Kind.

Indurated Ochry Wad.

Festes Ochrige Wad, *Hausmann.*

Manganese oxidé brunâtre concretionné, *Hauy*, Tabl. p. 111.—Festes Ochriges Wad, *Haus.* Handb. b. i. s. 297.

External Characters.

Its colours are clove-brown, and soot-brown.

It occurs massive, globular, small reniform, botryoidal, and stalactitic.

It is dull.

The fracture is fine earthy.

It sometimes occurs in curved lamellar concretions.

It is opaque.

It is soft, passing into very soft.

It becomes shining and darker in the streak.

It soils strongly.

It feels meagre.

It is light.

Geognostic and Geographic Situations.

It is associated with scaly brown iron-ore, at Iberg, and other places in the Hartz.

Second Kind.

Pulverulent Ochry Wad.

Loses Wad, *Hausmann.*

Black Wad, *Kirw.* vol. ii. p. 293.—Manganese oxydé noire brunâtre pulverulent, *Hauy,* Tabl. p. 111.—Loses Wad, *Haus.* Handb. b. i. s. 298.

External Characters.

Its colour is clove-brown, soot-brown, and blackish-brown.

It occurs massive, investing other minerals, and either in a completely pulverulent form, or slightly cohering.

Chemical Characters.

When dry, and mixed with one-fourth of its weight of linseed oil, and gently heated, it inflames *. At 95° of Wedgwood, it melts to a slag, and at 144°, into a perfect glass. It dissolves in borax, to which it communicates a violet-blue colour.

Constituent

* This effect is probably produced by the partial decomposition of the metallic oxide: its oxygen uniting with the oil, and, in consequence of the degree of heat present, causing combustion.—*Kid.*

Constituent Parts.

Oxide of Manganese,	43	Brown Oxide of Manganese, -	68.00
Oxide of Iron,	43	Oxide of Iron,	6.50
Lead, (accidental,	4.5	Water, -	17.50
Mica, (accidental),	5	Carbon, -	1.00
	——	Barytes, -	1.00
	Wedgwood.	Silica, - -	8.00
			102
		Klaproth, Beit. b. iii. s. 311.	

Geognostic and Geographic Situations.

It occurs at Leadhills in Scotland; and in Derbyshire and Devonshire: at Iberg and other places in the Hartz.

Third Subspecies.

Dendritic Wad.

Dendritisches Wad, *Hausmann.*

Manganese oxydé noire brunâtre ramuleux, *Hauy,* Tabl. p. 111. —Dendritisches Wad, *Haus.* Handb. b. i. s. 298.

External Characters.

Its colour is soot-brown and brownish-black.

It occurs in the dendritic external shape, which is either superimposed, or imbedded in other minerals.

It is dull.

Geognostic

Geognostic and Geographic Situations.

It occurs superimposed, or imbedded in basalt, clay, slate, hornstone, and other minerals. It occurs at the Dransfeld, near Göttingen.

4. Sulphuret of Manganese.

Manganglanz, *Karsten.*

Schwarzerz, *Müller v. Reichenstein,* Phys. arb. d. eintr. Fr. i. Wien. 1. Jahrg. 2. Quart. s. 86. *Id. Reuss,* b. ii. 4. s. 446. —Braunsteinkies, *Leonhard,* Tabel. s. 70.—Manganglanz, *Karsten,* Tabel. s. 72.—Manganese sulphuré, *Hauy,* Tabl. p. 111.—Sulphuret of Manganese, *Aikin,* p. 62.

External Characters.

Its colour on the fresh fracture is dark steel-grey, which approaches to iron-black, but on exposure, becomes tarnished of a brownish-black colour.

It occurs massive.

Its lustre on the fresh fracture is splendent; on the tarnished surface shining and metallic.

The fracture is imperfect foliated, which inclines to fine-grained uneven.

It is opaque.

It is soft, passing into semi-hard.

It is rather sectile.

Its streak is greenish.

Specific gravity, 3.950.

Chemical Characters.

Before the blowpipe, it gives out sulphur, and tinges borax violet-blue.

Constituent

Constituent Parts.

Oxide of Manganese,	82.00	85
Sulphur, - - -	11.50	15
Carbonic Acid, - -	5.00	
	98	100
	Klaproth, Beit. b. iii. s. 42.	*Vauquelin*, Annal. d. Mus. vi. s. 405.

Geognostic and Geographic Situations.

It is found in Cornwall; and at Nagyag in Transylvania, along with ores of tellurium, blende, copper-pyrites, compact red manganese-ore, and brown-spar.

5. Phosphate of Manganese.

Eisenpecherz, *Werner*.

Fer phosphaté, *Broch.* t. ii. p. 533.—Manganese phosphaté, *Brong.* t. ii. p. 112.—Phosphormangan, *Karsten*, Tabel. s. 72. —Manganese phosphaté ferrifere, *Hauy*, Tabl. p. 111.—Triplit, *Haus.* Handb. b. iii. s. 1079.—Phosphate of Manganese, *Aikin*, p. 72.

External Characters.

Its colour is brownish-black, sometimes inclining to clove-brown.

It occurs massive.

Internally it is shining, glistening, or glimmering, and the lustre is resinous.

The principal fracture is concealed foliated, with a threefold rectangular cleavage, in which two of the cleavages

ages are more distinct than the third: the cross fracture is imperfect flat conchoidal, passing into even, and fine grained uneven.

It is opaque in the mass, but semi-transparent in splinters.

It scratches glass.

Its streak is clove-brown.

Specific gravity, 3.4390, *Vauquelin.* 3.767, 3.775, *Ullmann.*

Chemical Characters.

It is readily fusible before the blowpipe into a black enamel.

Constituent Parts.

Oxide of Manganese, -	42
Oxide of Iron, - -	31
Phosphoric Acid, - -	27
	100

Vauquelin, Journ. d. Min. N. 64. p. 299.

Geognostic and Geographic Situations.

It occurs in a coarse granular granite at Limoges in France.

6. Red Manganese-Ore.

Rother Braunstein, *Werner.*

This species is divided into two subspecies, viz. Foliated Red Manganese-ore, and Compact Red Manganese-ore.

First

First Subspecies.

Foliated Red Manganese-Ore.

Blättricher Rotherbraunstein.

Rosenrod syrsatt Manganes, *Hisinger*, in Afhandling i Fys. Kem. och. Min. 1. 105.—Blättricher Rothstein, *Haus.* Handb. b. i. s. 302.—Manganspath, *Werner*, Vorles. MS. Vivian?

External Characters.

Its colour is rose-red, slightly inclining to flesh-red.
It occurs massive, and disseminated.
Internally it is shining.
The fracture is foliated.
It occurs in fine granular distinct concretions.
It is slightly translucent.
It is so hard as to scratch glass.
Specific gravity, 3.6.

Constituent Parts.

Oxide of Manganese, -	52.60
Silica, - - - -	39.60
Oxide of Iron, - -	4.60
Lime, - - - -	1.50
Volatile ingredients, - -	2.75
	101.5

Berzelius, in Afh. i Fys. och. Min. 1. 110.

Geognostic and Geographic Situations.

It occurs in beds of specular iron-ore or iron-glance and magnetic iron-stone, along with compact garnet and calcareous-spar, in the gneiss hills at Langbanshytta, in Wermeland in Sweden; and also in Siberia.

Observations.

Observations.

1. Its fracture, lustre, and distinct concretions, distinguish it from the second subspecies.

2. Mr Vivian, a pupil of Werner, sent me from London the description of a mineral under the name *Manganspath*, which appears to belong to this species.

3. Mr Vivian also transmitted to me the description of a mineral under the name *Piemontesicher Braunstein*, or Manganese of Piedmont, and which Werner arranges in his system as a distinct species. It is said to have a red colour, to be crystallised in prisms; the fracture is radiated; and it is hard and heavy. It is probably the mineral analysed by Napione, the constituents of which are as follows:—Ferruginous Oxide of Manganese, 45.281: Silica, 26.125: Lime, 23: Alumina, 0.781: Water and Carbonic Acid, 3.—Vid. Hauy, Traité, t. iv. p. 248.

Second Subspecies.

Compact Red Manganese Ore.

Dichter Roth Braunstein.

Dichtes Rothbraunstein, *Reuss*, b. iv. s. 470.—Rothstein, *Mohs*, b. ii. s. 122.—Rothbraunstein, *Leonhard*, Tabel. s. 70.—Manganese lithoide, *Brong*. t. ii. p. 110.—Roth Manganerz, *Karsten*, Tabel. s. 72.—Manganese oxydé carbonaté, *Hauy*, Tabl. p. 111. (in part).—Dichter Rothstein, *Haus.* Handb. b. i. s. 302.—White Manganese, *Aikin*, p. 61.

External Characters.

Its colour is rose-red, which passes into brownish-red, reddish-

reddish-brown, yellowish-brown, and sometimes slightly incl n s to violet-blue.

It occurs massive and disseminated.

Internally it is dull.

The fracture is even, which passes into splintery or conchoidal, and sometimes into uneven.

It is faintly translucent, or only translucent on the edges.

It is hard, inclining to semi-hard.

It is brittle.

It is easily frangible.

Specific gravity, 3.233, Kapnik, *Kirwan*.

Chemical Characters.

It is infusible before the blowpipe, but becomes black by ignition.

Constituent Parts.

	Siberia.
Oxide of Manganese, -	61
Silica, - - - -	30
Oxide of Iron, - -	5
Alumina, - - -	2
	98

Lampadius, in Pract. Chem. Abh. b. ii. s. 209.

Geognostic and Geographic Situations.

It occurs at Kapnik in Transylvania, in veins, along with quartz, black copper-ore, sulphuret of manganese, blende, galena or leadglance, calcareous-spar, and brown-spar; also at Langbanshytta, in Wermeland in Sweden; and Catharinenburg in Siberia.

ORDER VIII.—TITANIUM.

THIS Order contains six species, viz. Menachanite, Iserine, Nigrine, Sphene, Rutile, and Octahedrite *.

1. Menachanite.

Menacan, *Werner*.

Menachanite, *Kirw.* vol. ii. p. 326.—Le Menakanite, *Broch.* t. ii. p. 468.—Titane oxidé ferrifere granuliforme, *Hauy,* t. iv. p. 306.—Manacan, *Reuss,* b. iv. s. 54. *Id. Lud.* b. i. s. 305. *Id. Mohs,* b. iii. s. 452. *Id. Leonhard,* Tabel. s. 81.—Titane Menakanite, *Brong.* t. ii. p. 99.—Manakan, *Karsten,* Tabel. s. 74.—Menachanite, *Kid,* vol. ii. p. 224.—Titane oxidé ferrifere, *Hauy,* Tabl. p. 116.—Titaneisenstein, *Haus.* Handb. b. i. s. 251.—Menachanite, *Aikin,* p. 67.

External Characters.

Its colour is greyish-black, inclining to iron-black.

It occurs only in very small flattish angular grains, which have a rough glimmering surface.

Internally it is glistening and glimmering, and the lustre is adamantine, passing into semi-metallic.

The fracture is imperfect foliated, approaching to slaty.

The fragments are indeterminate angular and sharp-edged

It is perfectly opaque.

It yields to the knife, but is not so hard as nigrine.

It is brittle.

It

* The best account in the English language of the minerals of this Order, is that of Dr Mitchell, in the Transactions of the Royal Irish Academy.

It retains its colour in the streak.
It is easily frangible.
It is heavy in a moderate degree.
Specific gravity, 4.427, *Gregor.* 4.270, *Lampadius.*

Physical Character.

It is attractible by the magnet, but in a much weaker degree than iron-sand or magnetic ironstone.

Chemical Characters.

It is infusible, without addition, before the blowpipe: it tinges borax of a greenish colour, which inclines to brown.

Constituent Parts.

	Cornwall.	Botany Bay.
Oxide of Iron, -	51.00	49
Oxide of Titanium,	45.25	40
Oxide of Manganese,	0.25	
Silica, - - -	3.50	11
	100.00	100
	Klaproth, Beit. b. ii. s. 231.	*Chenevix*, in Nicholson's Journ. vol. v. p. 132.

Geognostic and Geographic Situations.

It is found, accompanied with fine quartz-sand, in the bed of a rivulet which enters the valley of Menaccan in Cornwall; on the shores of the Island of Providence in America; the vicinity of Richmond in Virginia; and at Botany Bay in New South Wales.

Observations.

It has been confounded with Iron-Sand, from which it may be readily distinguished by its fracture, lustre, and inferior hardness.

2. Iserine.

Iserin, *Werner.*

L'Iserine, *Broch.* t. ii. p. 478.—Iserin, *Reuss,* b. iv. s. 598.—*Id. Lud.* b. i. s. 306.—Iser-Titan, *Suck.* 2ter th. s. 489.—Iserin, *Mohs,* b. iii s. 450. *Id. Leonhard,* Tabel. s. 81. *Id. Karsten,* Tabel. s. 74.—Titane oxydé ferrifere, *Hauy,* Tabl. p. 116.—Titaneisenstein, *Haus.* b. i. s. 251.—Iserine, *Aikin,* p. 67.

External Characters.

Its colour is iron-black, inclining to brownish-black.

It occurs in small, seldom middle-sized, obtuse angular, grains, and in rolled pieces, with a somewhat rough, strongly glimmering surface.

Internally it is glistening, and the lustre is semi-metallic.

The fracture is more or less perfect conchoidal.

The fragments are indeterminate angular and sharp-edged.

It is completely opaque.

It is hard.

It is brittle.

It retains its colour in the streak.

It is heavy in a moderate degree.

Specific gravity, 4.5.

Chemical

Chemical Characters.

Before the blowpipe, it melts into a blackish-brown coloured glass, which is slightly attracted by the magnet. The mineral acids have no sensible effect on it; but the acid of sugar extracts a portion of the titanium.

Constituent Parts.

Oxide of Titanium,	-	48
Oxide of Iron,	- -	48
Oxide of Uranium,	- -	4
		100

Thomson, in Edin. Phil. Trans. for 1807.

Geognostic and Geographic Situations.

It occurs in granitic sand, along with iron-sand, in the bed of the river Don in Aberdeenshire *. On the Continent of Europe, it has been hitherto found only in the lofty Riesengebirge, near the origin of the stream called the Iser, disseminated in granite-sand. It is still uncertain whether it belongs to the granite or flœtz-trap formation: from its affinity with iron-sand, and its occurring in the neighbourhood of the basalt hill, the Buchberg, it is suspected to be an inmate of the flœtz-trap formation †.

 Observations.

* Vid. Dr Thomson's paper on the Black Sand of the river Don in Aberdeenshire.—Ed. Phil. Trans.

† The Buchberg is the highest basalt hill in Germany, being 2921 feet bove the level of the sea, and the highest basalt except that small portion lodged in the cavity of the Schneegrube, which is situated near 4000 feet above the level of the sea. The hill itself is elevated 500 feet above the level of the stream named Iser, that waters its granite base, and at some distance

Observations.

1. It bears a very great resemblance to *Iron-Sand*, into which it even passes. It is distinguished from Iron-sand by the shade of brown in its colour, by its superior external, and inferior internal lustre, by its lower specific gravity, and very strikingly by its being very slightly attractable by even a powerful magnet.

2. It is distinguished from *Nigrine* and *Menachanite*, by fracture and lustre.

3. Nigrine.

Nigrin, *Werner*.

Titane oxydé ferrifere, *Hauy,* t. iv. p. 303. (in part).—Nigrin, *Reuss,* b. iv. s. 593. *Id. Lud.* b. i. s. 306.—Nigrin-titan, *Suck.* 2ter th. s. 487.—Nigrin, *Bert.* s. 516. *Id. Mohs,* b. iii. s. 454. *Id. Leonhard,* Tabel. s. 82.—Titane nigrin, *Brong.* t. ii. p. 99.—Nigrin, *Karsten,* Tabel. s. 74.—Eisentitan, *Haus.* Handb. b. i. s. 319.—Nigrine, *Aikin,* p. 67.

External Characters.

Its colour is dark brownish-black, inclining to velvet-black.

It occurs in larger and smaller angular grains, and rolled pieces.

Externally

---

distance below which the Iserine is found. Whilst travelling through Silesia with that excellent and truly philosophical mineralogist, my amiable, and ever to be regretted friend, the late Dr Mitchell, we ascended to the Buchberg, with the view of ascertaining more particularly the geognostic situation of the Iserine; but after a very careful examination, we could discover it neither in the granite nor basalt, but only loose in the granitic sand.

Externally it is moderately glistening.

Internally the principal fracture is shining; the cross fracture is moderately glistening; and the lustre is intermediate between semi-metallic and adamantine.

The principal fracture is imperfect and straight foliated, with several cleavages; the cross fracture is flat and imperfect conchoidal.

The fragments are indeterminate angular, and sharp-edged.

It is opaque.

It is semi-hard.

It is brittle.

It is easily frangible.

It yields a yellowish-grey streak.

It is heavy in a moderate degree.

Specific gravity, 4.445,—4.740, *Klaproth*. 4.605, *Esmark*. 4.543, *Lampadius*. 4.673, *Lowitz*. 3.700, *Vauquelin* and *Hecht*.

Physical Character.

It is not attracted by the magnet.

Chemical Characters.

It is infusible, without addition, before the blowpipe: with borax, it melts to a transparent hyacinth-red globule: it yields its titanium readily to acid of sugar.

Constituent Parts.

	Transylvania.	Uralian Mountains.	Transylvania.
Oxide of Titanium,	84	53	87
Oxide of Iron, -	14	47	9
Oxide of Manganese,	2		3
	100	100	99
	Klaproth, Beit. b. ii. s. 238.	*Lowitz*, in Crell's Ann. for 1799.	*Lampadius*, Pr. Chem. Abh. b. iii. s. 246.

Geognostic and Geographic Situations.

Europe.—At Ohlapian in Transylvania, it is found in alluvial hills, consisting of yellow sand, intermixed with fragments of granite, gneiss, and mica-slate, and from which gold is extracted by working. It comes to us usually intermixed with grains of precious garnet, rutile, kyanite, and common sand. It is also found at Bodenmais in Bavaria.

Asia.—It occurs in granite in the Uralian Mountains; and in the Island of Ceylon, where it is dispersed through alluvial land, with iron-sand, hyacinths, and zircons.

Observations.

1. Its name is derived from its black colour.

2. It is easily distinguished from *Menachanite*, by its stronger lustre, superior hardness, the colour of its streak, and by its not being in the least magnetic; which latter character also distinguishes it from *Iron-Sand.*

3. As it is usually found along with fragments of dark-coloured *Rutile*, they have been frequently confounded under

under the same name; but the red colour, perfect foliated fracture, twofold rectangular cleavage, and cubical fragments of Rutile, distinguish it at once from Nigrine.

4. It has some resemblance to *Wolfram*, but it is harder, lighter, and has a higher lustre, and darker colours.

4. Sphene.

Sphen, *Hauy & Karsten.*

This species is divided into two subspecies, viz. Common Sphene and Foliated Sphene.

First Subspecies.

Common Sphene.

Gemeiner Sphen, *Karsten.*

Braun Mänakerz, *Werner.*

Brauner Titanit, *Schumacher's* Verz. s. 114.—Titane Siliceo-calcaire, *Hauy,* t. iv. p. 307.; also Sphene, *Hauy,* t. iii. p. 114.—Gemeiner Titanit, *Reuss,* b. iv. s. 584.—Gemeiner Titanit. *Suck.* 2ter th. s. 481.—Brunon, *Mohs,* b. iii. s. 465.—*Id.* Titanit, *Leonhard,* Tabel. s. 82.—Gemeiner Sphen, *Karsten,* Tabel. s. 74.—Titane-siliceo-calcaire, *Hauy,* Tabel. p. 116.—Gemeiner Sphen, *Hausmann Handb.* b. ii. s. 613.—Sphene, *Aikin,* p. 68.

External Characters.

Its colours are reddish, yellowish, and blackish-brown, also grass-green, pistachio-green, asparagus-green, olive-green, yellowish-grey, and greenish-white.

It

It occurs seldom massive and disseminated, more frequently crystallised, and in the following figures.

1. Oblique four-sided prism, acutely bevelled on the extremities, the bevelling planes set on the obtuse lateral edges *, Fig. 215. Sometimes the acute angles are bevelled, and the bevelling planes set on the edges formed by the meeting of the bevelling planes on the extremities and the lateral planes †. Fig. 216.
2. Broad six-sided prism, with two opposite broader, and four opposite smaller lateral planes, and acuminated with four planes.
3. Rectangular four-sided prism, which is either bevelled on the extremities, or acuminated with four planes, which are set on the lateral planes. This prism is formed when all the edges of the oblique prism are truncated, and the truncating planes meet together, entirely obliterating the oblique planes.
4. Elongated octahedron, in which the apices are bevelled.
5. The preceding figure so very flat or obtuse, that it has a lenticular form.

Sometimes two crystals unite, forming either a furrowed or canaliculated twin-crystal ‡, or a compressed rectangularly cruciform one.

The crystals are large, middle-sized, small, and very small.

The

* Titan siliceo-calcaire ditetraedre, Hauy.

† Titan siliceo-calcaire uniternaire, Hauy.

‡ The above figure is the Rayonnanté en forme de gouttiere of Saussure; Sphene canaliculé of Hauy.

The surface of the crystals is generally smooth; rarely longitudinally streaked, as in the octahedron.

Internally it is shining or glistening, and the lustre is adamantine, sometimes inclining to resinous, sometimes to vitreous.

The fracture is imperfect conchoidal, which inclines to uneven; also concealed foliated.

It sometimes occurs in granular distinct concretions.

It alternates from opaque to translucent.

Its streak is greyish or yellowish-white.

It scratches glass.

It is brittle.

It is easily frangible.

Specific gravity, 3.480, *Schumacher*. 3.510, *Klaproth*.

Chemical Characters.

Before the blow-pipe it is fusible with difficulty into a blackish-brown enamel; with borax it yields a grey-slag; with phosphat of soda a green globule.

Chemical Composition.

	Passau.	Salzburg.	St Gothard.
Oxide of Titanium,	33	46	33.3
Silica, - -	35	36	28.0
Lime, - -	33	16	32.2
Water, - -	0	1	0
	101	99	93.5
	Klaproth, Beit. b. i. s. 251.	*Ibid.* b. v. s. 344.	*Cordier* in Jour. de Mines, N. 73. 70.

Geognostic

Geognostic and Geographic Situations.

Europe.—It occurs in small and very small crystals, imbedded in the syenite of the Criffle and other hills in Galloway; in syenite on the south side of Loch-Ness; in the same rock, in the mountains around the King's House, and in the syenite mountains that extend from that dreary and desolate track towards Inverouran; in the syenite of Ben-Nevis *; the granite of Aberdeen †; in the syenite of Culloden in Inverness-shire ‡; and in flœtz-trap rocks in Mid-Lothian. It is also found in the iron mines of Arendal in Norway; Bovkhult in West Gothland, in primitive limestone; in the granite rocks of Trollhätta in Sweden; in the syenite of Passau on the Inn; and in that of Moravia; in the granite or syenite of Nantes in France; in clinkstone in the Sanadoire in the Department of Puy de Dôme; and in the volcanic rocks on the borders of the Rhine: and on St Gothard and at Salzburgh it is associated with chlorite; which mineral is either disseminated through the crystals, or incrusts them.

America.—It is found imbedded in a compound of hornblende, felspar, and graphite, in the vicinity of Lake St George, and in granular foliated limestone at Kingsbridge, island of New-York ||.

Africa.—In the antique syenite of Egypt.

Second

---

* Greenough.

† Mr Mackenzie *junior* of Applecross.

‡ Macculloch.

|| Bruce's American Min. Journal, p. 239,—244.

Second Subspecies.

Foliated Sphene.

Schaaliger Sphen, *Karsten.*

Gelb Mänakerz, *Werner.*

Gelber Titanite, *Schumacher,* Verz. s. 46.—Titan siliceo-calcaire, *Hauy,* t. iv. p. 307.; also Sphene, *Id.* t. iii. p. 114.—Spathiger Titanit, *Reuss,* b. iv. s. 590. *Id. Suck.* 2ter th. s. 485. *Id. Bert.* s. 280.—Titanit, *Leonhard,* Tabel. s. 82.—Schaaliger Sphen, *Karsten,* Tabel. s. 74.—Titan siliceo-calcaire, *Hauy,* Tabl. p. 116.—Spathiger Sphen, *Haus.* Handb. b. ii. s. 614.—Sphene, *Aikin,* p. 68.

External Characters.

Its colours are pea yellow, straw-yellow, cream-yellow, and wax-yellow, which inclines to brown.

It occurs massive, and crystallised in the same figures as the preceding subspecies.

The lustre on the foliated fracture is splendent or shining, on the imperfect conchoidal and uneven fractures, only shining or glistening, and is intermediate between vitreous and resinous.

The principal fracture is foliated with a double cleavage, parallel with the lateral planes of the oblique four-sided prism. The cross fracture imperfect conchoidal, inclining to uneven.

It occurs in lamellar distinct concretions, which are straight or curved.

It is translucent, or only translucent on the edges.

In

In other characters it agrees with the preceding subspecies.

Its chemical characters and composition are the same as that of common sphene.

Geognostic and Geographic Situation.

Europe.—It is found at Arendal in Norway, in beds of magnetic ironstone, subordinate to gneiss. In these beds, it is associated with common sphene, epidote or pistacite, hornblende, augite, scapolite, felspar, quartz, calcareous-spar, and garnet.

America.—In primitive limestone at Newton in New Jersey; in an aggregate of felspar, hornblende, and graphite, near Ticonderago, where it was discovered by an intelligent and patriotic gentleman Colonel Gibbs; in Staten Island, in greenstone and hornblende rocks; in the vicinity of Peekshill, New-York, in syenite; and in veins of green stone or syenite that traverse granite at Wantage, Sussex county, New Jersey *.

Observations.

1. Some varieties of sphene, particularly the figure N° 3. so nearly resemble Zircon, that they have been confounded with it by mineralogists.

2. The name Sphene given to this species by Hauy, refers to the wedge-like shape of some of its crystals, particularly the figure N° 4.

5. Rutile.

---

* Bruce's American Mineralogical Journal, p. 239,—242.

5. Rutile.

Rutil, *Werner*.

Rother Schorl, *Klapr.* b. i. s. 233.—Schorl crystallizé opaque rouge, *De Born,* t. i. p. 168.—Titanite, *Kirw.* vol. ii. p. 329. —Sagenite, ou Schorl rouge, *Saussure,* t. iv. § 1894.—Oxide rouge de Titanium, *Lam.* t. i. p. 414.—Crispite, *Id.* t. ii. p. 233. —Titane oxidé, *Hauy,* t. iv. p. 296.—Le Ruthile, *Broch.* t. ii. p. 470.—Titanschorl, *Reuss,* b. iv. s. 569. *Id. Lud.* b. i. s. 305. *Id. Suck.* 2ter th. s. 476. *Id. Bert.* s. 514. *Id. Mohs,* b. i. s. 455.—Titane oxydé, *Lucas,* p. 180.—Rutile, *Leonhard,* Tabel. s. 82.—Titane ruthile, *Brong.* t. ii. p. 97.—Titane oxidé, *Brard,* p. 383.—Rutill, *Karsten,* Tabel. s. 74. *Id. Haus.* s. 111.—Native Oxide of Titanium, *Kid,* vol. ii. p. 222.—Titane oxidé, *Hauy,* Tabl. p. 115.—Rutil, *Haus.* Handb. b. i. s. 319.—Titanite, *Aikin,* p. 66.

External Characters.

Its colour is dark blood-red, of various degrees of intensity, which passes into light hyacinth-red and dark brownish-red.

It occurs massive, disseminated, in membranes; and crystallised in the following figures.

1. Rectangular four-sided prism.
2. Rather oblique four-sided prism.
3. Six-sided prism, which is sometimes acuminated with six planes, which are set on the lateral planes, and these are rarely convex *. Fig. 217.

4. Eight-

* Vid. Bruce's American Mineralogical Journal, p. 288. where the above figure is described.

4. Eight-sided prism.
5. Sometimes two prisms are joined by their terminal planes, under a very obtuse angle, thus forming a kind of twin crystal *. Fig. 218.
6. It occurs sometimes in capillary and acicular crystals.

The crystals are occasionally curved ; have frequent transverse rents, and are sometimes apparently broken entirely across, the ends removed to some distance from one another, and the interstice filled up with the substance of which the matrix consists.

The crystals are usually small and very small, seldom middle-sized ; the capillary crystals are frequently scopiformly aggregated ; often reticulated, and the interstices have the shape of equilateral triangles †.

The crystals are longitudinally streaked.

Externally it is shining and glistening.

Internally the lustre is intermediate between adamantine and semi-metallic, and is splendent in the foliated, but only shining or glistening in the conchoidal or uneven fractures.

In some varieties the principal fracture is foliated, with a fourfold cleavage, and the cross fracture conchoidal ; in other varieties it is imperfect and small conchoidal, inclining more or less to uneven.

The fragments are cubical.

It sometimes occurs in thick lamellar, and also in granular distinct concretions.

It yields a yellowish-grey or orange-yellow coloured streak.

It

* Titanc oxidé geniculé, Hauy.

† The reticulated variety was named by Saussure Sagenite, and Schorl frieoté.

It is translucent in a slight degree.

It scratches glass.

It is brittle.

It is easily frangible.

Specific gravity, 4.180, *Klaproth* ; 4.246, *La Metherie* ; 4.1025, *Hauy*.

Chemical Characters.

Without addition, or even with phosphoric salts, it is infusible before the blow-pipe ; with borax or alkali, it affords a hyacinth-red transparent glass.

Constituent Parts.

According to the analysis of Klaproth, it is pure oxide of titanium, slightly intermixed with oxide of iron. Ekeberg and Vauquelin found in a variety from Westmannland a small portion of oxide of chrome *.

Geognostic Situation.

It is found imbedded, in veins and in drusy cavities, in granite, syenite, gneiss, mica-slate, limestone, chlorite-slate, and hornblende-slate.

Geographic Situation.

Europe.—It occurs in the granite of Cairngorm, the limestone of Rannoch, and in the rocks of Ben Gloe, where it was discovered by Dr Macculloch ; also at Craig Cailleach near Killin, imbedded in quartz, and near to Beddgelert in Caernarvonshire. In Norway it occurs

* Annales du Mus. t. vi. p. 98. It is the Titane oxidé chromifere of Hauy.

occurs at Arendal, in a vein of granite which traverses gneiss, and in a rock of the same kind at Aschaffenburg. On St Gothard it is met with in those drusy cavities that so often occur in granite, resting on the rock-crystal, adularia, and foliated chlorite, with which they are lined. In the country of Salzburg, it is imbedded in tremolite, and in Hungary in common quartz, and in rock-crystal, which lies in nests in mica-slate. It is found near St Grieux, and at Allemont in France; at Buitrago in Spain, in veins in gneiss along with schorl; at Boinik in Hungary, and in Transylvania.

Asia.—It is found at the town of Sarapulka, twelve wersts from Mursinska in Siberia.

America.—It occurs in veins of an aggregated rock of felspar, quartz, mica and granular foliated limestone, which traverse primitive limestone in the island of New York: also in primitive limestone on the Hudson River; in large quantity imbedded in quartz, in the vicinity of Richmond in Virginia; in quartz in hornblende-slate, at Worthington in Massachussets; in quartz from the neighbourhood of Baltimore; and in quartz near the Schuyler copper-mines in Bergen county, New Jersey. It is also met with in South Carolina; in quartz in the county of Delaware in Pennsylvania, and in the back part of North Carolina, where it is said to occur in great abundance *.

6. Octahedrite.

* Bruce's American Mineralogical Journal, p. 235,—238.

6. Octahedrite.

Octaedrit, *Werner*.

Schorl bleu, *Romé de Lisle*, t. ii. p. 406.—Schorl octaedre rectangulaire, *Bournon*, Journ. de Phys. 1787, Mai.—Oisanite, *Lam.* t. ii. p. 269.—Octaedrite, *Saussure*, Voyages dans les Alpes, t. vii. p. 139. § 1901.—Anatase, *Hauy*, t. iii. p. 129. 136. *Id. Broch.* t. ii. p. 548. *Id. Reuss*, b. iv. s. 580.—Pyramiden Manak, *Lud.* b. ii. s. 191.—Anatase-titan, *Such.* 2ter th. s. 480.—Anatas, *Mohs*, b. iii. s. 462. *Id. Leonhard*, Tabel. s. 82.—Titane-Anatase, *Brong.* t. ii. p. 101.—Anatas, *Karsten*, Tabel. s. 74. *Id. Haus.* s. 111.—Octahedral Titanite, *Kid*, vol. ii. p. 223 —Titan anatase, *Hauy*, Tabl. p. 116.—Anatas, *Haus.* Handb. b. i. s. 322.—Octohedrite, *Aikin*, p. 68.

External Characters.

Its colour passes from indigo-blue, through many shades, to dark reddish-brown, clove-brown, and yellowish-brown.

It has been hitherto found only crystallised, and the following are the figures which it assumes.

1. Lengthened octahedron *.
2. The preceding figure truncated on the extremities †.
3. Octahedron acuminated on the extremities with four planes, and the acuminating planes set on the lateral planes ‡.

 4. Octahedron,

* Anatase primitif, Hauy.

† Anatase basé, Hauy.

‡ Anatase dioctaedre, Hauy.

4. Octahedron acuminated with eight planes, of which two and two are set on each lateral plane ‡.

The crystals are small and very small, and are usually superimposed.

The surface is transversely streaked, and is splendent, and the lustre is semi-metallic.

Internally it is also splendent, and the lustre is adamantine.

The fracture is foliated.

The fragments are indeterminate angular and sharp-edged.

It is strongly translucent or semitransparent, or passing to transparent.

It scratches glass.

It is brittle.

Specific gravity, 3.8571, *Hauy*.

Chemical Characters.

It is infusible before the blowpipe. When melted with borax, a reddish-brown coloured glass is formed. When this glass is brought to the extremity of the flame, the reddish-brown colour changes into blue, and becomes opaque. If the action of the blowpipe be still continued, it at length becomes white. In a higher temperature, the reddish-brown colour again appears; and according as the temperature is altered, the appearance and disappearance of the colours can be produced.

Constituent

* Anatase prominulé, Hauy.

Constituent Parts.

According to the experiments of Vauquelin, it is an oxide of titanium.

Geognostic and Geographic Situations.

It is a rare mineral. In Dauphiny it occurs in veins in primitive rocks along with felspar, axinite, rock-crystal, and chlorite: it has also been met with in drusy cavities in transition limestone in Hadeland in Norway.

ORDER IX.—LEAD.

This Order contains the following species, viz. Galena or Lead-glance, Blue Lead-ore, Bournonite or Antimonial Lead ore, Cobaltic Lead-ore, Native Minium, White Lead-ore, Black Lead ore, Corneous Lead-ore or Muriate of Lead, Phosphate of Lead, Arsenate of Lead, Sulphate of Lead or Lead-vitriol, Yellow Lead-ore or Molybdate of Lead, and Red Lead-ore or Chromate of Lead.

1. Galena or Lead-Glance.

Bleiglanz, *Werner.*

This species is divided into two subspecies: 1. Common Galena or Lead-glance. 2. Compact Galena or Lead-glance.

First

First Subspecies.

Common Galena or Lead-Glance.

Gemeiner Bleiglanz, *Werner*.

Id. Werner, Pabst. b. i. s. 97. *Id. Wid.* s. 841.—Common Galena, *Kirw.* vol. ii. p. 216.—Gemeiner Bleiglanz, *Emm.* b. ii. s. 369.—Plomb sulphuré, Galene, *Lam.* t. i. p. 289.-292.—Plomb sulphuré, *Hauy*, t. iii. p. 456.—La Galene commune, *Broch.* t. ii. p. 295.—Gemeiner Bleiglanz, *Reuss*, b. iv. s. 174. *Id. Lud.* b. i. s. 258. *Id. Suck.* 2ter th. s. 306. *Id. Bert.* s. 445. *Id. Mohs*, b. iii. s. 469.—Blættriges Bleiglanz, *Hab.* s. 126.—Plomb sulphuré, *Lucas*, p. 114.—Gemeiner Bleiglanz, *Leonhard*, Tabel. s. 70.—Plomb sulphuré laminaire, *Brong.* t. ii. p. 195. *Id. Brard*, p. 265.—Bleiglanz, *Karsten*, Tabel. s. 68. *Id. Haus.* s. 74.—Sulphuret of Lead, *Kid*, vol. ii. p. 130.—Plomb sulphuré, *Hauy*, Tabl. p. 79.—Galena, *Aikin*, p. 44.

External Characters.

Its colour is fresh lead-grey of different degrees of intensity. On the surface, it sometimes shows a tempered-steel or iridescent tarnish.

It occurs most frequently massive, and disseminated; also in membranes, in angular pieces, in grains, reticulated, specular, corroded, and amorphous, seldom fused-like and cylindrical; frequently crystallised:

1. *Cube* *, Fig. 219. which is the fundamental figure, exhibits the following varieties:

 a. In

* Plomb sulphuré primitif, Hauy.

a. In which the faces are either straight or convex, or concave.

b. Truncated on all the angles *, Fig. 220.

c. Truncated on all the angles and edges †, Fig. 221.

d. Bevelled on all the edges, Fig. 222.

2. *Octahedron* ‡, Fig. 223. which exhibits the following varieties:

a. Cuneiform or elongated, Fig. 224.

b. The edges rounded, and the faces concave.

c. Truncated on all the edges ||, Fig. 225.

d. Truncated on all the angles, and the edges of the truncations truncated §, Fig. 226.

e. In which all the angles and edges are truncated at the same time.

f. All the angles truncated, and the edges bevelled ¶, Fig. 227.

g. All the edges bevelled, and the edges of the bevelment, and also the angles, truncated **, fig. 228

3. Rectangular four-sided prism, acuminated on both extremities with four planes, which are set sometimes on the lateral planes, sometimes on the lateral

* Plomb sulphuré cub-octaedre, Hauy.

† Plomb sulphuré triforme, Hauy.

‡ Plomb sulphuré octaedre, Hauy.

|| Plomb sulphuré pantogene, Hauy.

§ Plomb sulphuré unibinaire, Hauy.

¶ Plomb sulphuré octotrigesima, Hauy.

** Plomb sulphuré pentacontaedre, Hauy.

teral edges, and the summits of the acuminations truncated, fig. 229.

4. Six-sided prism, acuminated with four planes, of which two are set on two opposite lateral planes, and two on opposite lateral edges; or all the acuminating planes are set on lateral edges.
5. Six-sided prism, acuminated with three planes, which are set on the alternate lateral edges.
6. Equiangular six-sided table, which is either perfect, or with bevelled terminal planes.
7. Three-sided table, in which the angles and terminal planes are bevelled, or in which the terminal planes and edges are bevelled.

The crystals are usually middle-sized, small, and very small; seldom large. They are generally superimposed, or in druses, but seldom imbedded.

The planes of the crystals are smooth, drusy, or rough.

Externally it alternates from specular splendent to glimmering, according as the surface is smooth, drusy, or rough.

Internally it alternates from specular splendent to glistening, and the lustre is metallic.

The fracture is more or less perfect foliated; generally straight, sometimes also floriform foliated, which latter sometimes passes into narrow or broad, and stellular diverging radiated. It also occurs scaly foliated. Cleavage threefold and rectangular, and parallel with the sides of a cube.

The fragments are cubical.

The massive varieties occur in granular distinct concretions, of all degrees of magnitude from large to very fine granular. The latter often passes into compact, consequently

consequently into the following subspecies. It occurs also in distinct concretions, which are straight, thin and thick lamellar, and very rarely columnar.

The streak is shining.

It is soft.

It is perfectly sectile.

It is uncommonly easily frangible.

It is uncommonly heavy.

Specific gravity, 7.220, *Muschenbröck* ; 7 290, *Gellert.* The crystallised, 7.5873, *Brisson.* The radiated, 5.500, *Lametherie* ; 6.565 to 7.786, *Watson.* From the Hartz, 7.447, *Kirwan.* From Kantenbach, 6.140, *Vauquelin.* From the Kirschwald, 6.820, *Vauquelin.* From Kampfstein, 7.100, *Vauquelin.* From Eckelsberg, from 7.300 to 7.600, *Vauquelin.*

Chemical Characters.

Before the blow pipe it flies into pieces, then melts, emitting a sulphureous odour, and a globule of metallic lead remains. When it is alternately heated and cooled, it at length disappears entirely; and if it be argentiferous, a minute globule of silver remains behind.

Constituent Parts.

According to Vauquelin, galena or lead-glance contains the following ingredients :

	From Kirschwald in Deux Ponts.	Kampfstein.	Eckles-berg.	Kanten-bach.
Lead,	54	69	68.69	64
Sulphur,	8	16	16.18	18
Calcareous-earth and silica,	38	15	16.13	18
	—	—	—	—

All those specimens appear impure, so that the analyses are not of so much value as those that follow :

Lead,

	Klausthal.				Durham.
Lead,	83.0		77	Lead,	85 13
Sulphur,	16.41		20	Sulphur,	13.02
Silver,	0.08		1	Iron,	0.50
Westrumb.	99.49	*Kirwan.*	98	*Thomson.*	98.65

Geognostic Situation.

It occurs in veins, beds, and imbedded masses, and is not confined to any particular class of rocks, for it occurs in primitive, transition, and flœtz mountains. In primitive rocks it is met with in beds, subordinate to gneiss and clay-slate, and associated with blende and iron-pyrites; and in veins in primitive limestone. It occurs in beds in greywacke; and in veins traversing that rock and transition clay-slate. It forms beds along with calamine in the older flœtz limestones; occurs in veins and imbedded portions, and disseminated in shell limestone and conglomerate; and in veins, and disseminated in limestone and sandstone, belonging to the coal formation.

Geographic Situation.

Europe.—At Leadhills, in Lanarkshire, it occurs in veins that traverse transition rocks along with white lead-ore, green lead ore, lead earth, sulphat of lead, calamine, ochry brown ironstone, brown hematite, iron-pyrites, sparry ironstone, azure copper-ore, manganese-ore, brown par, calcareous-spar, heavy-spar, and mountain-cork. The same formation extends into the upper part of Dumfries-shire, where it forms the mines of Wanlockhead. It occurs along with fluor-spar, in veins that traverse granite at Monaltrie, in Aberdeenshire;

Aberdeenshire; in the old lead-mines of Clifton at Tyndrum, already described; in veins that traverse gneiss, along with heavy-spar and calcareous-spar, at Strontian in Argylshire; in veins that traverse sandstone, along with heavy-spar and calcareous-spar, at Cumberhead in Lanarkshire; in small veins or disseminated in the grey sandstone of the coal formation in the Lothians and Fifeshire; in veins traversing limestone, in the island of Isla; veins traversing gneiss, in the isle of Coll; and in conglomerate rocks near Stromness in Orkney. It forms large veins in shell limestone, in Northumberland, Durham, Derbyshire, Flintshire, Somersetshire, and in slate in Shropshire, and most of the counties of Wales *. In the mines of Derbyshire, the galena or lead glance is associated with white lead-ore, or green lead-ore, heavy-spar, calcareous-spar, and fluor-spar, and some ores of zinc and iron. Fluor-spar is one of the most common veinstones in that country; and there are some places in which the veins are entirely filled up with fluor-spar. It occurs in veins in primitive limestone at Sala, and in beds, along with copper-pyrites, iron-pyrites, and blende, at Fahlun, in Sweden. In veins that traverse transition rocks, or in beds subordinate to these, in the Hartz, and in veins in gneiss, in the Saxon Erzgebirge. Disseminated, and in nests, in shell limestone, as at Kulf, near Bruggen, in the *ci-devant* kingdom of Westphalia, and in the vicinity of Göttingen. At Mus in Bohemia, in veins that traverse clay-slate, associated with white, black, and green lead-ores, blende, copper-pyrites, malachite, iron-pyrites,

* Aikin's Manuel, 2d Ed. p. 108.

iron-pyrites, quartz, and heavy-spar. At Prizbram, also in Bohemia, where it occurs in veins that traverse clay-slate, it is accompanied with black and green lead ores, blende, white silver-ore, native silver, sparry ironstone, grey and white antimony ores, iron-pyrites, heavy-spar, calcareous-spar, and quartz. At Bleyberg, in Carinthia, it is accompanied with white, yellow, and black lead-ores, calamine, yellow and brown blende, and mountain-cork. At Offenbanya in Transylvania, it is associated with grey copper-ore, grey antimony-ore, iron-pyrites, and brown blende, in a bed of granular limestone; at Nagyag, with red antimony-ore, and amethyst, in clay-porphyry. At Querbach in Silesia it is accompanied with black blende, cobalt-glance, magnetic ironstone, iron-pyrites, copper-pyrites, arsenical-pyrites, garnet, and calcareous-spar; and at Altenberg in the same country, along with iron-pyrites, copper-pyrites, arsenical-pyrites, copper-glance or vitreous copper-ore, calcareous-spar and quartz. Besides the countries above enumerated, the following may be added as affording galena or lead-glance: Swabia, Bavaria, the Tyrol, Salzburg, Upper Austria, the Bannat, France (in which the most considerable mines are those of Pompæan, Poullaouen, and Huelgoët), Italy and Spain.

Asia.—This ore does not occur so abundantly in Asia as in Europe; it is met with at Irkutzk, Kolywan, and in the Uralian districts. Lead-ore, (I presume galena or lead-glance), is found at Dessouly in Higher Hindostan, about fifty coss east of Sirinagur; and we are informed by Captain Turner, that at a place situated nearly two days journey from Tessoolumboo in Thibet, there is a mine of this ore. In Lower India, this ore has been met with in small quantities at Jangumrauzpillay, in the

the Cumtum district. The greater part, however, of the lead met with in the Peninsula of India comes from Siam *, Araccan, and occasionally from the Burmah dominions †; it is also found at Omon in Arabia ‡.

America.—It occurs in the United States; thus it forms a large vein in granite and gneiss near Northampton in Massachusets, and the veinstones are quartz, heavy-spar, and calcareous-spar. It is met with as far north as Greenland, where it is associated with cryolite, brown-spar, sparry ironstone, and iron pyrites.

It abounds in a flœtz limestone in the north-eastern parts of New Spain, particularly in the district of Zimapan; in the kingdom of New Leon; and in the province of New Santander. In these districts the galena contains a small portion of silver, and is generally worked more for its silver than its lead. Mines of this ore also occur in Chili in South America.

Uses.

Nearly all the lead of commerce is obtained from this ore. In order to obtain the lead the ore is first roasted, in order to drive off the sulphur, and then mixed with the necessary quantity of coke, charcoal, or peat, and reduced in a common furnace. The lead which remains after the operation of roasting, is in an oxydated state; the inflammable matter, with which it is mixed in the furnace, decomposes the metallic oxide; and combining with

* Elmore's Guide to the Indian Trade, p. 309.

† Oriental Repertory, vol. I. p. 117.

‡ Ainslie's Materia Medica of Hindostan, p. 56.

with its oxygen, flies off in the form of carbonic acid gas, while the lead is reduced to the metallic state, and sinks to the bottom of the furnace. Almost all the varieties of galena or lead-glance contain a greater or less portion of silver. The silver, after the reduction of the lead, may be separated by the process of cupellation; but in the greater number of instances, the quantity of silver is so inconsiderable, as not to repay the expence of labour; and hence the lead of commerce almost always contains a minute portion of silver. Galena is also used for glazing pottery.

Observations.

1. Some dark-coloured varieties of galena might, with a superficial observer, be confounded with *blende*, or sulphuret of zinc; but the lustre of the zinc-ore is destroyed by scratching the surface with a knife, which is not the case with the galena or lead-glance. If both are breathed upon, the galena recovers its lustre in a moment; the blende very slowly. Galena is distinguished from *graphite* by its colour, greater specific gravity, and by the comparatively faint trace it leaves on paper. The same marks will serve to distinguish it from *molybdena*, which possesses, besides a foliated fracture, a considerable degree of flexibility.

2. Galena, which occurs in beds, is said to contain less silver than that found in veins.

Second

Second Subspecies.

Compact Galena or Lead-Glance.

Bleischweif, *Werner.*

Plumbum Plumbago, *Wall.* t. ii. p. 305.—Bleischweif, *Werner*, Pabst. b. i. s. 114. *Id. Wid.* s. 845.—Galene compacte, *De Born*, t. ii. p. 355.—Compact Galena, *Kirw.* vol. ii. p. 218.—Bleischweif, *Emm.* b. ii. s. 377.—Plomb sulphuré compacte, *Hauy*, t. iii. p. 461.—La Galene compacte, *Broch.* t. ii. p. 301. —Bleischweif, *Reuss*, b. iv. s. 188. *Id. Lud.* b. i. s. 259—Dichter Bleiglanz, *Suck.* 2ter th. s. 312.—Bleischweif, *Bert.* s. 447. *Id. Mohs.* b. iii. s. 486. *Id. Hab.* s. 127.—Dichtes Bleiglanz, *Leonhard*, Tabel. s. 71.—Plomb sulphuré compacte, *Brong.* t. ii. p. 195.—Bleischweif, *Karsten*, Tabel. s. 68. *Id. Haus.* s 74.—Plomb sulphuré compacte, *Hauy*, Tabl. p. 80. —Compact Galena, *Aikin*, p. 45.

External Characters.

Its colour is fresh lead-grey, somewhat lighter thanthat of the preceding subspecies ; sometimes it slightly inclines to steel-grey.

It occurs massive, disseminated, and specular.

The specular variety is smooth, shining or splendent, externally.

Internally it is only strongly glimmering, and the lustre is metallic.

The fracture is even, which in some varieties passes into flat conchoidal.

The fragments are indeterminate angular, not particularly sharp edged.

It seldom occurs in thin and curved lamellar concretions.

The

The streak is shining, inclining to splendent.

It is not so easily frangible as the preceding subspecies; but agrees with it in the remaining characters.

Constituent Parts.

It is a compound of Sulphuret of Lead and Sulphuret of Antimony, and a small portion of Silver.

Geognostic Situation.

It occurs in veins, and is usually accompanied with common galena or lead-glance. It is worthy of remark, that when the two subspecies occur together, the compact always forms the sides of the vein, and this probably owing to its having been in a less perfect state of solution. It is also accompanied with black blende, common iron-pyrites, copper-pyrites, quartz, and heavy-spar.

Geographic Situation.

Europe.—It is found at Leadhills in Lanarkshire; in Derbyshire; Sahlberg in Westermannland; in the Hartz; Freyberg and Gersdorf in Saxony; Rauschenberg in Upper Bavaria; Weiding in the Upper Palatinate; Leogang in Salzburg; Servoz, in the valley of Chamouni in Switzerland.

Asia.—Siberia.

Observations.

1. It seldom occurs pure, but is generally intermixed with common galena or lead-glance. A mixture of this kind is described under the name *Galena striata*, by some of the older mineralogists; *Plumbum stibiatum*, Lin.; *Galena plumbi antimonialis*, Waller. Syst. Min. t. ii. p. 305;

p. 305.; *Plomb sulphuré strie*, Hauy; *Stripmalm* of the Swedes.

2. The specular variety is known in Derbyshire under the name *Slikensides*, a term somewhat expressive of its smooth form. It occurs lining the walls of very narrow rents. It has a most remarkable property, that when the rock in which it is contained is struck with a hammer, a crackling noise is heard, which is generally followed by an explosion of the rock, in the direction and neighbourhood of the vein. The cause of this singular effect has not been satisfactorily accounted for.

2. Blue Lead-Ore.

Blau Bleierz, *Werner*.

Plumbi nigri crystallis regularibus, *Waller.* t. ii. p. 309. ?—Blau Bleierz, *Werner*, Pabst. b. i. s. 115. *Id. Wid.* s. 847.—Blue Lead-ore, *Kirw.* vol. ii. p. 220.—Blau Bleierde, *Emm.* b. ii. s. 380.—Plomb noire, *Hauy*, t. iii. p. 497.—La Mine de Plomb bleue, *Broch.* t. ii. p. 303.—Blaubleierz, *Reuss*, b. i. s. 209. *Id. Lud.* b. i. s. 260. *Id. Suck.* 2ter th. s. 322. *Id. Bert.* s. 453. *Id. Mohs*, b. iii. s. 487. *Id. Leonhard*, Tabel. s. 71.—Plomb noire, *Brong.* t. ii. p. 199.—Blaubleierz, *Karsten*, Tabel. s. 68.—Plomb sulphuré epigene prismatique, *Hauy*, Tabl. p. 83.—Blue Lead, *Aikin*, p. 46.

External Characters.

Its colour is intermediate between dark indigo-blue and lead-grey.

It occurs massive, and crystallised in perfect six-sided prisms, which are usually small, low, sometimes bulging, and with a rough and dull surface.

Internally

Internally it is feebly glimmering, and the lustre is metallic.

The fracture is even; sometimes it approaches to small and flat conchoidal.

The fragments are indeterminate angular.

The streak is shining and metallic.

It is soft, inclining to very soft.

It is sectile.

It is easily frangible.

It is heavy, bordering on uncommonly heavy.

Specific gravity, 5.461, *Gellert*.

Chemical Characters.

It melts easily before the blowpipe, emitting a pungent sulphureous vapour, and is reduced to the metallic state.

Constituent Parts.

Hauy is of opinion, that it is Phosphate of Lead, partly converted into Sulphuret of Lead; while Hausmann and and Karsten consider it as a mechanical mixture of Phosphate of Lead and Sulphuret of Lead.

Geognostic and Geographic Situations.

It occurs in veins, accompanied with black lead-ore, brown lead-ore, white lead-ore, malachite, radiated azure copper-ore, quartz, fluor-spar, and heavy-spar.

It is a rare fossil, having hitherto been found only at Zschoppau in Saxony, and Huelgöet in France.

3. Bournonite, or Antimonial Lead-Ore.

Spiessglanzbleierz, *Karsten.*

Schwarz Spiessglaserz, *Werner.*

Hatchet and *Bournon,* in Phil. Trans. for 1804.—Spiessglanzblei, *Karsten,* Tabel. s. 68.—Plomb sulphure antimonifere, *Hauy,* Tabl..p. 80.—Triple sulphuré d'Antimoine, Plomb et Cuivre, Endellione, *Bournon,* Catalogue Mineralogique, p. 409. —Spiessglanzbleierz, *Haus.* Handb. b. i. s. 173.—Triple Sulphuret of Lead, *Aikin,* p. 45.

External Characters.

Its colour is intermediate between lead-grey and steel-grey.

It generally occurs massive and disseminated, and sometimes crystallised in the form of a cube, variously modified by truncations on the edges and angles.

Externally it is shining and metallic.

Internally the lustre is intermediate between glistening and glimmering, and is metallic.

The fracture is coarse-grained uneven.

It is opaque.

It yields easily to the knife.

It becomes more shining in the streak.

It is brittle.

It is easily frangible.

Specific gravity, 5.700, *Hatchett.*

Chemical Characters.

Before the blowpipe, it generally splits and decrepitates, then melts, emitting a white and sulphureous vapour;

pour; after which, there remains a crust of sulphureted lead, inclosing a globule of copper.

Constituent Parts.

	Huel Boys, near Endellion, Cornwall.	Cornwall.	Klausthal.
Lead, -	42.62	39.00	42.50
Antimony,	24.23	28.50	19.75
Copper, -	12.80	13.50	11.75
Iron, -	1.20	1.00	5.00
Sulphur, -	17.00	16.00	18.00
	Hatchett, Phil. Trans. 1804, i. 63.	*Klaproth*, Beit. b. iv. s. 90.	Id. s. 86.

According to an estimate of Mr Smithson, this ore contains in 100 parts the following compounds:

Sulphuret of Lead or Galena,	Lead,	41.08	47.41
	Sulphur,	6.33	
Sulphuret of Antimony,	Antimony,	25.67	34.23
	Sulphur,	8.56	
Sulphuret of Copper,	Copper,	12.80	16.00
	Sulphur,	3.20	
Sulphuret of Iron, -	Iron,	1.20	2.60
	Sulphur,	1.40	
			100.24

Smithson, Phil. Trans. for 1808, P. i. p. 55. &c.

Geognostic and Geographic Situations.

Europe.—It is found near Endellion in Cornwall, along with grey antimony-ore, and brown blende. On the Continent, it is met with at Ratisbon, associated with brown blende, grey copper-ore, galena or lead-glance,

 and

and common iron-pyrites; also in Saxony; in the Hartz, accompanied with galena or lead-glance, sparry iron-ore, and heavy-spar, in veins that traverse grey-wacke and grey-wacke-slate.

Asia.—In Siberia, along with quartz, malachite, galena or lead-glance, and calcareous-spar.

America.—In Peru, associated with copper and iron pyrites.

Observations.

It is named *Bournonite*, in honour of Bournon, who first described it.

4. Cobaltic Lead-Ore.

Kobaltbleierz, *Hausmann.*

Kobaltbleierz, *Haus.* s. 75. *Id. Haus.* Handb. b. i. s. 183.—Kobalt-Bleiglanz, Nordeutsch. Beitr. z. Berg. und Hüttenk. iii. s. 120.

External Characters.

Its colour is fresh lead-grey

It occurs fine and minutely disseminated, and in extremely minute crystals, which are aggregated in a moss-like form.

Its lustre is shining and metallic.

It is small and fine scaly foliated.

It occurs in fine granular distinct concretions.

It is opaque.

It is soft.

It is sectile.

It soils feebly.

Chemical

Chemical Characters.

Before the blowpipe, it splits into small pieces; and communicates a smalt-blue colour to glass of borax.

Geognostic and Geographic Situations.

It occurs in small quantity in a vein in transition rocks, in the mine of Lorenz near Clausthal in the Hartz.

Observations.

This ore was first discovered by M. Bauersach of Zellerfeld. An ore of this kind is mentioned by Proust as occurring in Catalonia *.

5. Native Minium, or Native Red Oxide of Lead.

Natürliche Mennige, } *Hausmann.*
Roth Bleioxyd, }

Smithson, in Nicholson's Journal, xvi. p. 127.—*Hänle,* in Magaz. d. Gesel. Natf. Fr. zu Berlin, iii. s. 235.—Plomb oxydé rouge, *Hauy,* Tabl. p. 80. Note 120.—Das rothe Bleioxyd, *Haus.* Handb. b. i. s. 351.—Native Minium, *Aikin,* p. 110. 2d edit.

External Characters.

Its colour is scarlet-red.

It occurs massive, amorphous, and pulverulent; but when examined by the lens, exhibits a crystalline structure, like that of galena, on which it generally rests.

* Proust, Journ. de Phys. lxiii. Nov. 1806.

Chemical Characters.

Before the blowpipe, on charcoal, it is first converted into litharge, and then into metallic lead.

Geognostic and Geographic Situations.

It is found in Grassington Moor, Craven; Grasshill Chapel, Wierdale, Yorkshire. On the Continent, it is found in the mine of Hausbaden, near Badenweiler, on galena, and associated with quartz.

Observations.

This mineral, in the opinion of Mr Smithson, is produced by the decay of galena or lead-glance; and he adduces in confirmation of this idea, the description of a specimen, which is galena in the centre, but native minium towards the surface.

6. White Lead-Ore.

Weiss-Bleierz, *Werner.*

It is divided into two subspecies, viz. Common White Lead-Ore, and Earthy Lead-ore.

First

First Subspecies.

Common White Lead-Ore, or Sparry White Lead-Ore.

Weiss-Bleierz, *Werner.*

Minera Plumbi alba spathosa, *Wall.* t. ii. p. 307.—Mine de Plomb blanche, *Romé de L.* t. iii. p. 380.—Weiss Bleyerz, *Wern.* Pabst. b. i. s. 118. *Id. Wid.* s. 852.—Plomb spathique blanc, *De Born,* t. ii. p. 368.—White Lead-ore, *Kirw.* vol. ii. p. 203.—Weiss Bleyerz, *Emm.* b. ii. s. 388.—Plomb blanc, *Lam.* t. i. p. 305.—Plomb carbonaté, *Hauy,* t. iii. p. 475.—La Mine de Plomb blanche, ou le Plomb blanc, *Broch.* t. ii. p. 309.—Weissbleierz, *Reuss,* b. iv. s. 245. *Id. Lud.* b. i. s. 261. *Id. Suck.* 2ter th. s. 326. *Id. Bert.* s. 459. *Id. Mohs,* b. iii. s. 493.—Kohlenstoffsaures Bleierz, *Hab.* s. 128.—Plomb carbonaté, *Lucas,* p. 117.—Weissbleierz, *Leonhard,* Tabel. s. 71.—Plomb carbonaté, *Brong.* t. ii. p. 198. *Id. Brard,* p. 268.—Lichter Bleispath, *Karsten,* Tabel. s. 68.——Spathiges Bleiweiss, *Haus.* s. 114.—Crystallised Carbonate of Lead, *Kid,* vol. ii. p. 136.—Plomb carbonate, *Hauy,* Tabl. p. 81.—Carbonate of Lead, *Aikin,* p. 46.

External Characters.

Its colour is snow-white, greyish-white, and yellowish-white; also yellowish-grey, cream-yellow, and clove-brown. It has sometimes a tempered steel tarnish. It is sometimes coloured externally yellow or brown, by yellow or brown ironstone; occasionally green, by earthy malachite, and blue, by earthy azure copper-ore.

It

It occurs massive, dissminated, and in membranes; but most commonly crystallised.

1. Very oblique four-sided prism, flatly bevelled on the extremities, the bevelling planes set on the obtuse lateral edges *. Fig. 230.
2. Unequiangular six-sided prism, in which the terminal edges are truncated. Fig. 231 †.
3. Unequiangular six-sided prism, acutely acuminated with six planes, which are set on the lateral planes ‡, Fig. 232. Sometimes the acumination ends in a line, and the prism is occasionally so broad that it appears like a bevelled six-sided table.
4. Acute double six-sided pyramid, which is either perfect, as in Fig. 233 ||, or is truncated on the common base.
5. Unequiangular six-sided prism, acuminated with four planes, two of which are set on the lateral planes, bounded by the obtuse lateral edges, but the other two are set on the acuter lateral edges. Fig. 234 §.
6. Rectangular four-sided prism, bevelled on the terminal planes. Sometimes the edge of the bevelment is several times truncated, so that it appears arched.
7. Rectangular four-sided prism, acuminated on the terminal planes, and the acuminating planes set on

* Plomb carbonaté octaedre, Hauy.

† Plomb carbonaté annulaire, Hauy.

‡ Plomb carbonaté trihexaedre, Hauy.

|| Plomb carbonaté bipyramidal, Hauy.

§ Plomb carbonaté sexoctonal, Hauy.

on the lateral planes. Sometimes the summits and edges of the acumination are truncated.

8. Long acicular and capillary crystals, which are columnarly aggregated.

9. It accurs also in twin and triple crystals.

The crystals are usually small and very small; seldom middle-sized.

The crystals occur sometimes single, more frequently columnarly and scopiformly, or promiscuously aggregated.

Externally, it alternates from specular splendent to glistening.

Internally, it alternates from highly splendent to glistening, and the lustre is adamantine, sometimes inclining to semimetallic, sometimes to resinous.

The fracture is small conchoidal, which sometimes passes into uneven and splintery; also concealed foliated: when the cleavage can be detected it appears five fold, four of the folia appearing to be parallel with the sides of an octahedron, and the fifth parallel with the common base of the octahedron.

The fragments are indeterminate angular.

It alternates from tran lucent to transparent; and it refracts double in a high degree.

It is soft.

It is brittle.

It is very easily frangible.

It is heavy.

Specific gravity, 6.480, from Leadhills, according to *Klaproth*; 7.2357, *Chevenix*; 6.0717,–6,5586, *Hauy*; 6.255, *Karsten*; 6.000, *Ullman*.

Chemicol

Chemical Characters.

It is insoluble in water. It dissolves with effervescence in muriatic and nitric acids. Before the blow-pipe it decrepitates, becomes yellow, then red, and is soon reduced to a metallic globule.

Constituent Parts.

	Leadhills.	Nertschinsk.	
		Transparent.	Translucent.
Oxide of lead,	82	84.5	73.50
Carbonic acid,	16	15.5	15.00
Silica, -	—	—	8.00
Alumina, -	—	—	} 2.66
Oxide of iron, -	—	—	} 2.66
Water, -	2		
	100	100.0	100.0
	Klaproth, Beit. b. 3. s. 168.	*John's* Chem. Unters. b. ii. s. 233. 236.	

Geognostic Situation.

It occurs in veins, and sometimes also in beds, in gneiss, mica-slate, and clay-slate, foliated granular limestone, greywacke, and greywacke slate, and flœtz limestone. The veins in which it is found are generally lead veins, which contain besides this ore, galena or lead-glance, green, black and yellow lead-ores, arseniate of lead, sulphate of lead, earthy lead-ore, copper and iron-pyrites, malachite, azure copper-ore, grey manganese-ore, copper-green, brown ironstone, sparry ironstone, native copper, white silver-ore, blende, and calamine; and the following vein-stones, vein-stones, heavy-spar, fluor-spar, calcareous-spar, quartz, and sometimes mountain-cork.

Geographic Situation.

Europe.—It occurs at Leadhills in Lanarkshire, in veins that traverse transition rocks, in which it is associated with galena or lead glance, earthy lead-ore, green lead-ore, lead-vitriol or sulphate of lead, sparry ironstone, iron-pyrites, brown-hematite, calamine, and azure copper-ore; and the vein-stones are quartz, lamellar heavy-spar, calcareous-spar, brown-spar, and mountain-cork. It is found also with galena or lead-glance at Allonhead and Teesdale in Durham; with the same ore, at Alston in Cumberland, and Snailback in Shropshire.

On the Continent, it is met with in several mines in the Hartz; also at Johanngeorgenstadt in Saxony; Prizbram in Bohemia; Tarnowitz in Silesia; Freiburg in the Breisgau; Schemnitz in Hungary; Bleiberg in Carinthia; Huelgoet and Poullavuen in Brittany; Saska and Dognatska in the Bannat; and in the Crimea.

Asia.—In several mines, particularly those of Gazimour in Siberia, where specimens of great beauty are found.

America.—It is met with in the mines of Chili.

Observations.

It is distinguished from *calcareous-spar* by its inferior hardness and greater specific gravity; and also by a simple chemical character,—its soon becoming black, when thrown into an alkaline sulphuret. It is distinguished faom *heavy-spar* or sulphate of barytes, by its superior specific

specific gravity, its effervescing and dissolving in acids, and becoming black in an alkaline sulphuret, neither of which characters are exhibited by heavy-spar: and it cannot be confounded with *tungsten*, because that mineral does not dissolve in the mineral acids, on the contrary, when pounded and thrown into nitric acid, it soon becomes yellow, and it does not become black on exposure to the action of alkaline sulphurets.

2. Next to galena or lead-glance, it is the most common ore of lead, but does not occur so abundantly as to be an object of importance to the metallurgist.

Second Subspecies.

Earthy Lead-Ore, or Lead-Earth.

Bleierde, *Werner.*

This subspecies is divided into two kinds, viz. Indurated Earthy Lead-ore, and Friable Earthy Lead-ore.

First

Indurated Earthy Lead-Ore.

First Kind.

Verhärtete Bleierde, *Werner*.

Id. Wid. s. 868.—Le Plomb terreux endurci, *Broch.* t. ii. p. 329. —Verhärtete gelbe und grau Bleyerde, *Reuss*, b. iv. s. 270. & 272.—Bleierde, *Lud.* b. i. s. 265. *Id. Suck.* 2ter th. s. 345. *Id. Bert.* s. 462. *Id. Mohs*, b. iii. s. 553. *Id. Leonhard*, Tabel. s. 70.—Plomb oxydé terreux, *Brong.* t. ii. p. 197. *Id. Brard*, p. 270.—Verhärtete Bleierde, *Karsten*, Tabel. s. 68.—Erdiches Bleiweiss, *Haus.* s. 114.—Earthy Carbonate of Lead, *Kid*, vol. ii. p. 138.—Plomb carbonaté terreux, *Hauy*, Tabl. p. 82.—Earthy Carbonate of Lead, *Aikin*, p. 47.

External Characters.

Its most frequent colour is yellowish-grey, from which it passes, on the one side, into straw-yellow and cream-yellow; on the other, into yellowish-brown. It occurs also smoke-grey, bluish-grey, and light-brownish red.

It occurs massive, disseminated, and corroded.

Internally it is glimmering, inclining to glistening; and the lustre is resinous *.

The fracture is small and fine-grained uneven, which passes on the one side into fine splintery, on the other into earthy.

The fragments are indeterminate angular.

Is usually opaque, or extremely faintly translucent on the edges.

It

* This lustre is accidental, and appears to be owing to intermixed white lead-ore or lead-vitriol.

It yields a brown-coloured streak.

It is soft, passing into very soft, even into friable, particularly the yellowish-grey, and yellow varieties.

Specific gravity, 5.579, *John*.

Chemical Characters.

It is very easily reduced before the blow-pipe; effervesces with acids, and becomes black with sulphuret of ammonia.

Constituent Parts.

	Tarnowitz.
Oxide of lead,	66.00
Carbonic-acid,	12.00
Water,	2.25
Silica,	10.50
Alumina,	4.75
Iron and oxide of manganese,	2.25
	97.75

John, Chem. Unt. B. 2. S. 229.

Geognostic Situation.

The yellow-coloured varieties occur in a bed in primitive limestone, accompanied with galena or lead-glance and other ores of lead, in the Bannat; the grey-coloured varieties occur sometimes in veins, sometimes in beds, and either in transition or flœtz rocks, and are usually accompanied with galena or lead-glance, white lead-ore, iron-pyrites, malachite, and quartz.

Geographic

Geographic Situation.

Europe.—It is found in the lead veins of Wanlock-head and Leadhills; also t Grassfield Mine near Nent-head in Durham, and in Derbyshire. On the Continent, it is met with at Andreasberg and Zellerfeld in the Hartz, Johanngeorgenstadt in Saxony; Tarnowitz in Silesia; Chentzen in Poland; in the country of Salzburg; and at Saska in the Bannat.

Asia.—Nertschink in Siberia.

Second Kind.

Friable Earthy Lead-Ore.

Zerreibliche Bleierde, *Werner.*

Le Plomb terreux friable, *Broch.* t. ii. p. 328.—Zerreibliche gelbe Bleierde, und zerreibliche grüne Bleierde, *Reuss,* b. iv. s. 268, 269. 271, 272. *Id. Mohs,* b. iii. s. 356. *Id. Leonhard,* Tabel. s. 73. *Id. Karsten,* Tabel. s. 68.—Zerreibliches Blei-weiss, *Haus.* s. 114.—Zerreibliche Bleierde, *Haus.* Handb. b. iii. s. 1110.

External Characters.

Its colour is yellowish-grey and straw-yellow, which sometimes approaches to sulphur-yellow and lemon-yellow.

It occurs massive, disseminated, and in crusts.

It is composed of dull fine earthy particles, which are feebly cohering.

It soils feebly.

It is meagre and rough to the feel.

It is heavy.

Geognostic Situation.

It occurs on the surface or in the hollows of other minerals, and is usually accompanied with galena or lead-glance and other ores of lead.

Geographic Situation.

Europe.—It is found at Wanlockhead and Leadhills; Zellerfeld in the Hartz; Zschopau, and also near Freyberg in the kingdom of Saxony; in the mountains of Kracau in Poland: and at La Croix in Lothringen in France.

Asia.—The mines of Nertschinsk and Beresowskoi in Siberia.

7. Black Lead-Ore.

Schwarz Bleierz, *Werner*.

Id. Werner, Pabst. b. i. s. 116. *Id. Wid.* s. 850.—Black Lead-Ore, *Kirw.* vol. ii. p. 221.—Schwarz Bleyerz, *Emm.* b. ii. s. 385.—La Mine de Plomb noire, *Broch.* t. ii. p. 307.—Schwarz Bleyerz, *Reuss,* b. iv. s. 241. *Id. Lud.* b. i. s. 261. *Id. Suck.* 2ter th. s. 324. *Id. Bert.* s. 461. *Id. Mohs,* b. iii. s. 495. *Id. Leonhard,* Tabel. s. 71.—Plomb noire, *Brong.* t. ii. p. 199.—Dunkler Bleispath, *Karsten,* Tabel. s. 68.—Plomb carbonaté noire, *Hauy,* Tabl. p. 82.—Bleischwärtze, *Haus.* Handb. b. iii. s. 1111.

External Characters.

Its colour is greyish-black of different degrees of intensity.

It

It occurs massive, disseminated, corroded, cellular, and seldom crystallised, in six-sided prisms.

The surface of the crystals is sometimes drusy, sometimes smooth, and sometimes longitudinally streaked.

Externally it is generally splendent, and sometimes shining.

Internally it is only shining, sometimes passing into glistening, and the lustre is adamantine, which sometimes inclines to resinous.

The fracture is small-grained uneven, which sometimes passes into imperfect conchoidal, and occasionally inclines to concealed foliated.

It alternates from translucent to opaque.

Its streak is whitish-grey.

It is soft.

It is rather brittle.

It is easily frangible.

It is heavy.

Constituent Parts.

Oxide of Lead,	79	78.5
Carbonic acid,	18	18.1
Carbon,	2	1.5
	99	

Lampadius, Handb. Zu. Chem. Anal.

Geognostic Situation.

It generally occurs in the upper part of veins, associated with white lead-ore and galena or lead-glance. Frequently this ore incrusts galena, and has resting upon it white lead-ore, and sometimes even green lead-ore. We often

often observe a nucleus of galena incrusted with black lead-ore, or black lead-ore forms a nucleus, which is incrusted with white lead-ore.

Geographic Situation.

Europe.—It occurs at Leadhills; at Fair Hill and Flow Edge, Durham. On the Continent, it is met with at Miess and Prizbram in Bohemia; Freyberg and Zschopau in Saxony; Schwarzleogang in Salzburg; Poullaouen in Lower Brittany in France.

Asia.—Schlangenberg in Siberia.

8. Corneous Lead-Ore, or Muriate of Lead.

Hornblei, *Werner.*

Plomb corné, ou Muriaté de Plomb natif, *Brosh.* t. ii. p. 330. 547, 548.—Hornbei, *Karsten,* Tab. 1. Ausg. s. 78—*Chenevix,* in Nicolson's Journal, vol. v. p. 219.—Hornblei, *Reuss,* Min. b. ii. s. 261. *Id. Lud.* b. ii. s. 187. *Id. Suck.* 2ter th s. 344. *Id. Bert.* s. 453. *Id. Leonhard,* Tabel. s. 73.—Plomb muriaté, *Brong.* t. ii. p. 203.—Hornblei, *Karsten,* Tabel 2ter Ausg. s. 68.—Muriate of Lead, *Kid,* vol. ii. p. 145. *Id. Aikin,* p. 47.

External Characters.

Its colours are greyish white, and yellowish grey, passing into pale wine-yellow.

It occurs crystallised in the following figures:

1. Rectangular four-sided prism, with square bases.
2. The preceding figure truncated on all the angles.

3. Rectangular

3. Rectangular four-sided prism truncated on the lateral edges.
4. The four-sided prism acutely acuminated on both extremities, with four planes, which are set on the lateral planes; sometimes the summits of the acuminations, and also the lateral edges, are truncated.
5. Four sided prism, flatly acuminated with four planes, which are set on the lateral planes; the summits of the acuminations, and sometimes also the lateral edges, are truncated.

Internally it is splendent, and the lustre is adamantine.

The principal fracture is foliated, with a threefold cleavage, the cleavages parallel to the planes of the four-sided prism; the cross fracture is conchoidal.

It is more or less transparent.

It is soft; rather softer than white lead-ore.

It is sectile.

It is easily frangible.

Specific gravity, 6.065, *Chenevix*.

Chemical Characters.

On exposure to the blow-pipe or charcoal, it melts into an orange coloured globule, and appears reticular externally, and of a white colour when solid; when again melted it becomes white; and on increase of the heat the acid flies off, and minute globules of lead remain behind.

Constituent Parts.

Oxide of lead, -	85.5
Muriatic acid, -	8.5
Carbonic acid and water, -	6.0
	100.0

Klaproth, Beit. b. iii. s. 144.

Geognostic

Geographic Situation.

Europe.—In Cromford Level, near Matlock in Derbyshire; and at Hausbaden, near Badweiler in Germany *.

America.—In the neighbourhood of Northampton in the United States †.

Observations.

It is a very rare mineral. A good many years ago, a few specimens of it, the only ones hitherto collected in England, were found in Cromford Level, which was soon afterwards filled with water, and the spot which afforded the specimens hid from view.

9. Phosphate of Lead.

This species is divided into three subspecies, viz. Common Phosphate of Lead, Fibrous Phosphate of Lead, and Conchoidal Phosphate of Lead.

First Subspecies.

Common Phosphate of Lead.

Gemeines Phosphorblei, *Karsten.*

It is divided into two kinds, viz. Green Phosphate of Lead, and Brown Phosphate of Lead.

First

* Leonhard's Taschenbuch for 1815, p. 338.

† Found by William Meade, M. D. as mentioned at p. 152. of Bruce's Mineralogical Journal.

First Kind.

Green Phosphate of Lead, or Green Lead-Ore.

Grün Bleierz, *Werner.*

Minera Plumbi viridis, *Wall.* t. ii. p. 308.—Grün Bleyerz, *Wern.* Pabst. b. i. s. 123. *Id. Wid.* s. 857.—Phosphorated Lead-ore, *Kirw.* vol. ii. p. 207.—Oxide de Plomb spathique verte, Phosphaté de Plomb, *De Born,* t. ii. p. 377.—Grün Bleyerz, *Emm.* b. ii. s. 394.—Plomb phosphaté, *Hauy,* t. iii. p. 490. —La Mine de Plomb verte, ou le Plòmb verte, *Broch.* t. ii. p. 314.—Grünbleierz, *Reuss,* b. iv. s. 216. *Id. Lud.* b. i. s. 262. *Id. Suck.* 2ter th. s. 331, *Id. Bert.* s. 455. *Id. Mohs,* b. iii. s. 517. *Id. Leonhard,* Tabel. s. 72.—Plomb phosphaté, *Brong.* t. ii. p. 200. *Id. Brard,* p. 271.—Gemeines Phosphorblei, (in part), *Karsten,* Tabel. s. 68.—Phosphate of Lead, *Kid,* vol. ii, p. 141.—Plomb phosphaté, *Hauy,* Tabl. p. 82. —Phosphate of Lead, *Aikin,* p. 48.

External Characters.

Its colour is grass green, which passes on the one side through pistachio-green, olive and siskin-green, into sulphur-yellow ; on the other side, through asparagus-green into greenish-white. Some varieties approach to leek-green. The olive and pistachio-green colours are the most common.

It seldom occurs massive, sometimes stalactitic, reniform and botryoidal ; but most commonly crystallised in the following figures :

1. Perfect and regular six-sided prism *. Fig. 235.
2. Six-sided prism, truncated on all the lateral edges, so as to form a twelve-sided prism †. Fig. 236.

 Six-

* Plomb phosphaté prismatique, Hauy.

† Plomb phosphaté peridodecaedre, Hauy.

3. Six-sided prism, flatly acuminated on the extremities with six planes, which are set on the lateral planes *. Fig. 237.
4. Regular six-sided prism, in which the terminal edges are truncated †. Fig. 238.
5. In acicular crystals, which are generally short and diverging.

The prisms are usually low, sometimes bulging, and hollow at their extremities.

The crystals are small and very small, seldom middle-sized, and are often scalarwise aggregated.

Externally it is smooth and shining, internally glistening, and the lustre is resinous.

The fracture is small-grained uneven, passing on the one hand into splintery, on the other into conchoidal.

The fragments are indeterminate angular, and blunt-edged.

It is more or less translucent, seldom nearly transparent, and is sometimes only translucent on the edges.

It is soft; it scratches white lead-ore.

It is brittle.

It is easily frangible.

It is uncommonly heavy.

Specific gravity, 6.560, from Wanlockhead, *Klaproth*. 6.270, Zschoppau, *Klaproth*. 6.9411, from the Breisgaw, according to *Hauy*.

Chemical Characters.

It dissolves in acids without effervescence. Before the blowpipe, on charcoal, it usually decrepitates, then melts, and on cooling, forms a polyhedral globule, the faces of which present concentric polygons: if this globule be pulverised,

* Plomb phosphaté trihexaedre, Hauy.

† Plomb phosphaté annulaire, Hauy.

pulverised, and mixed with borax, and again heated, a milk-white opaque enamel is partly formed; on continuance of the heat, the globule effervesces, and at length becomes perfectly transparent, the lower part of it being studded with globules of metallic lead.

Constituent Parts.

	Zschoppau.	Hoffsgrund.	Wanlockhead.
Oxide of Lead,	78.40	77.10	80.00
Phosphoric Acid,	18.37	19.00	18.00
Muriatic Acid,	1.70	1.54	1.62
Oxide of Iron,	0.10	0.10	a trace
	98.57	97.74	99.96

Klaproth, Beit. b. iii. s. 153.—161.

Geognostic and Geographic Situations.

Europe.—It occurs along with galena or lead-glance, and other ores of lead, at Leadhills and Wanlockhead. In England, it is met with at Alston in Cumberland, Allonhead, Grasshill, and Teesdale, in Durham, and Nithisdale in Yorkshire. On the Continent, it is found in several of the mines in the Hartz; also at Zschoppau in Saxony; Prizbram in Bohemia; Hofsgrund in the Breisgau; and Erlenbach in Alsace.

Asia.—In the lead-mines in Siberia; and in those of the Beresof.

Observations.

1. Green lead-ore, when it has a very pale greenish-white colour, is apt to be confounded with White Lead-ore; but we can distinguish them by the following characters:—1. The prisms of green lead-ore are generally equiangular, but those of white lead-ore are unequiangular.

angular. 2. Its lustre is resinous, but that of white lead-ore is adamantine. 3. It is harder than white lead-ore. 4. Its crystals are often scalarwise aggregated. which is never the case with white lead-ore. 5. Its prisms are generally shorter than those of white lead-ore; and this mineral does not effervesce with acids, which is the case with white lead-ore, and is not reduced to the metallic state before the blowpipe, without addition.

2. M. Klaproth having discovered phosphate of lead, or green lead-ore, of a greyish-white colour, proposes it as an objection to the naming of minerals from their colours. It must be remembered, however, that the name does not imply the constant occurrence of a green colour; it only intimates, that the green colour is the most striking feature in the external aspect of the mineral, and that it occurs more frequently than any other member of the colour-suite.

Second Kind.

Brown Phosphate of Lead, or Brown Lead-Ore.

Braun Bleyerz, *Werner.*

Id. Wern. Pabst. b. i. s. 115. *Id Wid.* s. 848.—Brown Lead-ore, *Kirw.* vol. ii. p. 222.—Braun Bleyerz, *Emm.* b. i. s. 383.—La Mine de Plomb brune, *Broch.* t. ii. p. 305.—Braun Bleierz, *Reuss,* b. i. s. 212. *Id. Lud.* b. i. s. 260. *Id. Suck.* 2ter th. s. 323. *Id. Bert.* s. 454. *Id. Mohs,* b. iii. s. 489. *Id. Leonhard,* Tabel. s. 71.—Gemeiner Phosphorblei, *Karsten,* Tabel. (in part) s. 68.—Plomb phosphaté, *Hauy,* Tabl. p. 82.—Gemeiner Pyromorphit, *Haus.* b. iii. s. 1090.—Brown Phosphate of Lead, *Aikin,* p. 48.

External Characters.

Its colour is hair-brown, of different degrees of intensity, sometimes

sometimes so pale that it inclines to grey, and sometimes it passes into clove-brown.

It occurs massive; and crystallised in the following figures:

1. Equiangular six-sided prism, which is sometimes bulging.
2. Preceding figure, in which the lateral planes are alternately broad and narrow, and sometimes the lateral edges are truncated.
3. Six-sided prism, converging towards both ends, and thus inclining to the pyramidal form.
4. Acute double three-sided pyramid, in which the lateral planes of the one are set on the lateral planes of the other, and in which the common basis is sometimes more or less deeply truncated. It originates from the bulging six-sided prism, in which the alternate lateral planes have disappeared.

The crystals are middle-sized, and small, are sometimes short and acicular, and singly imbedded, or scopiformly or globularly aggregated.

The surface of the crystals is sometimes blackish or yellowish brown, and rough.

Internally, it is glistening, and the lustre is resinous.

The fracture is small and fine-grained uneven, and sometimes passes into small splintery.

The fragments are indeterminate angular.

The crystallised varieties sometimes occur in thin columnar distinct concretions, in which the surfaces are longitudinally streaked and shining.

It is feebly translucent.

It is soft.

The streak is greyish-white.

It

It is rather brittle.

It is easily frangible.

Specific gravity, 6.974, *Wiedeman*; 6.600 from Huelgoët, according to *Klaproth*; 6.909 from Huelgoët, *Hauy*.

Chemical Characters.

It melts pretty easily before the blow-pipe without being reduced, and during cooling shoots into acicular crystals. It does not effervesce with nitric acid, but is soluble in it.

Constituent Parts.

From Huelgoët in Brittany,

Oxide of Lead,	78.58
Phosphoric Acid,	19.73
Muriatic acid,	1.65
	99.96

Klaproth, Beit. B. iii. s. 157.

Geognostic Situation.

It occurs in veins that traverse gneiss, clay-slate, and porphyry. The veins generally contain lead and silver ores, also native silver, iron and copper pyrites, malachite, blende, ochry ironstone, heavy-spar, and quartz.

Geographic Situation.

Europe.—It is found at Miess in Bohemia: near Schemnitz in Hungary; Saska in the Bannat; Zschoppau in Saxony; Huelgoët and Poullaoen in Lower Brittany.

America.—Zimapan in Mexico.

Second

Second Subspecies.

Fibrous Phosphate of Lead.

Fasriges Phosphorblei, *Karsten.*

Fasriges Phosphorblei, *Karsten,* Tabel. s. 69 & 99.—Traubenerz, *Karsten,* in Journ. f. Chem. Phys. u. Min. iv. 3. 394.—Plomb phosphaté arsenié, *Hauy,* t. iii. p. 496.—Fasriges Traubenblei, *Haus.* Handb. b. iii. s. 1095.

External Characters.

Its colours are reddish-brown, or pistachio-green, with a yellowish-grey crust.

It occurs in the botryoidal external shape, which has a rough surface.

Externally it is dull.

Internally it alternates from glimmering to shining; the first being pearly, the second adamantine.

The fracture passes from fibrous into radiated.

It occurs in concentric lamellar distinct concretions.

It is opaque.

It becomes lighter in the streak.

It is soft

It is rather sectile.

Specific gravity 6.5; *Karsten.*

Constituent

Constituent Parts.

	From Auvergne.
Oxide of Lead, -	76.00
Phosphoric Acid, -	13.00
Arsenic Acid, -	7.00
Muriatic Acid, -	1.75
Water, - - -	1.75
	100

Klaproth, Beit. b. v. s. 204.

Geographic Situation.

It occurs at Pont Gibaud in Auvergne; Huelgoët in Brittany; Zschopau in Saxony; Hofsgrund in the Breisgau; and in Andalusia.

Third Subspecies.

Conchoidal Phosphate of Lead.

Muschliches Phosphorblei, *Karsten.*

Muschliches Phosphorblei, *Karsten,* Tabel. s. 68. *Id. Karsten,* in N. Journ. d Chem. b. iii. s. 60.—Muschliches Traubenblei, *Haus.* Handb. b. iii. s. 1094.

External Characters.

Its colour is orange-yellow, which passes on the one side into honey-yellow, wax-yellow, lemon-yellow, and sulphur-yellow, and on the other into aurora-red.

It occurs disseminated, stalactitic, reniform, small-botryoidal; and crystallised in the following figures:

1. Perfect

1. Perfect six-sided prism, generally bulging, and almost globular.
2. The same figure, with straight planes, flatly acuminated on the extremities, with six rather convex planes; the acuminating planes set on the lateral planes.
3. Double six-sided pyramid; the lateral planes often convex, and the planes of the one pyramid resting on those of the other. The edge of the common base, and the summits of the pyramids, sometimes truncated.

The crystals are sometimes single, sometimes aggregal, in the rose, bud, or globular form.

Internally, the lustre is shining and resinous.

The fracture is conchoidal.

It is translucent.

The streak is yellowish-white.

It is soft, inclining to semihard.

It is brittle.

Specific gravity 7.261; *Karsten.*

Constituent Parts.

Oxide of Lead,	76.8	77.5
Phosphoric Acid,	9.0	7.5
Arsenic Acid,	4.0	12.5
Muriatic Acid,	7.0	1.5
Water,	1.5	
	98.3	99
	Laugier, Ann. d. Mus. t. vi. p. 171.	*Rose*, Journ. d. Chem. & Phys. t. ii. 229.

Geognostic

Geognostic and Geographic Situations.

It occurs in a metalliferous bed, in transition rocks; along with green lead-ore, white lead-ore, muriate of lead, galena or lead-glance, lamellar heavy-spar, fluor-spar, quartz, and lithomarge at Haus-Baden near Badweiler; also at Johanngeorgenstadt in Saxony, in a vein along with silver-glance or sulphureted silver-ore, and red silver-ore. It occurs in Huel-Unity in the parish of Gwennap in Cornwall.

10. Arseniate of Lead.

Bleiblüthe, *Hausmann.*

This species is divided into three subspecies, viz. Reniform Arseniate of Lead, Filamentous Arseniate of Lead, and Earthy Arseniate of Lead.

First Subspecies.

Reniform Arseniate of Lead.

Bleiniere, *Hausmann.*

Bleiniere, *Reuss,* b. ii. 4. s. 225. *Id. Leonhard,* Tabel. s. 78. —Plomb arsenié, & Plomb reniforme, *Brong.* t. ii. p. 202. —Bleiniere, *Karsten,* Tabel. s. 68.—Plomb arsenié concretionné-mamelonné et compacte, *Hauy,* Tabl. p. 80.—Bleiniere, *Haus.* Handb. b. iii. s. 1097.—Reniform Arseniate of Lead, *Aikin,* p. 50.

External Characters.

Its colours on the fresh fracture are reddish-brown and brownish-red; externally ochre-yellow, and straw-yellow. It

It occurs reniform and tuberose.

Internally it is shining and resinous.

The fracture is conchoidal, sometimes inclining to even or uneven.

It occurs in curved lamellar concretions.

It is opaque.

It is soft.

It is brittle.

Specific gravity 3.933, *Karsten.*

Chemical Characters.

It is insoluble in water. Before the blow-pipe on charcoal it gives out arsenical vapours, and is more or less perfectly reduced.

It colours glass of borax lemon-yellow.

Constituent Parts.

Oxide of Lead, -	35.00
Arsenic acid, -	25.00
Water, - -	10.00
Oxide of iron, -	14.00
Silver, - -	1.15
Silica, - -	7.00
Alumina, - -	2.60
	95.15

Bindheim in Beob. u. Endeck. d. Berl. Ges. Natf. Fr. iv. s. 374.

Geographic Situation.

It has been hitherto found only in one mine near Nertschinsk in Siberia.

Second Subspecies.

Filamentous Arseniate of Lead.

Flockenerz, *Karsten.*

Plomb arsenié filamenteux, *Hauy*, t. iii. p. 465.—Flockenerz, *Karsten*, Tabel. s. 68.—Flockige Bleiblüthe, *Haus.* Handb. b. iii. s. 1098.—Arseniate of Lead, *Aikin*, p. 49.

External Characters.

Its colours are grass-green, wine-yellow, wax-yellow, and lemon-yellow.

It occurs either in small acicular six-sixed prisms, which are collected into flakes, or in very delicate capillary silky fibres, which are slightly flexible, and easily frangible.

Geographic Situation.

It occurs at Saint Prix in the department of the Saone and Loire in France.

Third Subspecies.

Earthy Arseniate of Lead.

Erdige Bleiblüthe, *Hausmann.*

Erdige Bleiblüthe, *Haus.* Handb. b. iii. s. 1098.—Plomb arsenié terreux, *Lucas*, t. ii. p. 315.

External Characters.

Its colour is yellow.

It

It occurs in crusts.
The fracture is earthy.
It is friable.

Geognostic and Geographic Situations.

It occurs along with the filamentous arseniate of lead, in veins of galena or lead-glance, and associated with violet blue fluor-spar, and quartz, at Saint Prix; and also in a vein of galena or lead-glance in the Hill of Herpie in Oisans, in which it is associated with white lead-ore, argentiferous grey copper-ore, malachite, azure copper ore, and quartz.

11. Sulphate of Lead, or Lead-Vitriol.

Blei Vitriol, *Werner*.

Id. Wid. s. 870.—Native Vitriol of Lead, *Kirw.* vol. ii. p. 211.—Natürlicher Bleyvitriol, *Emm.* b. ii. s. 413. & b. iii. s. 366.—Sulphate de Plomb, *Lam.* t. i. p. 211.—Le Vitriol de Plomb natif, *Broch.* t. ii. p. 325.—Plomb sulphaté, *Hauy,* t. iii. p. 503—Bleivitriol, *Reuss,* b. iv. s. 264. *Id. Lud.* b. i. s. 264. *Id Suck.* 2ter th. s. 32. *Id. Bert.* s. 452. *Id. Mohs,* b. iii. s. 547.—Plomb sulphaté, *Lucas,* p 121.—Bleivitriol, *Leonhard,* Tabel. s. 73.—Plomb sulphaté, *Brong.* t. ii. p. 200. *Id. Brard,* p. 275—Bleivitriol, *Karsten,* Tabel. s. 68. *Id. Haus.* s. 140.—Sulphate of Lead, *Kid,* vol. ii. p. 140.—Plomb sulphaté, *Hauy,* Tabl. p. 83.—Sulphate of Lead, *Aikin,* p. 49.

External Characters.

Its colours are yellowish-grey and greyish-white, which

which sometimes pass into smoke and ash grey; the lighter varieties incline very much to white.

It occurs massive, disseminated; and crystallised in the following figures:

1. Very acute and unequiangular double four-sided pyramid, or an irregular octahedron, in which the adjacent faces of each pyramid are inclined to each other at an angle of 109° 18′, and the corresponding faces of the two pyramids at an angle of 78° 28′ *, fig. 239.
2. Elongated octahedron, or oblique four-sided prism, with lateral edges of 109° 18′ and 70° 42′, acutely bevelled on the extremities, and the bevelling planes set on the acute lateral edges †, fig. 240.
3. The preceding figure truncated on the obtuse lateral edges ‡, fig. 241. Sometimes the truncating planes are so large, that the crystals have the appearance of a six-sided table, with four short and two long opposite terminal planes. When the original faces of the figure are obliterated by means of the truncating planes, there is formed
4. An oblique four-sided prism, with lateral edges of 78° 28 and 101° 32′, and the edges formed by the bevelling planes of fig. 240. with the truncating planes on the obtuse lateral edges, more or less deeply truncated, which gives rise to a four-planed acumination.

5. Octahedron,

---

* Plomb sulphaté primitif, Hauy.

† Plomb sulphaté primitif, var. Hauy.

‡ Plomb sulphaté semi-prisme, Hauy.

5. Octahedron, in which two opposite planes are larger than the others, so that it terminates in a line.
6. Octahedron, in which the common base is truncated.
7. Octahedron, in which the angles on the common base are bevelled, and the bevelling planes set on the lateral edges: Sometimes the edges of the bevelment are truncated.

The crystals are small and very small, very rarely middle-sized. They are sometimes single, seldom grouped together, or cross each other in various directions.

Externally it is shining, and the lustre is adamantine, sometimes inclining to vitreo-resinous; and internally it is splendent, and the lustre is adamantine.

The fracture is conchoidal, sometimes passing into unven.

It occurs in angulo-granular distinct concretions.

It alternates from transparent to translucent.

It is so soft as to yield to the nail; it is softer than white lead-ore.

Its streak is greyish white.

It is rather brittle.

It is very easily frangible.

Specific gravity, 6.300, from Anglesea, *Klaproth.* 6.714, Zellerfeld, *Jordan.*

Chemical Characters.

It decrepitates before the blowpipe, then molts, and is soon reduced to the metallic state.

 Constituent

Constituent Parts.

	From Anglesea.	Wanlockhead.
Oxide of Lead,	71.0	70.50
Sulphuric Acid,	24.8	25.75
Water of crystallisation, -	2.0	2.25
Oxide of Iron, -	1.0	
	98.8	98.5

Klaproth, Beit. b. iii. s. 164. & 166.

	Zellerfeld.
Oxide of Lead,	72.9146
Sulphuric Acid,	26.0191
Water, -	0.1242
Oxide of Iron,	0.1151
Oxide of Manganese,	0.1654
Intermixed Silica and Alumina, -	a trace
	99.7992

Stromeyer, Gott. Gel. anz. 812. 204.

Geognostic and Geographic Situations.

Europe.—It occurs in veins along with galena or lead-glance, and other ores of lead, at Wanlockhead in Dumfries-shire, and Leadhills in Lanarkshire; at Pary's Mine in Anglesea, and Penzance in Cornwall. On the Continent, it is met with at Zellerfeld in the Hartz, in veins that traverse clay-slate and greywacke, associated with quartz, calcareous-spar, brown-spar, heavy-spar, brown ironstone, copper-green, azure copper-ore, green lead-ore, and white lead-ore; and in lead mines in Andalusia in Spain

Asia.—Siberia.

America.—In the neigbourhood of Northampton in the United States *.

12. Yellow

* Vid. Bruce's American Mineralogical Journal, p. 150.

12. Yellow Lead-Ore, or Molybdate of Lead.

Gelbes Bleierz, *Werner*.

Id. Werner, Pabst. b. i. s. 127. *Id. Wid.* s. 864.—Oxide de Plomb spathique jaune, *De Born,* t. ii. p. 379.—Yellow Lead-spar, *Kirw.* vol. ii. p. 212.—Gelbes Bleyerz, *Emm.* b. ii. s. 403.—Plomb molybdaté, *Hauy,* t. iii. p. 498.—La Mine de Plomb jaune, ou le Plomb jaune, *Broch.* t. ii p 322. —Gelb Bleierz, *Reuss,* b. iv. s. 286. *Id. Lud.* b. i. s. 264. *Id. Suck.* 2ter th. s. 340. *Id. Mohs,* b. iii. s. 535.—Plomb molybdaté, *Lucas,* p. 120.—Gelb Bleierz, *Leonhard,* Tabel. s. 72.—Plomb molybdaté, *Brong.* t. ii. p. 205. *Id. Brard,* p. 274.—Gelbleierz, *Karsten,* Tabel. s. 68.—Bleigelb, *Haus.* s. 140.—Plomb molybdaté, *Hauy,* Tabl. p. 83.—Molybdate of Lead, *Kid,* vol. ii. p. 139. *Id. Aikin,* p. 50.

External Characters.

Its most frequent colour is wax-yellow ; from which it passes, on the one side, into lemon-yellow and orange-yellow ; on the other side, into yellowish-brown and yellowish-grey : sometimes of a colour which is intermediate between yellowish-white and greyish-white.

It occurs massive, in crusts, cellular ; and crystallised in the following figures :

1. Octahedron, with equilateral triangular planes *, fig. 242.
2. Octahedron truncated on the angles †, fig. 243.

* Plomb molybdaté primitif, Hauy.

† Plomb molybdaté epointé, Hauy.

3. The octahedron, so deeply truncated on the summits and the common base, that the original faces disappear, when there is formed a rectangular parallelopiped, which is either tabular, or in the form of a cuboidal prism *, as in fig. 244.
4. The octahedron, so deeply truncated on all the angles and on the common base, that the original faces disappear, when there is formed a regular eight-sided table †, fig. 245. which is sometimes so thick as to appear as an eight-sided prism.
5. Octahedron truncated on the summits and common base, and bevelled on the angles of the common base, which gives rise to the rectangular four-sided table, bevelled on the terminal edges ‡, fig. 246.
6. Octahedron truncated on the lateral edges, which gives rise to the double eight sided pyramid. When this figure is deeply truncated on the summits, there is formed
7. A regular eight-sided table, bevelled on the terminal planes.

The tables are usually broad and thin, and alternate from small to very small, but are seldom middle sized. They are sometimes grown together, and frequently intersect one another.

Externally it is shining and smooth.

Internally it is shining or glistening, and resinous in the foliated varieties; but splendent and adamantine in those which are conchoidal.

The

* Plomb molybdaté bis-unitaire, Hauy.

† Plomb molybdaté tri-unitaire, Hauy.

‡ Plomb molybdaté perioctogone, Hauy.

The fracture is imperfect, or concealed foliated, and mall conchoidal.

The fragments are indeterminate angular, and rather sharp-edged.

It is translucent, sometimes inclining to transparent.

It is soft.

It is rather brittle.

It is easily frangible.

It is heavy.

Specific gravity, 5.092. *Hatchett.*

Chemical Characters.

It decrepitates before the blow pipe, and then melts into a dark-greyish coloured mass, in which the globules of reduced lead are dispersed. With borax it forms a brownish-yellow globule; but when in small proportion, and heated by the interior flame, it occasionally produces a glass, which is greenish-blue, and sometimes deep-blue.

Constituent Parts.

Klaproth was the first who made us acquainted with the chemical composition of this ore; but we are indebted to our celebrated countryman Hatchett for the most complete and accurate analysis of it.

Oxide of Lead,	64.42	Oxide of Lead,	58.40
Molybdic Acid,	34.25	Molybdic Acid,	38.0
	——	Oxide of Iron,	2.08
	98.67	Silica,	0.28
Klaproth, Beit. b. ii. s. 275.			——
			96.66
		Hatchett, Phil. Trans. for 1796.	

Geognostic

Geognostic and Geographic Situations.

Europe.—It occurs at Bleiberg in Carinthia, in a compact limestone, which is much traversed by veins of calcareous-spar, and is associated with galena or lead-glance, white, black and green lead-ores, calamine, malachite, calcareous-spar and fluor-spar; also on the Maukeriz near Brixlegg in the Tyrol, along with brown ironstone and red copper-cre *; at Annaberg in Austria, and Rezbanya in Transylvania.

America—In compact limestone at Zimapan in Mexico; and near Northampton in the United States †.

13. Red Lead-Ore, or Chromate of Lead.

Roth Bleierz, *Werner*.

Minera Plumbi rubra, *Wall.* t. ii. p. 309.—Rothes Bleierz, *Werner*, Pabst. b i. s. 127. *Id. Wid.* s. 861.—Red Lead-spar, *Kirw.* vol. ii. p. 214.—Oxide de Plomb spathique rouge, *De Born*, t. ii. p. 376.—Rothes Bleyerz, *Emm.* b. ii. s. 399.—Oxide rouge de Plomb, *Lam.* t. i. p. 287.—Plomb chromaté, *Hauy*, t. iii. p. 476.—La Mine de Plomb rouge, ou le Plomb rouge, *Broch.* t. ii. p. 318.—Rothbleierz, *Reuss*, b. iv. s. 228. *Id. Lud.* b. i. s. 263. *Id. Mohs*, b. iii. s. 527.—Plomb chromaté, *Lucas*, p. 116.—Rothbleierz, *Leonhard*, Tabel. s. 62.—Plomb chromé, *Brong.* t. ii. p. 205. *Id. Brard*, p. 267—Roth Bleierz, *Karsten*, Tabel. s. 68.—Kallochrom, *Haus.* s. 139.—Chromate of Lead, *Kid*, vol. ii. p. 143.—Plomb chromaté, *Hauy*, Tabl. p. 81.—Chromate of Lead, *Aikin*, p. 51.

External Characters.

Its colour is hyacinth-red, sometimes inclining to aurora or morning red.

It

* Leonhard's Taschenbuch, b. vii. s. 517.

† Bruce's American Mineralogical Journal, p. 151.

It rarely occurs massive or disseminated; more frequently crystallised, and in the following figures:

1. Rectangular four-sided prism, in which the terminal planes are set on obliquely, which is the fundamental figure *.
2. The preceding figure, in which the lateral edges are truncated.
3. N° 1. in which two opposite lateral edges are bevelled. When these bevelling planes become of equal size with the original planes, an eight-sided prism is formed; and when the original planes entirely disappear, there is formed
4. An oblique four-sided prism.
5. Rectangular four-sided prism bevelled on the extremities; the bevelling planes set on the two opposite lateral edges; or obliquely bevelled on the extremities, and the bevelling planes set on two opposite lateral planes.
6. Four-sided prism, obliquely acuminated with four planes, which are set on the lateral planes.

Sometimes, in the secondary figures, one or other of the lateral edges is truncated by an elongated trapezoidal plane.

The crystals are middle-sized; the surface of the crystals is usually smooth, and sometimes longitudinally streaked.

Externally it is splendent. Internally it is splendent on the foliated fracture; but only shining on the uneven or conchoidal fractures, and is adamantine, inclining to resinous.

The

* The primitive form, according to Hauy, is a rectangular four-sided prism, in which the terminal planes are set on obliquely.

The principal fracture is foliated, and the cross fracture uneven or small conchoidal.

The fragments are indeterminate angular, and rather blunt-edged.

It is translucent, passing into semi-transparent, and even sometimes inclining to transparent.

Its streak is orange-yellow, passing into lemon-yellow.

It is soft.

It is intermediate between brittle and sectile.

It is easily frangible.

It is heavy, passing into uncommonly heavy.

Specific gravity, 5.750, *Bindheim*; 6.0269, *Brisson.*

Chemical Characters.

Before the blowpipe it crackles and melts into a grey slag. With borax it is partly reduced. It does not effervesce with acids.

Constituent Parts.

Oxide of Lead,	-	-	63.96
Chromic Acid,	-	-	36.40
			100.36

Vauquelin, Journ. d. Min. n. 34. 737.

Geognostic and Geographic Situations.

It occurs in a rock composed of quartz and silver-white talc, in which are also contained decomposed crystals of auriferous iron-pyrites, in one of the gold-mines at Beresof in Siberia. It is occasionally associated with green

green phosphate of lead, sometimes crystallised in six-sided prisms, or in the earthy form; and also with an olive-green-coloured mineral, and white lead-ore.

Use.

In Siberia a very costly and beautiful orange-yellow colour is prepared from it, which is used by painters.

Observations.

1. Red lead ore may be distinguished from *red orpiment*, by the form of its crystals, its superior weight, and its not emitting a garlic smell upon exposure to the blowpipe, and by its accompanying minerals; from *Red silver-ore*, by its inferior specific gravity, and its yellow-coloured streak; and from *Cinnabar*, by the colour of its streak, and also by its being reduced to the metallic state before the blowpipe, whereas cinnabar, by the same means, is entirely volatilised.

2. The green phosphate of lead, which usually accompanies red lead-ore, has been sometimes erroneously considered as chromate of lead. Another mineral is sometimes associated with the red lead-ore, which Hausmann suspects to be a chromate of lead. Its colours are dark-yellowish or liver-brown. It occurs in small botryoidal and stalactitic forms, generally invested with a siskin-green earthy crust. Internally it is feebly glistening, or glimmering and resinous. The fracture is flat conchoidal or uneven. It occurs in curved lamellar concretions. It is opaque It is soft. Its streak is siskin-green. Hausmann conjectures that it is allied to the brown lead-ore brought from Zimapan by Humboldt, and

and which, according to Collet-Descotils, contains in 100 parts 74.2 Oxide of Lead; 3.5 Oxide of Iron; 16 Chromic Acid; and 1.5 Muriatic Acid. *Journ. de Phys.* t. lxii. p. 38. The Siberian lead-ore is also sometimes associated with yellowish green crystals, which Vauquelin considers to be a compound of oxides of lead and chrome. Other green-coloured crystals, having the same form as the red lead ore, are considered as compounds of oxide of lead and chrome, and derived from the red lead-ore.

LEAD MINES.

The greatest lead mines in the world are those of Great Britain; the next in magnitude are the French, Westphalian, Austrian, Spanish, Prussian; and the least considerable are those of Saxony, Russia, and Bavaria.

TABLE

TABLE of the Annual Quantity of Lead raised from the Earth in different Countries.

	In Quintals.
1. Great Britain, - -	250,000
2. France, - - -	60,000
3. Kingdom of Westphalia in 1808,	59,771
4. Austria, including Bohemia, Gallicia, Hungary, Transilvania, Stiria, Carinthia, Carniola, Salzburg, Moravia, and Austria, - - -	45,809
5. Spanish European Mines, - -	32,000
6. Prussia, after the Treaty of Tilsit, -	12,992
7. Kingdom of Saxony in 1808, -	10,000
8. Russia, - - - - - -	10,000
9. Bavaria, including the Tyrol, -	400
Total,	480,972

ORDER X.

ORDER X.—ZINC.

This Order contains the following species, Red Zinc-ore or Red Oxide of Zinc, Electric Calamine, Common Calamine, and Blende.

1. Red Zinc-Ore, or Red Oxide of Zinc.

Red Oxide of Zinc, *Bruce.*

Red Zinc-ore, *Bruce's* American Mineralogical Journal, p. 96.

External Characters.

Its colours are blood-red and aurora red.

It occurs massive, disseminated.

Internally, on the fresh fracture it is shining; after long exposure to the air it becomes dull, and even covered with a pearly crust.

The principal fracture is foliated, the cross fracture conchoidal.

It is translucent on the edges, or opaque.

It is easily scratched by the knife.

It is brittle.

It affords a streak which is brownish-yellow, approaching to orange.

Specific gravity, 6.220.

Chemical Characters.

It is soluble in the mineral acids. It is infusible without addition before the blowpipe. With sub-borate of soda,

soda, melts with a transparent yellow bead. When exposed to the united flames of oxygen and hydrogen, it sublimes, attended with a brilliant white light. When pounded, mixed with potash, and exposed to heat, it fuses into an emerald-green mass, which, on solution, affords to water the same colour. On the addition of a few drops of nitric, sulphuric, or muriatic acids, the green coloured fluid is immediately changed into a rose-red.

Constituent Parts.

Zinc, - -	76
Oxygen, - -	16
Oxides of Manganese and Iron,	8
	100

Bruce, American Min. Journ. p. 99.

Geognostic and Geographic Situation.

This mineral has been hitherto found only in North America, where it occurs in several of the iron-mines in Sussex County, New-Jersey; as at the Franklin, Stirling, and Rutgers mines, and near Sparta. In some instances it is imbedded in foliated granular limestone; while in others, it serves as a basis in which magnetic ironstone occurs, either in crystals or grains.

At Franklin, it also assumes a micaceous form, and is imbedded in a whitish oxide of zinc, which is often, in the same specimen, found adhering to the black oxide of iron.

Uses.

This species occurs abundantly in the United States of America, and promises to be a valuable acquisition to that country.

country. Dr Bruce, to whom we are indebted for every thing we know of this mineral, remarks: "The recently discovered property of the malleability of zinc, at a temperature of 300° of Fahrenheit, has greatly enhanced its value, and raised it to a high rank among the useful metals. The inconvenience arising from its brittleness, being removed, this metal is now applied to many of the purposes for which copper has been hitherto used. As the demand for metallic zinc must necessarily increase as its application to the arts becomes more general, the red zinc-ore will prove a source from which this metal may be procured in abundance; and a series of experiments sufficiently shews the ease with which it may be separated from the ore. In the manufacture of brass, this ore possesses advantages over those generally used; as, without previous preparation of ustulation, &c. it affords with copper a compound possessing a high degree of malleability, a fine colour, and every requisite of the best kind of brass, such as is used in the finest and most delicate workmanship, equal in every respect to that made from the reduced metal, or, as it is more generally termed, *spelter*. This mineral may also be advantageously employed in the manufacture of sulphate of zinc, or white vitriol of commerce. Experiments also prove, that the oxide or flowers of zinc may, without much difficulty, be obtained from this ore. The oxide of zinc has of late been recommended as a substitute for white-lead as a pigment, over which it possesses some advantages, as it is not liable to change, and in its preparation is not subject to those deleterious consequences so frequently attendant on all the preparations of lead."

Observations.

1. It is distinguished from *Red Silver-ore*, by its infusibility before the blowpipe: from *Red Copper-ore*, by its superior specific gravity, and its solution in acids, which is colourless, whereas that of the red copper-ore is of a bright green: from *Red Lead-ore*, by its infusibility before the blowpipe, the red lead-ore melting into a blackish slag: *Red Orpiment*, with which it might be confounded, is distinguished from it by its volatility before the blowpipe, and giving out a blue flame, and a strong garlic smell; and its solubility in the mineral acids distinguishes it from *Rutile*, which is insoluble.

2. This interesting mineral was first discovered and described by Dr Bruce in the American Mineralogical Journal,—a valuable work, the publication of which, we regret to remark, has been discontinued.

3. The red colour of this ore is conjectured to be owing to the oxide of iron and manganese it contains.

2. Electric Calamine, or Siliceous Oxide of Zinc.

Electrical Calamine, *Smithson.*

Zinc oxydé, *Hauy*, t. iv. p. 159. (in part)—Blättricher Galmei, *Reuss*, b. ii. 4. s. 349. (in part).—Electrical Calamine, *Smithson*, Phil. Trans. p. i. for 1803.—Spathiger Galmei, *Leonhard*, Tabel. s. 72.—Zinc Calamine, *Brong.* t. ii. p. 136.—Zinkglaserz, *Karsten*, Tabel. s. 70.—Zinc oxydé, *Hauy*, Tabl. p. 102.—Zinkglas, *Haus.* Handb. b. i. s. 343.—Electric Calamine, *Aikin*, p. 54.

External Characters.

Its colours are greyish, bluish, and yellowish white;

 and

and seldom apple-green. Externally it is sometimes brownish and blackish.

It occurs massive, stalactitic, reniform, botryoidal, and crystallised in the following figures:

1. Six-sided prism.
2. Flat six-sided prism, bevelled on the terminal planes; the bevelling planes set on the broader lateral planes.
3. Acute octahedron, sometimes perfect, sometimes truncated on the summits.
4. Acute octahedron, acuminated on both extremities, with four planes, which are set on the lateral planes, and sometimes the summits are truncated.

The crystals are small; and either solitary, or scopiformly aggregated.

Internally, the lustre is shining, glistening, and pearly.

The fracture is foliated, diverging radiated, or fibrous.

It alternates from transparent to translucent on the edges.

It yields to the knife, but is much harder than common calamine.

Specific gravity, 3.4.

Physical Character.

When gently heated, it is strongly electric.

Chemical Characters.

It loses, according to Pelletier, about 12 *per cent.* by ignition; it is soluble in muriatic acid without effervescence; and the solution gelatinises on cooling.

Constituent

Constituent Parts.

	Wanlockhead.	Freyburg.	Rezbanya.	Raibel in Carinthia.	England.
Oxide of Zinc,	66.00	38.00	68.30	69.250	75.00
Silica, -	33.00	50.00	25.00	30.750	25.00
Water, -		12.00	4.40		
	99	100	97.70	100.00	100.00
	Klaproth, in Crell's Annalen for 1788, l. s. 398.	*Pelletier*, in Mem. & Obs. de Chim. t. i. p. 60.	*Smithson*, in Phil. Trans. part i. for 1803.	*John*, Chem. Unter. b. iii. s. 302. & 303.	

Geognostic Situation.

It occurs in metalliferous veins, principally along with ores of lead, in transition rocks, as greywacke, greywacke-slate, and clay-slate; also in flœtz limestone.

Geographic Situation.

It occurs in the lead-mines at Wanlockhead; also in Leicestershire and Flintshire. On the Continent, it is met with at Rezbania in Hungary, Bleiberg in Carinthia, Freyburg in the Breisgau.

3. Calamine.

Galmei, *Werner.*

This species is divided into three subspecies, viz. Sparry Calamine, Compact Calamine, and Earthy Calamine.

 First

First Subspecies.

Sparry Calamine.

Späthiger Galmei, *Karsten.*

Minera Zinci vitrea, *Waller.* Syst. Min. t. ii. p. 215. (in part).—Blättricher Galmei, *Reuss,* b. iv. s. 349. (in part).—Spathiger Galmei, *Karsten,* Tabel. s. 70.—Zinc carbonaté, var. 1, 2. *Hauy,* Tabl. p. 103.—Edler Galmei, or Zinkspath, *Haus.* Handb. b. i. s. 345.—Calamine, *Aikin,* p. 54.

External Characters.

Its colours are greyish-white, yellowish-white, bluish-grey, greenish-grey, siskin-green, apple-green, reddish-brown, and clove-brown.

It occurs massive, botryoidal, reniform, stalactitic, tubular, cellular, and crystallised, in the following figures:

1. Obtuse rhomboid.
2. Acute rhomboid.
3. Long four-sided table, which is either perfect, or bevelled on the terminal planes; and the angles of the bevelment sometimes more or less deeply truncated.

The crystals are small.

Internally it is shining and pearly.

The fracture is foliated or radiated.

The foliated varieties occur in granular concretions; but the radiated in curved lamellar concretions.

It alternates from semitransparent to opaque.

It yields easily to the knife.

Specific gravity, 4.300.

Chemical

Chemical Characters.

It dissolves with effervescence in muriatic acid; it is infusible; loses about 34 *per cent.* by ignition.

Constituent Parts.

	Derbyshire.	Somersetshire.
Oxide of Zinc,	65.2	64.8
Carbonic Acid,	34.8	35.2
	100	100

Smithson, in Phil. Trans. P. I. for 1803.

Second Subspecies.

Compact Calamine.

Gemeiner Galmei, *Karsten.*

Lapis calaminaris, *Wall.* Syst. Min. t. ii. p. 216.—Gemeiner Galmei, *Reuss,* b. iv. s. 345. (in part).—Gemeiner Galmei, *Karsten,* Tabel. s. 70.—Zinc carbonaté, var. 3, 4. *Hauy,* Tabl. p. 103.—Gemeiner Galmei, *Haus.* Handb. b. i. s. 347.

External Characters.

Its colours are yellowish, ash-greenish, and smoke-grey; also cream-yellow, straw-yellow, and yellowish-brown.

It occurs massive, disseminated, corroded, reniform, stalactitic, and cellular. Rarely in supposititious crystals, or incrusting other crystals.

Internally it is dull, or very feebly glimmering and resinous.

 The

The fracture is coarse-grained uneven, fine splintery, even, and flat conchoidal.

It sometimes occurs in concentric curved lamellar concretions.

It is opaque.

Chemical Character.

The same as in the preceding subspecies.

Third Subspecies.

Earthy Calamine.

Zinkblüthe, *Karsten.*

Zinkblüthe, *Karsten,* Tabel. s. 70.—Zinc carbonaté, *Hauy,* Tabl. p. 103. (in part).—Zinkblüthe, *Haus.* Handb. b. i. s. 348.—Earthy Calamine, *Aikin,* p. 54.

External Characters.

Its colours are snow-white, greyish-white, and yellowish-white; sometimes with a yellowish-brown exterior.

It occurs massive, disseminated, botryoidal, flat reniform, and with impressions.

Internally it is dull.

The fracture is fine earthy.

It is opaque.

It yields to the nail.

It adheres to the tongue.

Specific gravity, 3.358.

Chemical Characters.

The same as in the first subspecies.

Constituent

Constituent Parts.

		Bleiberg in Carinthia.
Oxide of Zinc,	-	71.4
Carbonic Acid,	-	13.5
Water, -	-	15.1
		100.0

Smithson in Phil. Trans.
P. I. for 1803.

Geognostic Situation of the Species.

It occurs in beds, nests, filling up or lining hollows, in transition limestone, and in flœtz limestone, and conglomerate rock ; also in veins. In these repositories it is generally associated with galena or lead-glance, and occasionally with copper-pyrites, copper-green, malachite, yellow and brown blende, sparry ironstone, ochry-brown ironstone, brown-spar, calcareous-spar, and quartz.

Geographic Situation of the Species.

Europe.—It occurs in the Mendip Hills, at Shipham, near Cross, Somersetshire ; at Allonhead in Durham ; at Holywell and elsewhere in Flintshire ; and in Derbyshire. On the Continent, it is met with at Raibel and Bleiberg in Carinthia ; Aachen ; Namur ; Chemnitz in Hungary ; Medziana Gora in Poland ; Beuthen and Tarnowitz in Silesia ; and Iserlohn in the Dutchy of Berg.

Asia.—Altai in Siberia.

Uses.

Calamine, when purified and roasted, is used for the fabrication of brass, which is a compound of zinc and copper ;

copper; and the pure metal is also employed for a variety of other purposes. The use of calamine in the composition of brass was known at a very early period; for it is mentioned by Aristotle, who also makes a distinction between the compound resulting from the mixture of copper and calamine, and that resulting from the mixture of copper and tin *.

4. Blende.

Blende, *Werner*.

This species is divided into three subspecies, viz. Yellow Blende; Brown Blende; and Black Blende.

First Subspecies.

Yellow Blende.

Gelbe Blende, *Werner*.

Id. Wern. Pabst. b. i. s. 188. *Id. Wid.* s. 898.—Yellow Blende, *Kirw.* vol. ii. p. 238.—Gelb Blende, *Emm.* b. ii. s. 443.—La Blende jaune, *Broch.* t. ii. p. 350.—Gelbe Blende, *Reuss,* b. iv. s. 326. *Id. Lud.* b. i. s. 273. *Id. Suck.* 2ter th. s. 367. *Id. Bert.* s. 464. *Id. Mohs,* b. iii. s. 557. *Id. Leonhard,* Tabel. s. 74.—Zinc sulphuré jaune, *Brong.* t. ii. p. 141.—Gelbe Blende, *Karsten,* Tabel. s. 70. *Id. Haus.* s. 77.—Phosphorescent Blende, *Aikin,* p. 55.

External Characters.

Its colour is dark wax-yellow, and sulphur-yellow; which

* Aristot. ed. Paris, 1654, vol. ii. p. 721.

which passes, on the one side, into grass-green, asparagus-green, oil-green, and olive-green; on the other, into hyacinth-red, aurora-red, brownish-red, even into reddish and yellowish-brown. All these colours incline more or less to green.

It occurs usually massive, disseminated, and crystallised, in several of the forms exhibited by brown-blende.

The crystals are middle-sized and small, but usually so much grown together, that it is difficult to determine their figure.

The crystals have a smooth surface.

Externally and internally it is shining and splendent, and the lustre is adamantine.

The fracture is more or less perfect and straight foliated, with a sixfold cleavage.

The fragments are dodecahedral, but on account of the distinct concretions can seldom be obtained perfect, and are therefore most commonly indeterminate angular, and sharp-edged.

It occurs in distinct concretions, which are large and coarse, but seldom small granular.

It is usually only translucent, but the lighter coloured varieties are semitransparent, inclining to transparent.

It is duplicating translucent.

It yields a yellowish-grey or yellowish-white streak.

It yields pretty easily to the knife.

It is brittle.

It is easily frangible.

It is heavy.

Specific gravity, 4.044 to 4.048, *Gellert*; 4.067, *Kirwan*; 4.103, *Karsten*.

Physical

Physical Character.

It becomes phosphorescent by friction; and, according to Bergmann, as powerfully under water as in the air.

Chemical Characters.

It decrepitates before the blowpipe, becomes grey, but is infusible either alone or with borax.

Constituent Parts.

	Yellow Blende from Scharfenberg.
Zinc, - -	64
Sulphur, - -	20
Iron, - -	5
Fluoric acid, -	4
Silica, - -	1
Water, - -	6
	100

Bergmann, Opuscul. t. II. p. 345.

Geognostic Situation.

It occurs in veins in primitive transition and flœtz rocks, where it is generally associated with galena or lead-glance.

Geographic Situation.

It occurs along with galena or lead-glance, copper-pyrites, copper-green, red cobalt ochre, and heavy-spar, in veins that traverse quartz-rock, at Clifton Mine, near Tyndrum in Perthshire; also in Flintshire. Very beautiful specimens are met with at Ratieborziz in Bohemia, where it is associated with galena or lead-glance, grey copper-per-ore, iron-pyrites, brown-spar, and quartz; and sometimes also with native silver, silver-glance or sulphureted silver-ore. It is also found at Scharfenberg in Saxony; Rammelsberg in the Hartz, in veins in transition rocks. The green varieties are found at Gumerud in Norway, associated with galena or lead-glance, smalt-blue apatite, in transition rocks; and it is accompanied with sulphuret of manganese and red manganese-ore, at Kapnic in Transilvania.

Second Subspecies.

Brown Blende.

Braun Blende, *Werner.*

This subspecies is divided into two kinds, viz. Foliated Brown Blende, and Fibrous Brown Blende.

First Kind.

Foliated Brown blende.

Blättrige Braune-Blende, *Werner.*

Id. Werner, Pabst. b. i. s. 191. *Id. Wid.* s. 896.—Brown Blende, *Kirw.* vol. ii. p. 239.—Braune Blende, *Emm.* b. ii. s. 447.——La Blende brune, *Broch.* t. ii. p. 353.—Braune Blende, *Reuss,* b. iv. s. 330. *Id. Lud.* b. i. s. 274. *Id. Suck.* 2ter th. s. 369. *Id. Bert.* s. 466. *Id. Mohs,* b. iii. s. 564. *Id. Leonhard,* Tabel s. 74.—Zinc sulphuré brun, *Brong.* t. ii. p. 141. —Braune Blende, *Karsten,* Tabel. s. 70. *Id. Haus.* s. 77.

External Characters.

Its colours are reddish and yellowish-brown; from yellowishlowish-brown it passes into blackish-brown, and from reddish-brown into hyacinth-red.

It is sometimes tarnished with variegated colours.

It occurs usually massive, and disseminated; often also crystallised:

1. Rhomboidal dodecahedron, which is the fundamental figure. It is either perfect or truncated on the alternate lateral edges and angles, with triangular planes.
2. Octahedron, which is sometimes elongated, and is either perfect, truncated on the edges or angles, or both at once; and sometimes bevelled on the edges.
3. Tetrahedron, which is either perfect or truncated on the angles.
4. Rectangular four sided prism, acuminated with four planes, which are set on the lateral edges.
5. Six-sided prism, acuminated with three planes, which are set on the alternate lateral planes.
6. Acicular crystals.
7. Twin crystals like those of spinel.

The crystals are small, very small, and middle-sized. Their lateral planes are generally convex.

Externally it is drusy and shining.

Internally it alternates from specular splendent to feebly glimmering, and the lustre is intermediate between resinous and adamantine.

The fracture is more or less perfect foliated, with a six-fold cleavage.

It occurs in granular distinct concretions of all degrees of magnitude, from large to extremely fine granular, which nearly passes into even.

It is more or less translucent, commonly strongly translucent on the edges. The extremely fine granular variety

variety is opaque. The large and coarse granular varieties are translucent, sometimes bordering on perfect transparent.

It yields a yellowish-grey and yellowish-brown streak.

It yields pretty easily to the knife.

It is brittle.

It is easily frangible.

Specific gravity, 3.770 to 4.048, *Gellert*; 3.963, *Kirwan*.

Constituent Parts.

From Sahlberg.		Allonheads, Northumberland.	
Zinc, -	44	Zinc, -	58.8
Iron, -	5	Sulphur,	23.5
Sulphur,	17	Iron, -	8.4
Silica,	24	Silica, -	7.0
Alumina,	5		
Water, -	5		97.7
	100		*Dr Thomson.*

Bergmann, Opusc. t. II. p. 332.

Blende, like all other ores, often contains what, chemically considered, may be viewed as accidental ingredients; thus the blende of Prizbram frequently contains silver; and that from Nagyag, manganese, lead, arsenic, and auriferous silver.

Geognostic Situation.

It occurs principally in veins and beds, in primitive and transition rocks; seldomer in flœtz rocks. It is associated with ores of different kinds, such as galena or lead-

lead-glance, copper-pyrites, iron-pyrites, grey copper-ore, and black silver-ore.

Geographic Situation.

It occurs in the Clifton lead-mine, near Tyndrum in Perthshire; in small veins along with galena, in the coal-fields around Edinburgh; at Cumberhead in Lanarkshire, along with galena or lead-glance; at Leadhills it is associated with galena, white lead-ore, sulphate of lead, sparry ironstone, iron-pyrites, brown hematite, azure copper-ore, electric calamine, and wad; the vein-stones are quartz, lamellar heavy-spar, calcareous-spar, brown spar, and mountain-cork. It is met with in all the lead-mines in England and Wales. On the Continent of Europe it forms a constant attendant of galena or lead-glance, whether it occurs in veins or beds; and it maintains the same relation in the lead-mines of Asia, Africa and America.

Second Kind.

Fibrous Brown Blende.

Fasrige Braune Blende, *Werner.*

Hepatisches Zinkerz, *Wiedenmann,* Min. s. 906.—Zink sulphuré compacte, *Broch.* t. ii. p. 359.—Schaalenblende, *Reuss,* b. iv. s. 342. *Id. Karsten,* Tabel. s. 70. *Id. Haus.* Handb. b. i. s. 233.—Fibrous Blende, *Aikin,* p. 56.

External Characters.

Its colour is reddish-brown.

It

It occurs massive and reniform.

It is glistening, passing into strongly glimmering, and is resinous.

The fracture is extremely delicate scopiform and stellular fibrous.

It occurs in large and coarse granular distinct concretions, which are intersected by curved lamellar concretions, bent in the direction of the external surface.

It is opaque, or very feebly translucent on the edges.

In other characters, it agrees with foliated brown blende.

Constituent Parts.

	From the Breisgau.
Zinc, - - - -	62
Iron, - - - -	3
Lead, - - - -	5
Arsenic, - - -	1
Sulphur, - - -	21
Alumina, - - -	2
Water, - - - -	4
	98

Hecht, in Journ. d. Min. t. xlix. N. 13.

Geognostic and Geographic Situations.

It occurs, along with galena or lead-glance and iron-pyrites, at Geroldseck in the Breisgau; and Raibel in Carinthia.

Third Subspecies.

Black Blende.

Schwarze-Blende, *Werner.*

Id. Werner, Pabst. b. i. s. 193. *Id. Wid.* s. 893.—Black Blende, *Kirw.* vol. ii. p. 241.—Schwarze Blende, *Emm.* b. ii. s. 451.—La Blende noire, *Broch.* t. ii. p. 357.—Schwarze Blende, *Reuss,* b. iv. s. 337. *Id. Lud.* b. i. s. 275. *Id. Suck.* 2ter th. s. 371. *Id. Bert.* s. 467. *Id. Mohs,* b. iii. s. 575. *Id. Leonhard,* Tabel. s. 74.—Zinc sulphuré noire, *Brong.* t. ii. p. 141. —Schwarze Blende, *Karsten,* Tabel. s. 70. *Id. Haus.* s. 77.

External Characters.

Its colour is intermediate between greyish and velvet black; sometimes it is brownish-black. When translucent, it appears blood-red. It is sometimes tarnished with variegated colours.

It occurs massive, disseminated, and crystallised in the same figures as brown blende.

The crystals are small; and so much grown together, that it is very difficult to ascertain their figure.

Internally it is shining, sometimes splendent, and the lustre is adamantine, inclining to metallic.

The fracture is foliated, with a six-fold cleavage, which in this subspecies is very indistinct.

The fragments are indeterminate angular, and rather sharp-edged.

It occurs in coarse, small and fine granular distinct concretions.

It

It is almost always opaque, excepting the brownish-black variety, which is translucent on the edges and angles.

The streak is intermediate between yellowish-grey and light yellowish-brown.

It is semi-hard.

It is brittle.

It is easily frangible.

Specific gravity, 3.967 to 3.930, *Gellert*. 4.1665, *Brisson*. 5.398, auriferous from Nagyag, according to *Von Müller*.

Constituent Parts.

	From Danemora.	Bowallon.
Zinc,	45	52
Iron, -	9	8
Lead,	6	-
Arsenic,	1	-
Copper,	-	4
Sulphur,	29	26
Silica, -	4	6
Water,	6	4
	100	100

Bergman, Opuscul t. iv. p. 329.

Oxide of Zinc,	53
Iron, -	12
Arsenic, -	5
Sulphur, -	26
Water, -	4
	100

Lampadius, Handb. z. Chem. Annal. d. Min. s. 282.

The black blende from Nagyag, besides zinc, iron, and manganese, contains a portion of auriferous silver. The lead and copper obtained from blende by Bergman, were probably derived from very minutely mixed galena or lead-glance and copper-pyrites, and the silica from the vein-stone.

Geognostic Situation.

It occurs in veins in gneiss, seldomer in grey-wacke. It is generally accompanied with copper-pyrites, arsenical-pyrites, iron-pyrites, magnetic ironstone, red silver-ore, white silver-ore, and galena. It is rarely associated with brown blende. Its accompanying vein-stones are calcareous-spar, brown-spar, and rarely asbestous actynolite, and garnet.

Geographic Situation.

It occurs in Sweden, Saxony, Hungary, and Transylvania..

Uses of the Species.

This ore is valued on account of the zinc which it affords. In order to obtain that metal from it, it is first roasted, to drive off the sulphur, and then ground with charcoal, and exposed to heat in a crucible, when the metal is reduced, and sublimes into a lute, so placed as to convey it into water, when it condenses in small drops.

Observations on the Species.

1. It is distinguished from *Tinstone*, by its inferior hardness: from *Galena* or *Lead-Glance*, by its grey-coloured dull streak; and it is distinguished from most other substances which it resembles, by exhaling a sulphureous odour, when either triturated in a mortar, or thrown into an acid.

2. Of all the subspecies, the brown is the most frequent and abundant.

3. It is named *Black Jack* by the miners in England; and is also known under the name *Pseudo-galena*.

ORDER IX

ORDER XI.—TIN.

This Order contains only three species, viz. Tin-pyrites, Tinstone, and Wood-tin or Cornish Tin-ore.

1. Tin-pyrites.

Zinnkies, *Werner.*

Id. Wid. s. 875.—Tin-pyrites, *Kirw.* vol. ii. p. 200.—Zinnkies, *Emm.* b. ii. s. 418.—Etaine sulphuré, *Lam.* t. i. p. 279. *Id. Hauy,* t. iv. p. 154.—La Pyrite d'Etaine, ou l'Etain pyriteux, *Broch.* t. ii. p. 332.—Zinnkies, *Reuss,* b. iv. s. 286. *Id. Lud.* b. i. s. 267. *Id. Suck.* 2ter th. s. 354. *Id. Bert.* s. 440. *Id. Mohs,* b. iii. s. 591.—Etain sulphuré, *Lucas,* p. 151.—Zinnkies, *Leonhard,* Tabel. s. 75.—Etain pyriteux, *Brong.* t. ii. p. 191.—Etain sulphuré, or Mussif natif, *Brard,* p. 337.—Zinnkies, *Karsten,* Tabel. s. 70. *Id. Haus.* s. 74.—Etain sulphuré, *Hauy,* Tabl. p. 102.—Sulphuret of Tin, *Kid,* vol. ii. p. 154.—Tin-pyrites, *Aikin,* p. 52.

External Characters.

Its colour is intermediate between steel-grey and brass-yellow, but usually more inclined to the first.

It occurs massive and disseminated.

Internally it is glistening, sometimes shining, and the lustre is metallic.

The fracture is small and coarse-grained uneven; sometimes inclining to small and imperfect conchoidal, and imperfect foliated.

 The

The fragments are intermediate angular, and blunt-edged.

It yields easily to the knife.

It is brittle.

It is easily frangible.

It is heavy.

Specific gravity, 4.350, *Klaproth*. 4.785, *La Metherie*.

It is not magnetic.

Chemical Characters.

Before the blowpipe it exhales a sulphureous odour, and melts easily, without being reduced, into a black scoria.

It communicates a yellow or green colour to borax.

Constituent Parts.

Tin, -	34	26.50
Copper, -	36	30.00
Iron, -	3	12.00
Sulphur,	25	30.50
Earthy Matter,	2	——
	——	99
	100	*Klaproth*, Beit. b. v. s. 230.
	Klaproth, Beit. b. ii. s. 257.—264.	

Geognostic and Geographic Situations.

It has been hitherto found only in Cornwall, as at St Agnes, Stenna Gwynn, Huel Rock, and Huel Scorier, associated with ores of copper and blende.

Observation.

It was formerly confounded with Magnetic Pyrites.

2, Tinstone.

2. Tinstone.

Zinnstein, *Werner.*

Stannum Arsenico et Ferro mineralisatum, *Wall.* t. ii. p. 319. *et seq.*—Zinnstein, *Werner,* Pabst. b. i. s. 171. *Id. Wid.* s. 880.—Common Tin-stone, *Kirw.* vol. ii. p. 197.—Etain vitreux, *De Born,* t. ii. p. 238.—Zinnkies, *Emm.* b. ii. s. 420. —Oxide d'Etain, *Lam.* t. i. p. 274.—Etain oxidé, *Hauy,* t. iv. p. 137.—La Pierre d'Etain, ou la Mine d'Etain commune, *Broch.* t. ii. p. 334.—Zinnstein, *Reuss,* b. iv. s. 288. *Id. Lud.* b. i. s. 267. *Id. Suck.* 2ter th. s. 354. *Id. Bert.* s. 441. *Id. Mohs,* b. iii. s. 596.—Etain oxidé, *Lucas,* p. 150.—Zinnstein, *Leonhard,* Tabel. s. 75.—Etain oxydé, *Brong.* t. ii. p. 189. *Id. Brard,* p. 335.—Zinnstein, *Karsten,* Tabel. s. 70. *Id. Haus.* s. 110.—Etain oxidé, *Hauy,* Tabl. p. 101 —Native Oxyd of Tin, *Kid,* vol. ii. p. 147.—Tinstone, *Aikin,* p. 51.

External Characters.

Its most common colour is blackish-brown; from which it passes, on the one side, into brownish-black and velvet-black; on the other side, into hair-brown and reddish-brown, from which it passes further into yellowish-green, yellowish-white, and greenish-white.

It occurs massive, disseminated, in rolled pieces, in grains, but most frequently crystallised, and in the following figures:

1. The fundamental figure is a flat octahedron, in which the common base is square, fig. 247.

This figure is rarely perfect, being usually more or less deeply truncated on the edges of the common base, and sometimes the edge of the base is bevelled, and the edge of the bevelment truncated. The angles on the common base, and also of the summits, are occasionally truncated.

2. Rectangular four-sided prism, acuminated with four planes, which are set on the lateral planes *, fig. 248.

3. The preceding figure, in which the lateral edges, and also those formed by the meeting of the acuminating planes, are truncated †, fig. 249. Sometimes in the same figure the edges formed by the meeting of the acuminating planes are truncated, and the lateral edges are bevelled, and the bevelling edges truncated ‡, fig. 250.

4. Rectangular four sided prism, acuminated with four planes, which are set on the lateral edges ||, fig. 251.

5. Rectangular four-sided prism, acuminated with four planes, which are set on the lateral edges, and the edges formed by the meeting of the acuminating and lateral planes truncated §, fig. 252. When these truncating planes on the edges become very large, an eight-planed acumination is formed, as is represented in the following figure.

6. Long rectangular four sided prism, acutely acuminated on both extremities with eight planes, of which two and two always meet together, under very obtuse angles, and are set on the lateral planes;

* Etain oxydé pyramidé, Hauy.

† Etain oxydé equivalent, Hauy.

‡ Etain oxydé sustractive, Hauy.

|| Etain oxydé dodecaedre, Hauy.

§ Ftain oxydé recurrent, Hauy.

planes; and again flatly acuminated with four planes, which are set on the obtuse edges of the first acumination *, fig. 253. The edges of the second acumination are sometimes truncated †.

7. Twin crystals of various descriptions; but of these the most frequent is that formed by the junction of two crystals, of the variety No. 2. which is represented in fig. 254 ‡.

The twin crystal here figured, is one of the most common forms of the species.

The surface of the crystals is usually smooth, seldom more or less strongly streaked, and it is commonly splendent.

Internally it is only shining and glistening, and the lustre is intermediate between resinous and adamantine, but more inclining to the latter.

The fracture is coarse and small-grained uneven, inclining to imperfect conchoidal; seldom imperfect foliated, and extremely seldom perfect foliated, and then it is highly splendent.

The fragments are indeterminately angular, and rather blunt-edged.

The massive varieties generally occur in coarse, small and fine granular distinct concretions.

It alternates from semi-transparent to opaque; the darker coloured varieties are opaque, the lighter translucent

* Etain oxydé opposite, Hauy.

† Etain oxydé distique, Hauy.

‡ The most complete account of the various crystallisation of tinstone we possess, is that by Mr William Phillips, in the second volume of the Transactions of the Geological Society of London. His memoir is accompanied with a series of beautiful plates, of which I could not avail myself in this work, because of their number and minuteness.

cent and semitransparent, often even inclining to transparent; the intermediate varieties are only translucent and translucent on the edges.

It yields a greyish-white streak.

It is hard.

It is easily frangible.

It is brittle.

It is uncommonly heavy.

Specific gravity, 6.300 to 6.989, *Gellert*; 6.750, *Brunnich*; 6.880, *Leysser*; 6.9009, the black, 6.9348, the red, *Brisson*; 5.845 to 6.970, *Klaproth*.

Chemical Characters.

Before the blowpipe it decrepitates, and becomes paler; when finely pounded, it is reducible on charcoal by the continued action of the blowpipe, to the metallic state. Acids dissolve the iron it contains, but only a very minute portion of the tin.

Constituent Parts.

	From Alternon.	Schlackenwald.	Ehrenfriedersdorf.
Tin, -	77.50	75.00	68
Iron, -	0.25	0.50	9
Oxygen, -	21.50	24.50	16
Silica, -	0.75	- -	7
	100	100	100

Klaproth, Beit. b. ii. s. 256. *Lampadius*.

Oxide of Tin,	94.50
Oxide of Iron,	1.00
Oxide of Manganese,	0.50
Silica, - -	1.00
Alumina, -	3.00
	100

John, Chem. Unters. b. ii. s. 242.

	Zinwald.
Oxide of Tin,	97.15
Oxide of Iron,	00.35
Alumina, -	2.50
	100

Kastner, Beit. zu Begrundung einer Wissench. Chem. b. s. 26.

Geognostic

Geognostic Situation.

It occurs disseminated, in beds, in imbedded masses, and veins, in granite, gneiss, mica-slate, clay-slate, and in an alluvial form, in what are in Cornwall named *Stream Works*. It is associated with wolfram, tungsten, molybdena, arsenical-pyrites, copper-pyrites, specular iron-ore, blende, rock-crystal, topaz, shorl, hornblende, chlorite, mica, steatite and fluor-spar; less frequently with calcareous-spar, heavy-spar, and with ores of lead, silver, and iron.

Geographic Situation.

Europe.—Tin is not found in many different countries, but when it does occur, it is generally in considerable quantity. There are only three principal tin districts in Europe. The first and most considerable is in Cornwall, where it occurs in veins, or disseminated in granite and slate, whether clay slate or chlorite-slate. It is sometimes raised in large blocks; for we are informed by Mr Phillips, that one block raised from the mine called Polberrow in St Agnes's, weighed 1200 lbs. and produced more than half that of pure metal. It is rarely found in massive portions, being generally crystallised; and it is worthy of notice, that all the varieties of form are not found indiscriminately in the same vein or set of veins, but appear rather to be distributed in different veins or sets of veins. Thus, according to Mr Phillips, the tin-mine of Pednandrae, near Redruth, affords scarcely any other form but that of a particular kind of twin crystal; the veins of Huel Fanny Mine, only three particular varieties of crystallisation; and the tin-mine of Polgooth near St Austle, only minute crystals of one particular

particular form. Of the same nature nearly is the observation that certain varieties of calcareous-spar are peculiar to Derbyshire, others to particular districts in Saxony, and some are only met in particular mines in France or Spain. Alluvial depositions of tinstone are met with of considerable extent and depth, in several parts of Cornwall; these are named *Stream Works*, because the tin is extracted from them by passing a stream of water across them. It is worthy of remark, that the only traces of gold hitherto met with in Cornwall are in the stream-works, where it is found generally detached, and sometimes accompanied with quartz *.

The second tin district is situated in the Erzgebirge, both on the Bohemian and Saxon sides of that mountain group. There it occurs disseminated on the granite, and in beds that alternate with that rock, where it is associated with wolfram, tungsten, common quartz, rock-crystal, mica, talc, fluor-spar, &c. in massive or crystallised forms. It often also occurs in veins in granite, gneiss, and mica-slate, and also in clay-slate. Alluvial deposites of this ore, resembling the stream-works of Cornwall, are also met with in these districts.

The third tin district is situated at Monte Rey in Gallicia in Spain, where the ore occurs in beds in mica-slate.

Very lately this ore has been found in small quantity in grains and crystals, in veins which traverse the granite hill of Puy-les Vignes, in the vicinity of St Leonhard, in the department of Haut-Vienne in France, where

* For other particulars in regard to the Tin of Cornwall, I refer to Mr Phillips' paper, in the second volume of Transactions of Geological Society.

where it is associated with wolfram, arsenical-pyrites, and martial arseniate of copper.

Asia.—It does not appear that tin has hitherto been met with in any of our possessions in India. It is found on the east coast of Sumatra, of Siam, and of Pegu; but it is principally imported into our Indian Empire as an article of commerce from Queda, Junk-Ceylon, Tavai in Lower Siam *, and the Island of Banca. The tin mines of Banca are said to be of great extent, and Mr Ellmore informs us, that there are exported from them annually no less than from forty to sixty thousand peculs of tin. It is said also to occur at a place five days journey from Nankin, in the province of Kianfu in China †.

America.—Tin-stone is found in Mexico, where it is extracted from alluvial deposites by means of washing, in the intendancy of Guanuaxuata and Zacatecas. It is said also to occur in Chili.

Uses.

It is worked as an ore of tin, and nearly all the tin of commerce is obtained from it.

Observations.

1. It is distinguished from *wolfram*, by its superior hardness, as it gives sparks with steel, whereas wolfram yields readily to the knife; and also by the streak, which is of a greyish-white colour, whereas that of wolfram is reddish-brown. It is distinguished from *blende* by its superior hardness, and its not emitting a sulphureous odour when

---

* Vid. Franklin's Tracts "On the Dominions of Ava," p. 64.

† Sage, in Journ. de Phys. t. 54. p. 113.

when triturated; from *garnet* by its resino-adamantine lustre, and higher specific gravity; and from *shorl* by colour, form, lustre, and higher specific gravity.

2. Its name is derived from the great proportion of tin it contains, and its unmetallic-like aspect.

3. Wood-Tin, or Cornish Tin-Ore.

Kornisch Zinnerz, *Werner.*

Mine d'Etain mamelonné, ou en Stalactites, *Romé de L.* t. iii. p. 428.—Kornisch Zinnerz, *Werner,* Pabst. b. i. s. 183. *Id. Wid.* s. 877.—Wood Tin-ore, *Kirw.* vol. ii. p. 198.—Etain limoneux, *De Born,* t. ii. p. 248.—Kornisch Zinnerz, *Emm.* b. ii. s. 427.—Mine d'Etain ferrugineux, *Lam.* t. i. p. 281.—Etain oxidé concretionné, *Hauy,* t. iv. p. 147.—La Mine d'Etain grenue, ou l'Etain grenu, *Broch.* t. ii. p. 340.—Holzzinnerz, *Reuss,* b. iv. s. 300. *Id. Lud.* b. i. s. 269. *Id. Suck.* 2ter th. s. 358.—Cornisch Zinnerz, *Bert.* s. 443. *Id. Mohs,* b. iii. s. 593.—Etain oxydé concretionné, *Lucas,* p. 150.—Holzzinnerz, *Leonhard,* Tabel. s. 75.—Etain oxidé concretionné, *Brong.* t. ii. p. 190. *Id. Brard,* p. 336.—Holzzinnerz, *Karsten,* Tabel. s. 70.—Fasriger Zinnstein, *Haus.* s. 110.—Etain oxidé concretionné, *Hauy,* Tabl. p. 102.—Wood-tin, *Aikin,* p. 52.

External Characters.

Its most common colour is hair-brown, of different degrees of intensity, which passes into wood-brown, yellowish-grey, and sometimes into reddish-brown. In single pieces it is oocasionally striped in a concentric manner.

It occurs in rolled pieces, which are generally wedge-shaped, also reniform, botryoidal, and globular.

Externally

Externally it is glistening.

Internally it is feebly glistening or glimmering, and the lustre is resinous, inclining to pearly.

The fracture is delicate, straight, scopiform, and stellular fibrous.

The fragments are wedge-shaped and splintery.

It usually occurs in large and coarse granular distinct concretions, which are intersected by curved and thin lamellar concretions, and the colour-delineation is in the direction of the latter.

It is opaque.

It is hard.

The streak is shining, and yellowish-brown.

It is brittle.

It is easily frangible.

It is uncommonly heavy.

Specific gravity, 5.800, *Brunnich*. 6.450, *Klaproth*.

Chemical Characters.

Before the blowpipe, it becomes brownish-red, and decrepitates, but is not fused, or reduced to the metallic state: when strongly heated in a charcoal crucible, it affords about 73 *per cent.* of metallic tin.

Constituent Parts.

			Mexican.
Oxide of Tin,	-	91	95
Oxide of Iron,	-	9	5
		100	100
		Vauquelin, N. Journ. d. Chem. b. v. s. 231.	*Collet-Descotils*, Ann. d. Chem. t. liii. p. 268.

Geognostic

Geognostic and Geographic Situations.

Europe.—It occurs loose, and in small quantities, along with stream tin, in alluvial deposites (stream-works), at Sithney, St Creet, Gossmoor, Pentowan, Gavrigan, St Mewan, St Columb, St Roach, and St Denis, in Cornwall.

America.—It is one of the most common ores of tin in Mexico. In that country, it is found at Guanuaxuato, (Goanachuato), in veins that traverse trap-porphyry, and is also met with in alluvial deposites *.

Observations.

It very much resembles *Brown Hematite*, but can be distinguished from it by its colour-suite, greater hardness, and higher specific gravity.

ORDER XII.

* Some time ago, Mr Mawe of London sent me a drawing of a mass of Mexican wood-tin, now in his possession, which weighs ten ounces and a half. It is the largest specimen of this ore I am acquainted with.

ORDER XII.—BISMUTH.

This Order contains the following species, viz. Native Bismuth, Bismuth-glance or Sulphureted Bismuth, Needle-ore or Plumbo-cupriferous Bismuth-ore, Cupriferous Bismuth-ore, and Bismuth-ochre.

1. Native Bismuth.

Gediegen Wissmuth, *Werner*.

Wismuthum nativum, *Wall.* t. ii. p. 205.—Gediegen Wismuth, *Werner,* Pabst. b. i. s. 183. *Id. Wid.* s. 887.—Native Bismuth, *Kirw.* vol. ii. p. 264.—Bismuth natif, *De Born,* t. ii. p. 214.—Gediegen Wismuth, *Emm.* b. ii. s. 434.—Bismuth natif, *Lam.* t. ii. p. 331. *Id. Hauy,* t. iv. p. 184. *Id. Broch.* t. ii. p. 343.—Gediegen Wismuth, *Reuss,* b. iv. s. 310 *Id. Suck.* 2ter th. s. 361. *Id. Lud.* b. i. s. 270. *Id. Bert.* s. 472. *Id. Mohs,* b. iii. s. 633.—Bismuth natif, *Lucas,* p. 156.—Gediegen Wismuth, *Leonhard,* Tabel. s. 77.—Bismuth natif, *Brong.* t. ii. p. 131. *Id. Brard,* p. 348.—Gediegen Wismuth, *Karsten,* Tabel. s. 70. *Id. Haus.* s. 70.—Bismuth natif, *Hauy,* Tabl. p. 105.—Native Bismuth, *Kid,* vol. ii. p. 212. *Id. Aikin,* p. 52.

External Characters.

Its colour is silver white, which inclines to red; on the fresh fracture it acquires a pavonine tarnish.

It occurs massive, disseminated, in leaves whose surface is plumosely streaked; reticulated; and crystallised, in the following figures:

1. Octahedron,

1. Octahedron, which is slightly oblique, and the lateral planes sometimes concave, and deeply transversely streaked.
2. Double three-sided pyramid, on which the lateral planes of the one are set on the lateral planes of the other: Or it may be described as an acute rhomboid, the alternate angles of which are 60° and 90°.
3. Four-sided tables, which are scalarwise aggregated.

Internally it is splendent, and the lustre is metallic.

The fracture is perfect foliated, with a threefold cleavage.

The fragments are indeterminate angular, and blunt-edged.

It occurs in distinct concretions, which are small and fine granular, and very seldom coarse granular.

It is soft.

It is sectile, passing into malleable.

It is rather difficultly frangible.

It is uncommonly heavy.

Specific gravity, 9.0202, *Brisson*; 9.570, *Kirwan*.

Chemical Characters.

It melts even by the flame of a candle; before the blowpipe it melts very quickly to a silver-white globule, which, by continuance of the heat, is volatilised, and deposites a white covering on the charcoal. It dissolves with effervescence in nitric acid; but if we add water to the solution, it is precipitated in the form of a white powder.

Geognostic

Geognostic Situation.

It occurs in veins in gneiss, mica-slate, and clay-slate. It is usually accompanied with ores of cobalt, particularly tin-white cobalt-ore, and grey cobalt-ore; also with copper nickel, bismuth ochre, iron-pyrites, sparry iron-ore, and brown blende; sometimes with native silver, and very seldom galena or lead-glance; the vein-stones are quartz, hornstone, calcareous-spar, brown spar, and heavy spar.

Geographic Situation.

Europe.—It is found at St Columb and Botallack, in Cornwall, but more frequently at Johanngeorgenstadt and Schneeberg, in the Kingdom of Saxony, than in any place in Enrope; it occurs also in considerable quantity at Joachimsthal in Bohemia; and in less abundance in the Black Forest (Schwarzwald) in Swabia. It has been also met with at Zalathna in Transilvania; Temeswar in the Bannat; at Biber in Hanau; St Saveur, and in the mines of Brittany in France; Dalecarlia and Nerike in Sweden; and Modum in Norway.

America.—It occurs at Huntington, parish of New Stratford, in the State of Connecticut, in a vein of quartz, along with common and magnetic pyrites, and galena or lead-glace *.

Uses.

It enters as an ingredient into the composition of types, pewter; is used as solder, in the construction of mirrors, and for the refining of gold and silver; its oxide is used as a white pigment, as an essential ingredient in a kind of salve, which is used for giving a black colour to the hair, and as an ingredent in sympathetic ink.

* Bruce's American Mineralogical Journal, p. 267.

Observations.

It is distinguished from bismuth-glance by its colour and regular external figures; the recticulated varieties, from recticulated native silver, by their colour and inferior malleability.

2. Bismuth-Glance or Sulphureted Bismuth.

Wismuth-Glanz, *Werner.*

Galena Wismuthi, *Wall.* t. ii. p. 206.—Minera Wismuthi cinerea-versicolor-martialis, *Id.* p. 207. and 208.—Wismuth-glanz, *Wern.* Pabst. b. i. s. 187. *Id Wid.* s. 890.—Sulphurated Bismuth, *Kirw.* vol. ii. p. 266.—Bismuth sulphuré, *De Born,* t. ii. p. 217.—Wismuth-glanz, *Emm.* b. ii. s. 438.—Bismuth sulphuré, *Lam.* t. ii. p. 333. *Id. Hauy,* t. iv. p. 190.—La Galena de Bismuth, ou le Bismuth sulphuré, *Broch.* t. ii. p. 346.—Wismuth-glanz, *Reuss,* b. iv. s. 314. *Id. Lud.* b. i. s. 271. *Id. Suck.* 2ter th. s. 363. *Id. Bert.* s. 473. *Id. Mohs,* b. iii. s. 631.—Bismuth sulphuré, *Lucas,* p. 157.—Wismuth-glanz, *Leonhard,* Tabel. s. 77.—Bismuth sulphuré, *Brong.* t. ii. p. 133. *Id. Brard,* p. 350.—Wismuth-glanz, *Karsten,* Tabel. s. 70. *Id. Haus.* s. 75.—Bismuth sulphuré, *Hauy,* Tabl. p. 105.—Sulphureted Bismuth, *Aikin,* p. 53.

External Characters.

Its colour is light lead-grey, more or less inclining to tin white.

Externally it is yellowish, or tarnished with variegated colours.

It occurs massive, disseminated, and in acicular and capillary crystals.

Internally

Internally the foliated fracture is splendent; the radiated fracture only shining.

The fracture is perfect foliated, and the cleavages are parallel to the sides and to the short diagonal of a slightly oblique rhomboïdal prism; the latter cleavages are very distinct, the others less so; sometimes also the fracture is narrow and promiscuous radiated.

The fragments are indeterminate angular.

The foliated varieties occur in large and coarse granular distinct concretions.

It is soft and very soft.

It soils.

It is brittle, inclining to sectile.

It is easily frangible.

It is heavy.

Specific gravity, 6.4672, *Brisson*; 6.131, *Kirwan*.

Chemical Characters.

It melts in the flame of a candle. It is volatilised before the blowpipe, and deposites on the charcoal a yellow crust, which becomes white on cooling.

Constituent Parts.

Bismuth,	- -	60
Sulphur,	- -	40
		100

Sage in Mem. de l'Acad. d. Sc. 1782, p. 307.

Geognostic Situation.

It occurs in veins, and is usually accompanied with native bismuth, grey cobalt-ore, cerite, sparry ironstone, arsenical-

arsenical-pyrites, copper-pyrites, tinstone, quartz, and fluor-spar.

Geographic Situation.

It is found in Herland-mine in Cornwall; at Joachimsthal and Schlackenwald in Bohemia; Johanngeorgenstadt and Altenberg in the kingdom of Saxony; and Bastnäs near Riddarhytta in Sweden.

Observations.

1. It is distinguished from grey antimony-ore by its lighter lead-grey colour, and its greater specific gravity.
2. It is a rare mineral.

3. Needle-Ore, or Plumbo-cupriferous sulphureted Bismuth-Ore.

Nadelerz, *Werner.*

Nadelerz, *Karsten,* Tabel. s. 70. *Id. Haus.* s. 75.—Bismuth sulphuré plumbo-cuprifere, *Hauy,* Tabl. p. 108.—Plumbo-cupriferous Sulphureted Bismuth, *Aikin,* p. 122. 2d edit.

External Characters.

Its colour is steel-grey, with a pale copper-red tarnish.

It occurs disseminated and crystallised in oblique four or six-sided prisms, in which the lateral planes are deeply longitudinally streaked. The crystals are long, often acicular, or grown together, frequently curved, and sometimes divided by cross rents.

Internally

Internally it is splendent in the foliated fracture; but only shining on the uneven fracture; and the lustre is metallic.

The principal fracture is foliated; the cross fracture is small grained uneven.

It is opaque.

It yields easily to the knife.

Specific gravity, 6.125, *John.*

Chemical Characters.

It is fusible before the blowpipe into a steel-grey globule; by continuance of the heat, it partly volatilises, and deposites on the charcoal a yellow powder, after which there remains a red globule, inclosing a grain of cupriferous metallic lead, which, when treated with glass of borax, communicates a bluish-green colour to it.

Constituent Parts.

Bismuth, - -	43.20
Lead - -	24.32
Copper, - -	12.10
Sulphur, - -	11.58
Nickel, - -	1.58
Tellurium? - -	1.32
Gold, - -	0.79
	94.89

John, n. Chem. Untersuchungen, s. 216.

Or we may estimate its ingredients in the following manner:

Sulphuret of Bismuth,	Bismuth,	43.20	50.76
	Sulphur,	7.56	
Sulphuret of Lead,	Lead,	24.32	28.07
	Sulphur,	3.75	
Sulphuret of Copper,	Copper,	12.10	15.13
	Sulphur,	3,03	
Nickel,	- -	- -	1.58
Tellurium?	- -	- -	1.32
Gold,	- -	- -	0.79
			97.65

Geognostic Situation.

It occurs imbedded in quartz, and is associated with galena or lead glance, and native gold. The crystals are sometimes invested with a greenish crust, which appears to be copper green, and sometimes with a yellow crust of bismuth-ochre.

Geographic Situation.

It occurs in the mines of Pyschminskoi and of Klintzefskoi near Beresof, in the district of Catharinenburg in Siberia.

Observations.

In the former edition of this work, the Needle ore, on the authority of Werner, was arranged as an ore of Chrome; but the late investigations of John have proved, that it belongs to the bismuth order. It was at one time considered as an auriferous ore of nickel; and Patrin, so early as the year 1786, approached very near to its true nature, for he describes it as a sulphuret of bismuth.

4. Cupreous

4. Cupreous Bismuth-Ore, or Cupriferous sulphureted Bismuth-Ore.

Kupferwismutherz, *Karsten.*

Kupferwismuth, *Karsten,* Tabel. s. 70. *Id. Haus.* s. 75. *Id. Klaproth,* Beit. b. iv. s. 91. *Id. Selb,* in d. Annal. der Wetterauischen Gesellsch. b. i. s. 40. *Id. Haus.* Handb. b. i. s. 189. —Cupriferous Sulphureted Bismuth, *Aikin,* p. 122. 2d edit.

External Characters.

Its colour is light lead-grey, which passes on the one side into steel-grey, and on the other into tin-white; and its tarnish is yellowish or reddish.

It occurs massive, disseminated, seldom in small scopiformly aggregated prisms.

Internally it is shining and metallic.

The fracture is fine grained uneven, and sometimes inclines to radiated.

It is sectile.

Constituent Parts.

Bismuth, - -	47.24
Copper, - -	34.66
Sulphur, - -	12.58
	94.48

Klaproth, Beit. b. iv. s. 96.

Or probably more correctly, according to estimation,

Sulphuret of Bismuth,	Bismuth,	47.240	55.507
	Sulphur,	8.267	
Sulphuret of Copper,	Copper,	34.660	43.325
	Sulphur,	8.665	
			98.832

Hausmann, Handb. b. i. s. 188.

Geognostic

Geognostic and Geographic Situations.

It occurs in veins in granite, along with native bismuth, copper-pyrites, and heavy-spar, in the mines named Neuglück, and Daniel at Gallenbach, near Wittichen, in Furstemberg.

Observation.

This rare ore was first discovered by Mr Selb.

5. Bismuth-Ochre.

Wismuthocker, *Werner*.

Ochra Wismuthi, *Wall.* t. ii. p. 209.—Wismuthocker, *Werner*, Pabst. b i. s. 188. *Id. Wid.* s. 891.—Bismuth-Ochre, *Kirw.* vol. ii. p. 265.—Ocre de Bismuth, *De Born*, t. ii. p. 194.—Wismuth-ocre, *Emm.* b. ii. s. 440.—Oxide de Bismuth, *Lam.* t. i. p. 332.—Bismuth oxidé, *Hauy*, t. iv. p. 194, 195.—L'Ocre de Bismuth, *Broch.* t. ii. p. 348.—Wismuth-Ochre, *Reuss*, b. iv. s. 318. *Id. Lud.* b. i. s. 272. *Id. Suck.* 2ter th. s. 364. *Id. Bert.* s. 474. *Id. Mohs*, b. iii. s. 662. *Id. Leonhard*, Tabel. s. 77.—Bismuth oxidé, *Brong.* t. ii. p. 134.—Wismuth-Ochre, *Karsten*, Tabel. s. 70.—Bismuth oxidé, *Hauy*, Tabl. p. 106.—Wismuth-Ocker, *Haus.* Handb. b. i. s. 337.—Native Oxide of Bismuth, *Kid*, vol. ii. p. 212.—Bismuth Ochre, *Aikin*, p. 52.

External Characters.

Its colour is straw-yellow, which sometimes passes into light yellowish-grey and ash-grey ; sometimes even verges on apple-green *.

It

* The apple-green varieties contain nickel.

It occurs massive and disseminated.

It is glimmering and glistening when the fracture is fine and small grained uneven; shining when it is foliated; and dull when it is earthy. Its lustre is adamantine.

The fracture is fine and small-grained uneven, which passes on the one side into foliated, on the other into earthy.

The fragments are indeterminate angular, and rather blunt-edged.

It is opaque.

It is soft and very soft, verging on friable.

It is rather brittle.

It is easily frangible.

It is heavy.

Specific gravity, 4.3711, *Brisson.*

Chemical Characters.

Before the blowpipe, on charcoal, it is easily reduced, but is also volatilised if the heat be continued. It dissolves with effervescence in acids, and the solution is decomposed by means of water, when a white precipitate is formed.

Constituent Parts.

Oxide of Bismuth, -	86.3
Oxide of Iron, -	5 2
Carbonic Acid, - -	4.1
Water, - - -	3.4
	99.0

Lampadius, Handb. z. Chem. Annal. s. 286.

Geognostic

Geognostic and Geographic Situations.

It occurs along with red cobalt-ochre, copper-nickel, grey copper-ore, copper-glance or vitreous copper-ore, azure copper-ore, sparry ironstone, quartz, and calcareous-spar.

It is found at St Agnes in Cornwall; Schneeberg and Johanngeorgenstadt in Saxony; and Joachimsthal in Bohemia; but is a rare mineral.

Observation.

It has been confounded with green iron-earth, from which it is well distinguished not only by its external aspect, but by its accompanying minerals.

ORDER XIII.

ORDER XIII.—TELLURIUM.

This Order contains four species, viz. Native Tellurium, Graphic Tellurium, Yellow Tellurium-ore, and Black Tellurium-ore.

1. Native Tellurium.

Gediegen Sylvan, *Werner.*

Tellure natif aurifere et ferrifere, *Hauy,* t. iv. p. 325.—Le Silvane natif, *Broch.* t. ii. p. 480.—Gediegen Tellur, *Reuss,* b. iv. s. 604. *Id. Lud.* b. i. s. 310. *Id. Suck.* 2ter th. s. 492.—Gediegen Sylvan, *Bert.* s. 520. *Id. Mohs,* b. iii. s. 57.—Tellure natif, *Lucas,* p. 185.—Gediegen Tellur, *Leonhard,* Tabel. s. 80.—Tellure natif, *Brong.* t. ii. p. 125. *Id. Brard,* p. 391.—Gediegen Tellur, *Karsten,* Tabel. s. 70. *Id. Haus.* s. 70.—Tellure natif auro-ferrifere, *Hauy,* Tabl. p. 119.—Gediegen Tellur, *Haus.* Handb. b. i. s. 129.—Native Tellurium, *Aikin,* p. 70.

External Characters.

Its colour is intermediate between tin white and silver-white, and sometimes inclines to pale steel-grey.

It occurs massive and disseminated; and, it is said, also crystallised in four-sided prisms acutely acuminated with four planes, which are set on the lateral planes.

Internally it is shining, and the lustre is metallic.

The fracture is foliated, or promiscuous radiated.

It

It occurs in small and fine granular distinct concretions.

It is semihard, approaching to soft.

It is rather brittle.

It is easily frangible.

It is heavy.

Specific gravity, 5.723, *Müller*; 5.730, *Kirwan*; 6.115, *Klaproth.*

Chemical Characters.

Before the blowpipe it melts as easily as lead, emits a thick white smoke, and burns with a light green colour, and a pungent acrid odour, like that of horse-radish. When exposed to a low heat, it is converted into a yellowish or blackish coloured oxide: by an increase of temperature, it melts into a dark-brown or blackish coloured glass, in which gold grains are interspersed: at a still higher heat, the oxide is entirely volatilised. In concentrated nitric acid, it is converted into a yellow oxide, and a small portion is dissolved, which is precipitated in yellow flakes, on the addition of water.

Constituent Parts.

Tellurium, - -	92.55
Iron, - -	7.20
Gold, - -	0.25
	100

Klaproth, Beit. b. iii. s. 8.

Geognostic Situation.

It occurs in veins in porphyry, along with iron-pyrites and quartz.

Geographic

Geographic Situation.

It has been hitherto only found at Facebay in Transilvania.

Observations.

1. It bears a very striking resemblance to Native Antimony, from which it is distinguished by its inferior hardness, its lower specific gravity, and its smaller granular concretions.

2. It is known in older works on Mineralogy under the names *Aurum problematicum*, *Aurum paradoxum*, and White Gold-ore.

3. The alchemists distinguished metallic substances by the same names that the ancients had applied to the planets,—a practice which has been adopted in the present instance; and the metal we are now describing was named, by Klaproth, the discoverer of it, from our own planet the Earth, or "Tellus," while Kirwan names it *Sylvan*, from the country, Transylvania, where it is found.

2. Graphic

2. Graphic Tellurium, or Graphic-Ore.

Schrifterz, *Werner*.

Weiss Golderz, *Wid.* s. 673.—Schrifterz, *Esmark*, N. Bergm. Journ. t. ii. p. 10.—Or blanc d'Offenbanya, ou graphique, Aurum graphicum, *De Born*, t. ii. p. 470.—Schrifterz, *Emm.* b. iii. s. 405.—Tellure natif graphique, *Hauy*, t. iv. p. 327.—Le Silvane graphique, *Broch.* t. ii. p. 482.—Schrifterz, *Reuss*, b. iv. s. 608. *Id. Lud.* b. i. s. 310. *Id. Suck.* 2ter th. s. 493. *Id. Mohs*, b. iii. s. 65.—Tellure natif aurifere et argentifere, *Lucas*, p. 186.—Schrifterz, *Leonhard*, Tabel. s. 80.—Tellur natif graphique, *Brong.* t. ii. p. 123.—Schrifterz, *Karsten*, Tabel. s. 70.—Schrift Tellur, *Haus.* s. 70.—Tellur natif auro-argentifere, *Hauy*, Tabl. p. 119.—Graphic Tellurium, *Aikin*, p. 70.

External Characters.

Its colour is steel-grey, which sometimes becomes white, yellow, or lead-grey, or variously tarnished by exposure to the air.

It occurs massive, disseminated, in leaves; and crystallised in the following figures:

1. Rectangular four-sided prism, acutely acuminated with four planes, which are set on the lateral edges. Frequently two opposite lateral planes are broader than the others, when the prism appears broad; and sometimes the lateral edges and acuminating edges are truncated.
2. Six sided prism, acuminated with four planes.

The crystals are small, and very small, and are generally arranged in rows on the surface of quartz.

Frequently

Frequently there are attached to the extremities of the prisms, others at right angles, giving to the whole row the appearance of a line of Persepolitan characters.

The planes of the crystals are smooth.

Externally it is splendent, and the lustre is metallic.

Internally it is glistening, and the lustre is metallic.

The fracture is fine grained uneven, and sometimes imperfect foliated.

The massive variety, which is very rare, occurs in granular distinct concretions.

It is soft, but not so soft as yellow tellurium-ore.

It is brittle.

It is easily frangible.

It affords a lead-grey streak.

It soils slightly.

It is heavy.

Specific gravity, 5.723, *Müller*.

Chemical Characters.

Before the blowpipe it burns with a green flame, and is volatilized.

Constituent Parts.

Tellurium,	-	-	60
Gold,	-	-	30
Silver,	-	-	10
			100

Klaproth, Beit. b. iii. s. 20.

Geognostic Situation.

It occurs in veins in porphyry, along with quartz, calcareous-

calcareous-spar, iron-pyrites, blende, and brass-yellow, native gold.

Geographic Situation.

It has been hitherto only found at Offenbanya in Transylvania.

Use.

It is worked as an ore of gold.

Observation.

Its name is derived from the particular appearance formed by the aggregation of the crystals.

3. Yellow Tellurium-Ore.

Weiss Sylvanerz, *Werner.*

Var. de Nagyagerz, *Wid.* s. 671.—Or gris jaunâtre, *De Born*, t. ii. p. 464.—Var. de Nagyagerz, *Emm.* b. ii. s. 121.—Mine jaune de Nagyag, *Journ. des Min.* No. 38. p. 150.—Gelberz, *Karst.* Tabel. s. 56.—Tellure natif aurifère et plombifere, *Hauy*, t. iv. p. 327.—La Silvane blanc, *Broch.* t. ii. p. 484.—Gelberz, *Reuss*, b. iv. s. 612.—Gelbtellurerz, *Lud.* b. i. s. 311. *Id. Suck.* 2ter th. s. 495.—Weiss-sylvanerz, *Bert.* s. 521. *Id. Mohs*, b. iii. s. 59.—Tellure aurifere et plombifere, *Lucas*, p. 186.—Weiss Tellurerz, *Leonhard*, Tabel. s. 80.—Tellure natif plombifere, *Brong.* t. ii. p. 124. *Id. Brard*, p. 392.—Gelberz, *Karsten*, Tabel. s. 70.—Weiss Tellur, *Haus.* s. 71.—Tellure natif auro-plombifere, *Hauy*, Tabl. p. 119.—Yellow Tellurium, *Aikin*, p. 71.

External Characters.

Its colour is silver-white, which inclines very much to yellow,

yellow, and sometime to reddish and ash-grey. It occasionally exhibits a yellow and green play of colour, which, however, is not of long duration, as the whole surface soon becomes of one tint of colour.

It occurs disseminated, less frequently massive, very rarely imperfectly reticulated; and seldom crystallised, in broad four-sided prisms, which are generally acicular.

The principal fracture is intermediate between splendent and shining; the cross fracture is glistening, and the lustre is metallic. The principal fracture is foliated, and the cross-fracture is small-grained uneven.

It is soft.

It is rather sectile.

It is uncommonly heavy.

Specific gravity, 10.678, *Müller*.

Constituent Parts.

Tellurium,	-	-	44.75
Gold,	-	-	26.75
Lead,	-	-	19.50
Silver,	-	-	8.50
Sulphur,	-	-	0.50
			100.

Klaproth, Beit. b. iii. s. 25.

Geognostic and Geographic Situations.

This ore occurs in small and very irregular veins in porphyry. The most frequent vein-stones are brown-spar and quartz; sometimes it is also associated with red manganese-ore, sulphuret of manganese, native arsenic, plumose antimony, and brass-yellow native gold. It has been hitherto found only at Nagyag in Transylvania.

 Use.

Use.

As it contains a considerable portion of gold and silver, it is worked on account of both these metals.

4. Black Tellurium-Ore.

Nagyagerz, *Werner*.

Id. Wid. s. 671.—Or gris lamelleux, *De Born*, t. ii. p. 463.—Blättererz, *Karst.* Tabel. 56.—Nagyagerz. *Emm* b. ii. s. 121.—Mine d'Or de Nagayag, *Lam.* t. i. p. 110.—Tellure natif aurifere et plombifere, *Hauy*, t. iv. p. 327.—La Mine de Nagyag, ou le Silvane lamelleux, *Broch.* t. ii. p. 486 —Blättererz, *Reuss*, b. iv. s. 615.—Nagyagerz, *Lud* b. i. s. 311.—Blätter Tellurerz, *Suck.* 2ter th. s. 497.—Blättererz, *Bert.* s. 522. *Id. Mohs*, b. iii. s. 70. *Id. Leonhard*, Tabel s. 80.—Tellur natif plombifere, var. feuilleté, *Brong.* t. ii p. 124—Blättererz, *Karsten*, Tabel s 70.—Blätter Tellur, *Haus.* s. 71.—Tellur natif auro-plombifere, lamellaire et lamelliforme, *Hauy*, Tabl. p. 119.—Black Tellurium, *Aikin*, p. 71.

External Characters.

Its colour is intermediate between iron-black and blackish lead-grey.

It occurs massive, in leaves, and crystallised in the following figures:

1. Oblique four-sided table.
2. Rectangular four sided table.
3. Six-sided table.
4. Eight sided table.
5. Acute octahedron, truncated on the summits.

Externally

Externally it is splendent and the lustre is metallic.

Internally it is shining.

The fracture is foliated, with a single cleavage, and the folia are curved.

The fragments are tabular.

It is very soft; it is the softest of the ores of tellurium.

It is sectile; it is the most sectile of the ores of tellurium.

It soils slightly.

The thin leaves and tables are common flexible.

It is uncommonly heavy.

Specific gravity, 8.919, *Müller*.

Chemical Characters.

It melts very easily before the blow-pipe; the sulphur and tellurium are soon volatilised, and a blackish-brown coloured globule remains, which on being melted with borax, affords an argentiferous gold globule; the slag which remains. tinges borax violet blue. It dissolves with effervescence in acids; the nitrico-muriatic acid extracts the gold from it.

Constituent Parts.

Tellurium, - -	32.2
Lead, - - -	54.0
Gold, - - -	9.0
Sulphur, - - -	3.0
Copper, -	1.3
Silver, - - -	0.5
	100

Klaproth, Beit. b. iii. s. 32.

 Geognostic

Geognostic and Geographic Situations.

It is generally associated with yellow tellurium-ore, in veins that traverse porphyry, and has been hitherto found only at Nagyag in Transylvania.

Use.

As it affords gold and silver, it is worked both as a gold and silver ore.

Observations.

1. A new ore of Tellurium has been lately found in Norway by Esmark.

2. The celebrated traveller Dr Clarke, in the fourth volume of his Travels, just published, gives a full account of the tellurium mines.

ORDER XIV.

ORDER XIV.—ANTIMONY.

This order contains six species, viz. Native Antimony, Grey Antimony-ore, Nickeliferous Grey Antimony-ore, Red Antimony-ore, White Antimony-ore, and Antimony-ochre.

1. Native Antimony.

Gediegen Spiesglas, *Werner*.

Swab, in d. Schriften der K. Schwed. Acad. 10. b. v. I. 1748, s. 100.—Regulus Antimonii nativus, *Wall.* t. ii. p. 196.—Gediegen Spiesglas, *Werner*, Pabst. b. i. s. 197. *Id. Wid.* s. 909.—Native Antimony, *Kirw.* vol. ii. p. 245.—Antimoine natif, *De Born*, t. ii. p. 137.—Gediegen Spiesglas, *Emm.* b. ii. s. 464.—Antimoine natif, *Hauy*, t. iv. p. 252 —L'Antimoine natif, *Broch.* t. ii. p. 369.—Gediegen Spiesglanz, *Reuss*, b. iv. s. 362. *Id. Lud.* b. i. s. 277. *Id. Suck.* 2ter th. s. 383. *Id. Bert.* s. 475. *Id. Mohs*, b. iii. s. 688.—Antimonie natif, *Lucas*, p. 171.—Gediegen Spiesglanz, *Leonhard*, Tabel. s. 78.—Antimoine natif, *Brong.* t. ii. p. 126. *Id. Brard*, p. 370.—Gediegen Spiesglanz, *Karsten*, Tabel. s. 72. *Id. Haus.* s. 70. —Native Antimony, *Kid*, vol. ii. p. 199.—Antimoine natif, *Hauy*, Tabl. p. 112.—Native Antimony, *Aikin*, p. 56.

External Characters.

Its colour is perfect tin-white. On the fresh fracture it sometimes becomes covered with a blackish or yellowish tarnish.

It occurs massive, disseminated, reniform, and also crystallised, in the following figures :

1. Octahedron.

1. Octahedron.
2. Rhomboidal dodecahedron.

On the fresh fracture it is splendent, and the lustre is metallic.

The fracture is perfect, usually straight, and sometimes also curved foliated, with a fourfold cleavage.

The fragments are sometimes rhomboidal, more usually indeterminate angular, and blunt-edged.

It occurs in coarse, small and fine granular distinct concretions; sometimes in thin and curved lamellar concretions.

It is soft, passing into semi-hard.

It is rather sectile.

It is easily frangible.

Specific gravity, 6,720, *Klaproth.*

Chemical Characters.

Before the blowpipe it melts easily, and volatilises in the form of a grey inodorous vapour; if the melted globule be allowed to cool slowly, it becomes covered with white brilliant acicular crystals. A very minute globule of silver generally remains after the antimony has been dissipated.

Constituent Parts.

	Andreasberg.
Antimony, - -	98.00
Silver, - - -	1.00
Iron, - - -	0.25
	99.25 *

Klaproth, Beit. b. iii. s. 172.

Geognostic

* It is sometimes alloyed with a small and variable proportion of arsenic; in consequence of which, its vapour, when exposed to the blowpipe, has an arsenical odour.

Geognostic and Geographic Situations.

Europe.—It is found in argentiferous veins in the gneiss mountain of Chalanches in Dauphiny, where it is accompanied with grey antimony-ore, white antimony-ore, red antimony ore, silver white cobalt-ore or cobalt-glance, and quartz; at Andreasberg in the Hartz, associated with red silver-ore, calcareous spar and quartz; at Sahlberg, in Westermannland in Sweden, disseminated in calcareous spar.

America.—At Cuencamé in Mexico.

Observations.

1. It is distinguished from all other minerals with which it might be confounded, by colour, fracture, hardness and weight.

2. It is a rare mineral.

2. Grey Antimony-Ore.

Grau Spiesglaserz, *Werner.*

This species is divided into two subspecies, viz. Common Grey Antimony-ore, and Plumose Antimony-ore.

First Subspecies.

Common Grey Antimony-Ore.

Gemeines Grau Spiesglaserz, *Werner.*

This subspecies is divided into three kinds, viz. Common, Foliated, and Radiated.

First

First Kind.

Compact Grey Antimony-Ore.

Dichtes Grauspiesglaserz, *Werner.*

Minera Antimonii solida, *Wall.* t. ii. p. 198.—Dichter Grauspiesglaserz, *Werner,* Pabst. b. i. s. 197. *Id. Wid.* s. 912.—Compact sulphurated Antimony, *Kirw.* vol. ii. p. 247.—Dichter Grauspiesglaserz, *Emm.* b. ii. s. 468.—L'Antimoine gris compacte, *Broch.* t. ii. p. 372.—Dichter Grauspiesglanzerz, *Reuss,* b. iv. s. 367. *Id. Lud.* b. i. s. 278. *Id. Suck.* 2ter th. s. 384. *Id. Bert.* s. 475. *Id. Mohs,* b. iii. s. 687. *Id. Leonhard,* Tabel. s. 79.—Antimoine sulphuré pure compacte, *Brong.* t. ii. p. 127.—Dichtes Grauspiesglanzerz, *Karsten,* Tabel. s. 72.—*Id. Haus.* s. 75.—Sulphuret of Antimony, *Kid,* vol. ii. p. 201.—Antimoine sulphuré compacte, *Hauy,* Tabl. p. 113.—Grey Antimony, *Aikin,* p. 56.

External Characters.

Its colour is light lead-grey, and it has sometimes a pavonine or steel-coloured tarnish.

It occurs massive, disseminated, and seldom in membranes.

Internally, it is shining and glistening, and the lustre is metallic.

The fracture is small and fine-grained uneven, which latter sometimes passes into even.

The fragments are indeterminate angular, and blunt-edged.

It is soft.

It is easily frangible.

It

It soils.

The lustre is increased in the streak.

It is heavy.

Specific gravity, 4.368, *Kirwan.*

Geographic Situation.

Europe.—It is found at Sahlberg in Sweden; Braunsdorf near Freyberg in Saxony; Hungary; Baireuth; Salzburg; and Auvergne in France.

Asia.—Siberia.

America.—Chili.

Second Kind.

Foliated Grey Antimony-Ore.

Blättriges Grauspiesglaserz, *Werner.*

Id. Wern. Pabst. b. i. s. 197.—Foliated sulphurated Antimony, *Kirw.* vol. ii. p. 248.—Blättriches Grauspiesglaserz, *Emm.* b. ii. s. 470.—L'Antimoine gris lamelleux, *Broch.* t. ii. p. 373. —Blättriches Grauspiesglanserz, *Reuss,* b. iv. s. 368. *Id. Lud.* b. i. s. 278. *Id. Suck.* 2ter th. s. 385. *Id. Bert.* s. 475. *Id. Mohs,* b. iii. s. 687. *Id. Leonhard,* Tabel. s. 79.—Antimoine sulphuré pure lamelleux, *Brong.* t. ii. p. 127.—Blättriches Grauspiesglaserz, *Karsten,* Tabel. s. 72. *Id. Haus.* s. 75.

External Characters.

The colour is the same as that of the preceding subspecies.

It occurs massive and disseminated.

Internally, it alternates from shining to splendent, and the lustre is metallic.

The

The fracture is foliated, which sometimes passes into broad radiated. It appears to have a single cleavage.

The fragments are indeterminate angular, and not particularly sharp-edged.

It occurs in coarse, small, fine, and usually longish granular distinct concretions.

It is soft.

It is not particularly brittle.

It is easily frangible.

It is heavy.

Specific gravity, 4.368, *Kirwan.*

Third Kind.

Radiated Grey Antimony-Ore.

Strahliches Grauspiesglaserz, *Werner.*

Id. Werner, Pabst. b. i. s. 198. *Id. Wid.* s. 914.—Striated sulphurated Antimony, *Kirw.* vol. ii. p. 249.—Strahliches Grauspiesglaserz, *Emm.* b. ii. s. 374.—L'Antimoine gris rayonné, *Broch* t. ii. p. 374.—Strahliches Grauspiesglaserz, *Reuss,* b. iv. s. 370. *Id. Lud.* b. ii. s. 279. *Id. Mohs,* b. iii. s. 690. *Id. Leonhard,* Tabel. s. 79.—Antimoine sulphuré pure rayonné, *Brong.* t. ii. p. 127.—Strahliges Graubraunsteinerz, *Karsten,* Tabel. s. 72. *Id. Haus.* s. 75.

External Characters.

The colour is light lead-grey, and it is sometimes tarnished with an azure-blue colour, or it exhibits the colours of tempered steel, or it is pavonine.

It occurs massive, disseminated, and crystallised in the following figures:

1. Oblique

1. Oblique four-sided prism, rather acutely acuminated with four planes, which are set on the lateral planes, fig. 255. Sometimes the obtuse lateral edges of the prism are truncated, sometimes bevelled, or even rounded off, so that the prism appears reed-shaped.
2. Oblique four-sided prism, flatly acuminated with four planes, which are set on the lateral planes, Fig. 256.
3. Oblique four-sided prism, rather acutely acuminated with four planes, which are set on the lateral planes; and this acumination flatly acuminated with four planes, which are set on the planes of the first acumination, Fig. 257.
4. Oblique four-sided prism, rather acutely acuminated with four planes, which are set on the lateral planes; and the angles formed by the meeting of the acuminating and lateral planes bevelled, Fig. 258.
5. Broad six-sided prism, rather acutely acuminated on the extremities with four planes, which are set on the narrow lateral planes, Fig 259.
6. Broad six-sided prism, flatly acuminated on both extremities with four planes, which are set on the narrow lateral planes, Fig. 260.
7. In acicular, and sometimes in capillary crystals.

The crystals usually intersect one another, or are scopiformly aggregated. Their surface is strongly longitudinally streaked, and usually shining.

Internally, it alternates from splendent to glistening, and the lustre is metallic.

The fracture is very broad, broad, or narrow, straight, and sometimes scopiform and stellular diverging radiated;

ted; occasionally promiscuous radiated. The very narrow radiated variety passes into fibrous, and the very broad variety into foliated.

The fragments are usually indeterminate angular, and not particularly sharp-edged; sometimes also splintery.

It sometimes occurs in thin and imperfect columnar concretions, and sometimes in coarse and small longish granular concretions.

It is soft.

It is rather brittle.

It is easily frangible.

It is heavy.

Specific gravity, 4.200, *Bergman*; 4.229, *Gellert*; 4.1327 to 4.5165, *Brisson*; 4.440, *Kirwan*.

Chemical Characters.

It melts by the mere flame of a candle; it is almost entirely dissipated before the blowpipe, in the form of a white vapour, with a sulphureous odour.

Constituent Parts.

Antimony,	74	75
Sulphur,	26	25
	100	100
	Bergman, Chem. Opusc. t. ii. p. 167.	*Proust*.

Geognostic Situations of the Foliated and Radiated Kinds.

These minerals occur in veins, and it is said sometimes also in beds, in primitive and transition mountains. The veins sometimes contain no other minerals besides antimony

timony and quartz; in other instances these minerals are associated with gold, or ores of silver, and more frequently with galena or lead-glance, grey copper-ore, iron-pyrites, arsenical-pyrites, blende, heavy-spar, and brown-spar.

Geographic Situation.

Europe.—It occurs at Glendinning in Dumfries-shire, in veins that traverse transition rocks, accompanied with fine granular brown blende, iron-pyrites, quartz, and calcareous-spar *; in Cornwall at St Stephen's, Padstow, and Huel Boys in Endellion; at Narverud and Hillebeck near Eger in Norway, along with common garnet; in veins in transition rocks in the Hartz; in veins that traverse gneiss in Massiac and Langle in Auvergne: at Braunsdorf in Saxony; in Bohemia, Silesia, Swabia, Salzburg, Tuscany, Sardinia, Corsica, Sicily, and Spain; also at Offenbanya in Hungary, in veins with galena, grey copper-ore, iron-pyrites, and brown-blende, in foliated granular limestone; at Felsobania in Transylvania, associated with grey copper-ore, plumose antimony, red orpiment, red antimony-ore, rose-red brown-spar, calcareous-spar, and quartz.

America.—It is found at Catorce and Los Pozuelos near Cuencamé in Mexico, and also in Louisiana †.

Second

* Jameson's Mineralogy of Dumfriesshire, p. 74.

† Bruce's American Mineralogical Journal, p. 125.

Second Subspecies.

Plumose Grey Antimony-Ore.

Federerz, *Werner.*

Minera Antimonii plumosa, *Wall.* t. ii. p. 197.—Federerz, *Wern.* Pabst. b. i. s. 201. *Id. Wid.* s. 916 —Plumose Antimonial ore, *Kirw.* vol. ii. p. 250.—Federerz, *Emm.* b. ii. s. 474.—L'Antimoine en plumes, *Broch.* t. ii. p 377.—Haarförmiges Grauspiesglanzerz, *Reuss,* b. iv. s. 375.—Federerz, *Lud* b. i. s. 280. *Id. Suck.* 2ter th. s. 339. *Id. Bert.* s. 478. *Id. Mohs,* b. iii. s. 702.—Haarförmiges Grauspiesglanzerz, *Leonhard,* Tabel. s. 79.—Antimoine sulphuré capillaire, *Brong.* t. ii. p. 127.—Haarförmiges Grauspiesglanzerz, *Karsten,* Tabel. s. 72.—Federerz, *Haus.* s. 75.—Antimome sulphuré capillaire, *Hauy,* Tabl. p. 113.—Plumose Grey Antimony, *Aikin,* p. 57.

External Characters.

Its colour is intermediate between blackish lead-grey and steel-grey. The lighter coloured varieties have sometimes a tempered steel coloured tarnish.

It occurs sometimes massive; most commonly, however, in thin capillary crystals, which are almost always promiscuously aggregated, and very much grown together.

Externally the crystals are glistening.

Internally it is glimmering, and the lustre is semi-metallic or metallic.

The fracture is delicate and promiscuous fibrous.

The fragments are indeterminate angular, and blunt-edged.

It

It is opaque.
It is very soft, passing into soft.
It is rather brittle.
It is easily frangible.

Chemical Characters.

Before the blowpipe, it melts into a black slag, after giving out a vapour, which, when condensed, appears in form of a white and yellow powder.

Constituent Parts.

According to Bergman, it is a compound of Antimony, Sulphur, Arsenic, Iron, and Silver.

Geognostic Situation.

It occurs most frequently in veins in primitive rocks, that contain ores of silver, particularly white silver-ore; also in antimony veins. It is usually accompanied with argentiferous arsenical-pyrites, native tellurium, and the ores, already mentioned as accompanying the other ores of this metal. A newer formation is met with in transition rocks, where it is associated with galena or lead-glance, grey copper-ore, sparry ironstone, and fluor-spar.

Geographic Situation.

Europe.—It occcurs at Andreasberg and Clausthal in the Hartz; Freyberg and Braunsdorf in the kingdom of Saxony; Rathhausberg in Gastein, and Schwarzleogang in Salzburg; Schemnitz in Hungary; Nagyag and Felsobanya in Transylvania.

America.—Mexico.

 3. Nickeliferous

3. Nickeliferous Grey Antimony-Ore, or Nickel-Antimonial Ore.

Antimoine sulphuré nickelifere, *Hauy.*

Id. Lucas, t. ii. p. 471. *Id. Vauquelin,* Annal. du Mus. t. xix. p. 52. Spiessglanzkies, *Haus.* Handb. b. i. s. 192.—Nickel Antimonerz, *John,* in Schweigger's Journal for 1814.

External Characters.

Its colour is steel grey, which passes on the one side into lead-grey, on the other into tin-white, and is tarnished with tempered-steel colours.

It occurs massive and disseminated; it is shining in the principal fracture, and glistening in the cross fracture.

The principal fracture is foliated, with a double cleavage. The fragments are cubical.

It is harder than grey antimony-ore.

It is brittle.

It is easily frangible.

Specific gravity, 5 65.

Chemical Characters.

On exposure to the blowpipe, it melts, emits a white vapour, having the smell of arsenic, part of which remains attached to the charcoal, to which it communicates a yellow colour. In proportion as the vapours are exhaled, the fusibility is diminished, until the remaining portion becomes infusible: the infusible portion appears as a small white easily frangible button, which proves that at least two metals enter into the composition of this ore.

It is partly soluble in nitric acid, to which it communicates a green colour, and deposites a white powder. It is almost entirely dissolved in muriatic acid. *Vauquelin.*

Constituent

Constituent Parts.

It is composed of antimony, nickel, arsenic, iron, lead, and sulphur: of these, the antimony is the most abundant, forming about the half of the ore; the next in quantity is the nickel; arsenic the third; sulphur the fourth; iron the fifth; and lead, but in very small quantity. It is probable that the antimony and sulphur form a particular combination, the arsenic and nickel another, which is mechanically mixed with the first, and that the lead and iron are combined with the sulphur. *Vauquelin.*

According to John, it contains antimony with arsenic, 61.68. Nickel, 23.33. Sulphur, 14.16. Silica with silver and lead, 0.83, Trace of iron.

Geognostic and Geographic Situations.

It occurs in veins near Freussberg, in the county of Sayn-Altenkirchen, in the principality of Nassau, along with sparry iron-ore, galena or lead-glance, and copper-pyrites.

4. Red Antimony-Ore.

Roth-spiesglaserz, *Werner.*

This species is divided into two subspecies, viz. Common Red Antimony-ore, and Tinder-ore.

First Subspecies.

Common Red Antimony-Ore.

Gemeines Rothspiesglaserz, *Werner.*

Minera Antimonii colorata, *Wall.* t. ii. p. 199.—Roth-spiesglaserz, *Wern.* Pabst. b. i. s. 202. *Id. Wid.* s. 918.—Red Antimonial ore, *Kirw.* vol. ii. p 250.—Rothspiesglaserz, *Emm.* b. ii. s. 477.—Antimoine rougeâtre, mineralisé par le Soufre, *Lam* t. i. p. 343.—Antimoine hydro-sulphuré, *Hauy,* t. iv. p. 276.—L'Antimoine rouge, *Broch.* t. ii. p. 379.—Rothspiesglanzerz, *Reuss,* b. iv. s. 379. *Id. Lud.* b. i. s. 281. *Id. Suck.* 2ter th. s. 390. *Id. Bert.* s. 480. *Id. Mohs,* b. iii. s. 706.—Antimoine hydro-sulphuré, *Lucas,* p. 174.—Rothspiesglanzerz, *Leonhard,* Tabel. s. 79.—Antimoine hydro-sulphuré capillaire, *Brong.* t. ii. p. 129.—Rothspiesglanzerz, *Karsten,* Tabel. s. 72. *Id. Haus.* s. 77.—Hydro-sulphuret of Antimony, *Kid,* vol. ii. p. 202.—Antimoine hydro-sulphuré, *Hauy,* Tabl. p. 113.—Red Antimony, *Aikin,* p. 57.

External Characters.

Its colour is cherry-red, and it has sometimes a tempered-steel coloured tarnish.

It occurs massive, disseminated, in membranes, but most frequently in delicate capillary crystals, which are sometimes promiscuous, sometimes scopiformly aggregated.

Externally and internally it is shining, and the lustre is nearly adamantine.

The fracture is delicate, scopiform, and stellular fibrous, which passes into narrow radiated.

The

The fragments are wedge-shaped and splintery.

It occurs in coarse, small, and longish granular distinct concretions, which sometimes approach the wedge-shaped columnar.

It is opaque.

The colour is not changed in the streak.

It is very soft, passing into friable.

It is not very brittle.

It is easily frangible.

Specific gravity, 4.090, *Klaproth.*

Chemical Characters.

It melts and evaporates before the blowpipe, giving out a sulphureous odour.

Constituent Parts.

From the mine called Neue Hoffnung Gottes at Braüns-dorf:

Antimony, - -	67.50
Oxygen, - - -	10.80
Sulphur, - - -	19.70
	98.00

Klaproth, Beit. b. iii. s. 182.

Geognostic Situation.

This rare mineral occurs in veins, in primitive rocks, generally along with native antimony, grey antimony-ore, and ores of arsenic, with quartz, iron-pyrites, and calcareous-spar,.

Geographic Situation.

It occurs at Braunsdorf in Saxony; Allemont in France;

France; in Tuscany; at Malaczka in Hungary; and and Felsobanya in Transylvania.

Second Subspecies.

Tinder-Ore.

Zundererz, *Werner.*

Zundererz, *Reuss,* b. iii s. 382. *Id. Leonhard,* Tabel. s. 79. *Id. Karsten,* Tabel. s. 72. *Id. Haus.* s. 77.

External Characters.

Its colours are muddy cherry-red, reddish brown, and sometimes it inclines to blackish-brown, and dark lead-grey.

It occurs in very delicate felt-like flexible leaves, which have sometimes a promiscuous fibrous texture.

Its lustre is feebly glimmering.

It is opaque.

It is friable.

It becomes shining when rubbed.

It soils strongly.

It is supernatant.

Chemical Characters.

Before the blowpipe, the antimony, lead and sulphur evaporate, and colour the charcoal white and yellow; the residuum melts into a black magnetic slag.

Constituent Parts.

According to an analysis by Link, this ore appears to contain in 100 parts, 33 of Oxide of Antimony, 40 of Oxide

Oxide of Iron, 16 of Lead, and 4 of Sulphur. It also contains a portion of Silver.—Vid. N. Journ. de Chem. v. s. 461.

Geognostic and Geographic Situations.

It occurs principally in the mines named Carolina and Dorothea at Clausthal, where it is associated with crystals of quartz, calcareous-spar, and lead-glance.

Use.

It is worked as an ore of silver.

5. White Antimony-Ore.

Weiss-spiesglaserz, *Werner.*

Id. Wern. Pabst. b. ii. s. 203. *Id Wid.* s. 920.—Muriated Antimony, *Kirw.* vol. ii. p. 251.—Muriate d'Antimoine, *De Born,* t. ii. p. 147.—Weiss-spiesglaserz, *Emm.* b. ii. s. 480.—Antimoine muriatique, *Lam.* t. i. p. 348.—Antimoine oxydé, *Hauy,* t. iv. p. 273.—Antimoine blanc, *Broch.* t. ii. p. 381.—Weiss-spiesglanzerz, *Reuss,* b. iv. s. 382. *Id. Lud.* b. i. s. 281. *Id. Suck.* 2ter th. s. 392. *Id. Bert.* s. 470. *Id. Mohs,* b. iii. s. 710.—Antimoine oxydé, *Lucas,* p. 173.—Weiss-spiesglanzerz, *Leonhard,* Tabel. s. 70.—Antimoine oxydé, *Brong.* t. ii. p. 128. *Id. Brard,* p. 374.—Weiss-spiesglanz, *Karsten,* Tabel. s. 72.—Antimoine oxydé, *Hauy,* Tabl. p. 113.—Spiessglanzweiss, *Haus.* Handb. b. i. s. 341.—White Antimony, *Aikin,* p. 58.

External Characters.

Its colours are snow-white, greyish-white, light ash-grey, and yellowish-white, which latter colour is the most common.

It seldom occurs massive, more frequently disseminated, in membranes; and crystallised in the following figures:

1. Rectangular four-sided prism.
2. Rectangular four-sided table.
3. Six sided prism.
4. Acicular and capillary crystals.

The tables are small and very small, usually adhering by their lateral planes, and sometimes, although seldom, manipularly aggregated, and often intersect each other, in such a manner as to form a cellular group.

The crystals are sometimes smooth, sometimes feebly longitudinally streaked, and splendent.

Internally it is shining, and the lustre is intermediate between pearly and adamantine.

The fracture is foliated, with a single cleavage; sometimes also narrow, scopiform, and stellular radiated.

The fragments are indeterminate angular, and not particularly sharp-edged.

It occurs in coarse and small granular distinct concretions; and also in thin columnar concretions.

It is translucent.

It is soft.

It is rather sectile.

It is heavy.

Chemical Characters.

Before the blowpipe it melts very easily, and is volatilised in the form of a white vapour.

Constituent

Constituent Parts.

	Allemont.
Oxide of Antimony,	86
Oxides of Antimony and Iron,	3
Silica, - - - - -	8
	98

Vauquelin, Hauy, t. iv. p. 274.

Geognostic and Geographic Situations.

It occurs in veins in primitive rocks, and is usually accompanied with the other ores of antimony.

At Prizbram in Bohemia, it occurs along with crystallised galena or lead-glance; and at Allemont, with native antimony, and grey and red antimony-ores. It has also been found at Malaxa in Hungary.

6. Antimony-Ochre.

Spiesglanzocker, *Werner*.

Spiessglanzocher, *Reuss*, b. iv. s. 388. *Id. Lud.* b. i. s. 282. *Id. Suck.* 2ter th. s. 394. *Id. Bert.* s. 478. *Id. Mohs*, b. iii. s. 713. *Id. Leonhard*, Tabel. s. 79. *Id. Karsten*, Tabel. s. 72.— Antimoine oxydé terreux, *Hauy*, Tabl. p. 113.—Spiessglanzocher, *Haus.* Handb. b. i. s. 339.—Antimonial Ochre, *Aikin*, p. 58.

External Characters.

Its colour is straw-yellow, of different degrees of intensity, which inclines on the one side into yellowish-grey, on the other into yellowish-brown.

It

It occurs massive, disseminated, spongy, and usually incrusting crystals of grey antimony-ore.

It is dull.

The fracture is earthy, and sometimes inclines to radiated.

It is soft, passing into very soft.

It is rather sectile.

It is heavy.

Chemical Characters.

Before the blowpipe, on charcoal, it becomes white, and evaporates without melting. With borax, it intumesces, and is partly reduced to the metallic state.

Geognostic and Geographic Situations.

It occurs always in veins, and accompanied with grey antimony-ore, and sometimes with red antimony-ore.

It is found at Dublowitz, near Saltschaw in Bohemia; Telkebanya in Hungary; Toplitz in Transylvania; Braunsdorf, in the kingdom of Saxony; on the Sonnenberg, near Mittersill in Salzburg; and in Siberia.

ORDER XV.

ORDER XV.—MOLYBDENA.

This Order contains but two species, viz. Molybdena, and Molybdena-ochre.

1. Molybdena, or Sulphuret of Molybdena.

Wasserblei, *Werner*.

Ferrum Molybdæna pura membranacea nitens, (in part), *Wall.* t. ii. p. 249.—Wasserblei, Molybdæna, *Scheele*, i. d. Abhand. d. Schwed. Acad. 1778, s. 238.—Wasserblei, *Werner*, Pabst. b. i. s. 221. *Id. Wid.* s. 962.—Molybdena, *Kirw.* vol. ii. p. 322.—Sulphure de Molybdene, *De Born*, t. ii. p. 119.—Wasserbley, *Emm.* b. ii. s. 541.—Molybdene sulphuré, *Lam.* t. i. p. 397. *Id. Hauy*, t. iv. p. 289.—Le Molybdene sulphuré, *Broch.* t. ii. p. 432.—Wasserblei, *Reuss*, b. iv. s. 478. *Id. Lud.* b. i. s. 295.—Molybdankies, *Suck.* 2ter th s. 437.—Wasserblei, *Bert.* s. 499. *Id. Mohs*, b. iii. s. 588.—Molybdene sulphuré, *Lucas*, p. 179.—Wasserblei, *Leonhard*, Tabel. s. 80.—Molybdene sulphuré, *Brong.* t. ii. p. 92. *Id. Brard*, p. 381.—Molybdän, *Karsten*, Tabel. s. 70.—Molybdänkies, *Haus.* s. 76.—Molybdena, *Kid*, vol. ii. p. 216.—Molybdene sulphuré, *Hauy*, Tabl. p. 114.—Molybdena, *Aikin*, p. 62.

External Characters.

Its colour is fresh lead grey.

It occurs usually massive, disseminated, in plates, and sometimes crystallised, in the following figures:

1. Regular

1. Regular six-sided table, fig. 261 *.
2. Very short six-sided prism flatly acuminated on both extremities, with six planes, which are set on the lateral planes, fig. 262 †.

The crystals are small and middle-sized, and always imbedded, or in druses.

Internally it is splendent, sometimes passing into shining, and the lustre is metallic.

The fracture is perfect foliated, with a single cleavage, which is parallel with the lateral planes of the table. almost always curved, and sometimes floriform foliated.

The fragments are indeterminate angular, and blunt-edged.

It occurs in large and coarse granular distinct concretions.

It is opaque.

It writes with a bluish-grey streak on paper, but with a greenish-grey streak on porcelain.

It retains its lustre in the streak.

It soils slightly.

It is very soft

It is easily frangible.

It splits easily.

In thin leaves it is flexible, but not elastic.

It is sectile, approaching to malleable,

It feels greasy.

It is heavy.

Specific gravity, 4.7385, *Brisson;* 4.048, *Kirwan;* 4.569, *Karsten:* 4.667, *Schumacher.*

Chemical

---

* Molybdene sulphuré prismatique, Hauy.

† Moybdene sulphuré trihexaedre, Hauy.

Chemical Characters.

It gives out a sulphureous odour before the blowpipe; and when urged by the utmost force of the heat, it gives out white vapour, and a pale-blue flame; it is soluble, with violent effervescence, in carbonate of soda.

Constituent Parts.

Molybdena,	-	60
Sulphur,	-	40
		100

Bucholz in Gehlen's Journ de Chem. u. Phys. b. iv. s. 603.

Geognostic Situation.

It occurs disseminated in granite, gneiss, mica-slate, and chlorite-slate, or in veins traversing these rocks, in which it is associated with wolfram, tungsten, tinstone, magnetic ironstone, arsenical pyrites, fluor-spar, topaz, quartz and heavy-spar.

Geographic Situation.

Europe.—It occurs imbedded in chlorite-slate along with actynolite in Glenelg in Inverness shire; in six-sided tables, in quartz in granite, on the mountain of Corybuy, at the head of Loch Creran; in granite at Shap in Westmoreland; also at Coldbeck in Cumberland, and Huel Gorland in Cornwall. In Norway and Sweden, it is found imbedded in granite and gneiss, in Bohemia and Saxony in veins accompanied with tinstone; in the *snow-pits* in the Riesengebirge in Silesia, disseminated in granite; in the country of Glatz, imbedded in gneiss and mica-slate; in porphyritic syenite near the copper-mines of Chessy in the

the department of the Rhone in France; and in granitic rocks in the Alps and the Vosges.

Asia.—Siberia.

America.—Greenland.

Observations.

This mineral has frequently been confounded with *graphite*; but the following characters sufficiently distinguish them from one another: The colour of graphite is steel-grey, inclining more or less to iron-black; whereas that of molybdena is lead grey: if both minerals are rubbed on a piece of white porcelain, it will be seen that the streak made by the graphite is of the same colour with the substance by which it was made; while that made by the molybdena is greenish grey: and graphite soils strongly, but molybdena only slightly.

2. Molybdena-Ochre.

Molybdänocher, *Karsten.*

Molybdänocher, *Karsten,* Tabel. s. 70. *Id. Haus.* Handb. b. i. s. 336.

External Characters.

Its colour is sulphur yellow, which passes on the one side into straw yellow and orange yellow, and on the other into siskin green.

It occurs disseminated, and incrusting molybdena.

It is friable.

It is dull.

Geographic Situation.

It is found investing and intermixed with molybdena, in the granite of Corybuy at Loch Creran; and also at Nummedalen in Norway.

ORDER XVI.

ORDER XVI.—COBALT.

This Order contains eight species, viz. Tin-white Cobalt-ore, Grey Cobalt-ore, Silver-white Cobalt-ore, Cobalt-pyrites, Black Cobalt ochre, Brown Cobalt-ochre, Yellow Cobalt-ochre, and Red Cobalt-ochre.

1. Tin-white Cobalt-Ore.

Weisser Speisskobald, *Werner.*

Weisserspeisskobalt, *Lud.* b. i. s. 283. *Id. Mohs,* b. iii. s. 647. *Id. Leonhard,* Tabel. s. 76. *Id. Karsten,* Tabel. s. 72.—Speisskobalt, *Haus.* s. 75. (in part).—Cobalt arsenical, var. Blanc argentine, *Hauy,* Tabl. p. 107.—Arsenical Cobalt, *Aikin,* p. 59. (in part.)

External Characters.

On the fresh fracture its colour is tin-white; but it tarnishes grey or blackish, rarely reddish or yellowish, and sometimes variegated.

It occurs massive, disseminated, in membranes, specular, globular, botryoidal, thin cylindrical, frequently reticulated, fruticose; and crystallised in the following figures:

1. Cube, in which the faces are sometimes spherical convex; and the edges and angles are sometimes more or less deeply truncated. This figure is frequently cracked or burst in determinate directions.

2. Flat

2. Flat octahedron, which is either perfect, or is more or less deeply truncated on all the angles.

The external surface of the crystal is smooth and splendent, and the lustre is metallic.

Internally it is glistening, and the lustre is metallic.

The fracture is coarse and fine grained uneven, seldom scopiform and stellular fibrous, and radiated.

The fragments are indeterminate angular, rather sharp-edged; in the varieties with a radiated fracture, wedge-shaped.

It occurs in coarse, small and fine granular distinct concretions; sometimes also in lamellar fortification-wise bent concretions.

It is semihard in a high degree.

It becomes shining in the streak.

It is brittle.

It is easily frangible.

It is uncommonly heavy.

Specific gravity, 7.379,—7.751, *Karsten;* 6.2173, *Kopp.*

Chemical Characters.

Before the blowpipe it gives out a copious arsenical vapour on the first impression of the heat; it melts only partially, and that with great difficulty, and is not attractable by the magnet; on the addition of borax it immediately melts into a grey metallic globule, colouring the borax of a deep blue.

It is considered to be a compound of cobalt, iron and arsenic.

Geognostic Situation.

It occurs in veins in granite, gneiss, mica-slate, and clay-slate; seldomer in transition clay-slate and limestone;

stone; more frequently in flœtz rocks, and in beds in primitive, and old flœtz rocks, as red sandstone and bituminous marl-slate. In primitive rocks, it is associated with copper-nickel, nickel ochre, native bismuth, bismuth-glance, black cobalt-ochre, ores of silver, and fluor-spar, calcareous-spar, brown-spar, lamellar heavy spar and quartz; in transition rocks it is accompanied with copper-pyrites and quartz; in flœtz rocks with the various ores of cobalt, different ores of copper, calcareous-spar and lamellar heavy-spar.

Geographic Situation.

It occurs at Huel Sparnon, Redruth and Dolcoath in Cornwall; at Schneeberg, Annaberg and Johanngeorgenstadt in Saxony; Joachimsthall in Bohemia; Thuringia; Hessia; Stiria; Crawitza in the Bannat.

2. Grey Cobalt-Ore.

Grauer Speiskobold, *Werner.*

Grauer Speisskobalt, *Lud* b. i. s. 284. *Id. Mohs,* b. iii. s. 644. *Id. Leonhard,* Tabel. s. 76. *Id. Karsten,* Tabel. s. 72.—Speisskobalt, *Haus.* s. 75. (in part) —Cobalt arsenical, var. gris-noirâtre subluisant, *Hauy,* Tabl. p. 107.—Arsenical Cobalt, *Aikin,* p. 59. (in part).

External Characters.

On the fresh fracture its colour is light steel-grey, which sometimes inclines to tin-white; but by exposure it gradually acquires a tempered-steel and greyish black tarnish.

 It

It occurs massive, disseminated, seldom small reniform, and small botryoidal, tubiform, and specular.

Externally it is generally dull and tarnished.

Internally strongly glimmering or glistening, and the lustre is metallic; but the specular variety is splendent.

The fracture is even, which sometimes passes into flat and large conchoidal, sometimes into fine-grained uneven.

The fragments are indeterminate angular, and pretty sharp-edged

It sometimes occurs in thick and curved lamellar distinct concretions; also in granular concretions.

It becomes shining in the streak, without change of colour.

It is semihard, but in a lower degree than the preceding species.

It is brittle.

It is easily frangible, and when struck, emits an arsenical odour.

It is heavy.

Specific gravity, 5.503, *Gellert*, Schneeberg; 5.309, *Kirwan*, Schneeberg; 5.511, Annaberg; 5.4306, *Kopp*. Riechelsdorf, 5.021, *Karsten*

Constituent Parts.

According to Klaproth, it contains 19.60 parts of Cobalt, with Iron and Arsenic. Vid. *Klaproth* in d. Beot. u. Entd. b. i. s. 182.

Geognostic Situation.

It occurs in veins in granite, gneiss, mica-slate and clay slate, associated with the tin-white cobalt-ore; but it is neither so frequent nor abundant.

Geographic

Geographic Situation.

It is found at Herland, and Dolcoath in Cornwall; Norway; Annaberg, Schneeberg, and Freyberg in the electorate of Saxony; Joachimsthall in Bohemia; Krobsdorf, Hindorf, and Kupferberg in Silesia: Wittichen in Swabia; Nassau; Salzburg; Allemont in France; Stiria; and Hungary.

3. Silver-white Cobalt-Ore, or Cobalt-glance.

Glanz-Kobold, *Werner*.

Minera Cobalti tessularis, *Wall.* Syst. Min. t. ii. p. 176.—Minera Cobalti crystallisata, *Waller.* Syst. Min. t. ii. p. 179. (in part).—Bright white Cobalt-ore, *Kirwan.* vol. ii. p. 273.—Le Cobalt eclatant, *Broch.* t. ii. p. 390.—Cobalt gris, *Hauy,* t. iv. p. 204.—Kobaltglanz, *Lud.* b. i. s. 284. *Id. Suck.* 2ter th. s. 400. *Id. Bert.* s. 482. *Id. Mohs,* b. iii. s. 639. *Id. Leonhard,* Tabel. s. 76.—Cobalt gris, *Lucas,* p 160. *Id. Brong.* t. ii. p. 116. *Id. Brard,* p. 354.—Glanzkobalt, *Karsten,* Tabel. s. 72.—Cobalt-glanz, *Haus.* s. 73.—Cobalt gris, *Hauy,* Tabl. p. 107.—Bright-white Cobalt, *Aikin,* p. 58.

External Characters.

Its colour is silver-white, slightly inclining to reddish. Sometimes it is tarnished yellowish or columbine.

It occurs commonly massive and disseminated; also reticulated; and crystallised in the following figures:

1. Cube, which is either perfect, or truncated on the angles.
2. Octahedron.

3. Cube, in which all the edges are truncated, and in such a manner, that each face supports two opposite truncating planes: It is the middle figure, between the cube and the pentagonal dodecahedron.
4. Pentagonal dodecahedron.
5. Middle figure between the dodecahedron and the icosahedron.
6. Icosahedron.

The surface is smooth or streaked.

Externally it is splendent.

Internally it is intermediate between shining and glistening, and the lustre is metallic.

The principal fracture is foliated; the cross fracture small conchoidal.

The fragments are indeterminate angular, and rather blunt-edged.

It sometimes occurs in granular concretions.

It is semihard in a low degree.

Its streak is of a grey colour.

It is not very brittle.

Specific gravity, 5.320 to 6.466, *Karsten*; 6.198, *Lowry*; 4.9411, *Kopp*.

Chemical Character.

Before the blowpipe it gives out an arsenical odour; and after being roasted, colours glass of borax smalt-blue.

Constituent

Constituent Parts.

		Tunnaberg.		Tunnaberg.
Cobalt,	-	44.00		36.66
Arsenic,	-	55.00		49.00
Sulphur,	-	0.50	Iron,	6.50
				5.66
		99.50		
		Klaproth, Beit. b. ii. s. 307.		*Tassaert* in Annal. d. Chim. xxviii. p. 82.

Geognostic Situation.

It occurs in primitive rocks, particularly in a quartzose mica-slate, and in gneiss, in imbedded masses, intermixed with the rock at their line of junction; also disseminated, and in imbedded crystals. It is associated with copper-pyrites, iron-pyrites, and red cobalt ochre.

Geographic Situation.

It occurs principally at Skutternd in the parish of Modum in Norway; at Tunnaberg in Sweden; and in small quantity at Queerbach in Silesia.

Uses.

This is the most common species of cobalt, and is that from which the cobalt of commerce is principally obtained. When roasted and melted in certain proportions with pounded quartz, it forms *small*, a compound which is highly useful in the manufacturing of porcelain and glass, and also for painting.

The other species of cobalt are employed for similar purposes.

 4. Cobalt-

4. Cobalt-Pyrites.

Cobaltkies, *Hausmann.*

Cobaltum pyriticosum, (ferro sulphurato mineralisatum), *Lin.* Syst. Nat. t. iii. p. 129.—Minera Cobalti sulphurea, *Waller.* Syst. Min. t. ii. p. 178.—Kobolt, med jern och suafelsyra, *Brandt,* in K. vet. Acad. Handl. 1746, p. 119.—Kobolt, med. forsvafladt jern, *Cronstedt,* Mineralogie, § 250.—Svafvelbunden Kobolt, *Hisinger,* in Afhandl. i Fysik, Kemi och Min iii. 316.—Kobaltkies, *Haus.* Entw. s 73. *Id. Haus.* Handb. b. i. s. 158.—Cobalt sulphuré, *Lucas,* t. ii. p. 516.

External Characters.

Its colour is pale steel-grey ; which by tarnishing approaches to copper-red.

It occurs massive, disseminated, and it is said also crystallised in a cubical form.

Its lustre is shining and metallic.

Its fracture is uneven, passing into imperfect conchoidal, and sometimes concealed foliated.

It is semihard.

Chemical Characters.

Before the blowpipe it emits a sulphureous odour, and after being roasted colours glass of borax smalt-blue.

Constituent Parts.

Cobalt,	-	-	43.20
Sulphur,	-	-	38.50
Copper,	-	-	14.40
Iron,	-	-	3.53

Hisinger in Afhandl. i Physik. Kemi och Min. iii. 321.

Geognostic

Geognostic and Geographic Situations.

It occurs along with copper-pyrites, and common actynolite, in a bed in gneiss, at Nya Bastnas at Riddarhyttan, in Sweden.

5. Black Cobalt-Ochre.

Schwarz Erdkobold, *Werner.*

This species contains two subspecies, viz. Earthy Black Cobalt-ochre, and Indurated Black Cobalt-ochre.

First Subspecies.

Earthy Black Cobalt-Ochre.

Schwarzer Kobold Mulm, *Werner.*

Id. Wern. Pabst. b. i. s. 205.—Zerreiblicher schwarzer Erdkobold, *Wid.* s. 933.—Loose Black Cobalt-ochre, *Kirw.* vol. ii. p. 275.—Schwarzer Kobold Mulm, *Emm.* b. ii. s. 498.—Le Cobalt terreux noire friable, *Broch.* t ii. p. 397.—Cobalt oxydé noire terreux, *Hauy,* t. iv. p. 215.—Zerreiblicher schwarzer Erdkobold, *Reuss,* b. iv. s. 411.—Schwarzer Kobaltmulm, *Lud.* b. i. s. 285.—Zerreiblicher schwarzer Erdkobold, *Leonhard,* Tabel. s. 76.—Cobalt oxydé terreux, *Brong.* t. ii. p. 118. —Lockere Kobaltschwärze, *Haus.* Handb. b. i. s. 332.

External Characters.

Its colour is intermediate between brownish and bluish-black; the bluish-black is essential, the brownish-black it derives from iron.

It is composed of dull dusty particles, which soil very little.

It is usually cohering, sometimes also loose.

The streak is shining.

It is meagre to the feel.

It is light.

Chemical Characters.

Before the blowpipe it yields a white smoke, which has an arsenical smell; it colours borax blue.

It dissolves in muriatic acid.

Second Subspecies.

Indurated Black Cobalt-Ochre.

Fester Schwarz Erdkobold, *Werner*.

Minera Cobalti scoriformis, *Wall.* Syst. Min. t. ii. p. 180.—Verhärteter schwarzer Erdkobold, *Wid.* s. 933.—Indurated Black Cobalt-ochre, *Kirw.* vol. ii. p. 275.—Verhärteter schwarzer Erdkobold, *Emm.* b. ii. s. 499.—Le Cobalt terreux noire endurci, *Broch.* t ii. p. 397.—Cobalt oxidé noire, var. 1.—3. *Hauy,* t. iv. p. 215.—Verhärteter schwarzer Erd-cobalt, *Reuss,* b. iv. s. 413. *Id. Lud.* b. i. s. 286. *Id, Mohs,* b. iii. s 665. *Id. Leonhard,* Tabel. s. 76.—Cobalt oxidé vitreux, *Brong.* t. ii. p. 118.—Fester Kobaltschwärze, *Haus.* Handb. b. i. s. 333.

External Characters.

Its colour is almost always bluish-blaek; seldom inclines to brownish-black.

It occurs massive, disseminated, as a coating small botryoidal,

botryoidal, small reniform, fruticose, corroded, and with impressions.

The surface is feebly glimmering.

Internally it is dull, or very feebly glimmering.

The fracture is fine earthy, sometimes passing into conchoidal.

The fragments are indeterminate angular, and blunt-edged.

It sometimes occurs in thin and curved lamellar distinct concretions.

The streak is shining and resinous.

It is very soft, approaching to soft.

It is sectile.

It is very easily frangible.

Specific gravity, 2.019 to 2.425 *Gellert*; 2.9287, *Kopp*.

Chemical Characters.

Before the blowpipe it gives an arsenical odour, and colours glass of borax smalt-blue.

Constituent Parts.

It is considered as black oxide of cobalt, with arsenic and oxide of iron.

Geognostic Situation.

Both subspecies usually occur together, and in the same kind of repository; but the first subspecies is the rarest. They are found sometimes in primitive mountains, but most frequently in flœtz mountains, where they are accompanied with ochry-brown ironstone, red, brown, and yellow cobalt-ochres, native-silver, several other

other ores of silver and of copper, and lamellar heavy-spar, calcareous-spar, and quartz.

Geographic Situation.

It is found at Alderly Edge, Cheshire, in red sandstone; in slate clay in the peninsula of Howth near Dublin; at Reigelsdorf in Hessia; Schneeberg, Kamsdorf, and Saalfeld in Saxony; Bohemia; Wittichen in Furstenberg, Alpirsbach in Wurtenberg in Swabia; in the Upper Palatinate; Fuyen in the Zillerthall in Salzburg; Kleinzell in Austria; Kitsbichel in the Tyrol; Allemont in France; and in the valley of Gistain in Spain.

Use.

It affords a most excellent blue colour; hence is highly valued as an ore of cobalt.

Observation.

In the principality of Nassau there is a black ore of cobalt which is an intimate mixture of black cobalt-ochre and quartz, and which when melted, changes its colour, becoming smalt blue.

6. Brown

6. Brown Cobalt-Ochre.

Brauner Erdkobold, *Werner.*

Ochra Cobalti lutea, *Waller.* Syst. Min. t. ii. p. 183.—Brauner Erdkobold, *Wern.* Pabst. b. i. s. 206. *Id. Wid.* s. 935.—Brown Cobalt-ochre, *Kirw.* vol. ii. p. 276.—Brauner Erdkobold, *Emm.* b. ii. s. 503.—Le Cobalt terreux brun, *Broch.* t. ii. p. 400.—Brauner Erdkobalt, *Reuss,* b. iv. s. 415. *Id. Lud.* b. i. s. 287. *Id. Suck.* 2ter th. s 406 *Id. Bert.* s. 488. *Id. Mohs,* b. iii s. 667. *Id. Leonhard,* Tabel. s. 76.—Cobalt oxidé brun, *Brong.* t. ii. p. 118.—Erdkobalt, *Haus.* Handb. b. i. s. 334. (in part).

External Characters.

Its colour is liver-brown, which sometimes passes into yellowish-brown, sometimes into brownish-black, and into yellowish grey and ash-grey.

The yellowish-grey makes the transition into the yellow cobalt-ochre, and the black into the black cobalt-ochre.

It occurs massive, and disseminated.

Internally it is dull.

The fracture is fine earthy.

The fragments are indeterminate angular, and blunt-edged.

The streak is shining and resinous.

It is very soft.

It is sectile.

It is very easily frangible.

Chemical

Chemical Characters.

Before the blowpipe it emits an arsenical odour, and communicates a blue colour to borax.

Constituent Parts.

It is considered to be a compound of Brown Oxide of Cobalt, Arsenic, and Oxide of Iron.

Geognostic Situation.

It appears to occur principally in flœtz mountains, and is generally accompanied with red and black cobalt-ochre, ochry-brown ironstone, and lamellar heavy-spar.

Geographic Situation.

It is found at Kamsdorf and Saalfeld in Saxony; Alpirsbach in Wurtemberg; and in the valley of Gistain in Spain.

7. Yellow Cobalt-Ochre.

Gelber Erdkobold, *Werner.*

Ochra Cobalti lutea et alba, *Wall.* t. ii. p. 183.—Gelber Erdkobold, *Wid.* s. 936.—Yellow Cobalt-ochre, *Kirw.* vol. ii. p. 277.—Gelber Erdkobold, *Emm.* b. ii. s. 504.—Le Cobalt terreux jaune, *Broch.* t. ii. p. 401.—Gelber Erdkobold, *Reuss,* b. iv. s. 417. *Id. Lud.* b. i. s. 287. *Id. Suck.* 2ter th. s. 407. *Id. Bert.* s. 488. *Id. Leonhard,* Tabel. s. 76.—Cobalt oxidé jaune, *Brong.* t. ii. p. 118.—Erdkobalt, *Haus.* Handb. b. i. s. 334. (in part).

External Characters.

The colour is muddy straw-yellow, which in some varieties

rieties passes through light yellowish grey into yellowish-white.

It occurs massive, disseminated, incrusting, and less frequently corroded.

It has frequently the appearance of being burst in different directions.

Internally it is dull.

The fracture is fine earthy.

The fragments is indeterminate angular, and blunt-edged.

The streak is shining.

It is soft, passing into friable.

It is rather sectile.

It is very easily frangible.

Specific gravity, 2.677, *Kirwan*, after having absorbed water.

Chemical Characters.

It emits an arsenical odour before the blowpipe, and colours borax blue.

Geognostic Situation.

It occurs in the same geognostic situation as the preceding species, and is almost always associated with earthy red cobalt ochre, and sometimes with radiated red cobalt ochre, nickel-ochre, iron-shot copper-green, and azure copper-ore.

Geographic Situation.

It occurs at Saalfeld in Thuringia; Kupferberg in Silesia; Wittichen in Furstenberg and Alpirsbach in Wurtemberg in Swabia; and Allemont in France.

Observations.

Observations.

The preceding ochres of cobalt are characterised by their lightness, softness, sectility, and their yielding a resinous streak.

8. Red Cobalt-Ochre.

Rother Erdkobold, *Werner.*

This species contains three subspecies, viz. Earthy Red Cobalt-ochre, Radiated Red Cobalt-ochre, and Slaggy Red Cobalt-ochre.

First Subspecies.

Earthy Red Cobalt-Ochre, or Cobalt-crust.

Koboldbeschlag, *Werner.*

Ochra Cobalti rubra, *Wall.* Syst. Min. t. ii. p 181.—Koboldbeschlag, *Wid.* s. 938.—Cobalt Incrustations, *Kirw.* vol. ii. p. 279.—Kobaldbeschlag, *Emm.* b. ii s. 509.—Le Cobalt terreux rouge, pulverulent, *Broch.* t. ii. p. 405.—Cobalt arseniaté pulverulent, *Hauy,* t. iv. p. 218.—Erdiger rother Erdkobold, *Reuss,* b. iv. s. 419.—Kobaldbeschlag, *Lud.* b. i. s. 287. *Id. Mohs,* b. iii. s. 671.—Cobalt arseniaté, *Lucas,* p. 161.—Erdiger rother Erdkobold, *Leonhard,* Tabel. s. 77.—Cobalt arseniaté pulverz, *Brong.* t. ii. p. 119. *Id. Brard,* p. 357.—Gemeine Kobaltblüthe, *Karsten,* Tabel. s. 72.—Arseniate of Cobalt, *Kid,* vol. ii. p. 211.—Cobalt arseniaté pulverulent, *Hauy,* Tabl. p. 108.—Erdige Kobaltbluthe, *Haus.* Handb. b. iii. s. 1125.—Red Cobalt, *Aikin,* p. 60.

External Characters.

Its colour is peach blossom-red, of different degrees of intensity, which sometimes inclines to crimson-red, sometimes

times verges on cochineal-red, and also passes into reddish white.

It seldom occurs massy, often in velvety drusy coatings, or disseminated, and also small botryoidal and small reniform.

It is very feebly glimmering, bordering on dull.

The fracture is fine earthy.

The fragments are indeterminate angular, and blunt-edged.

It is very easily frangible.

It is very soft, bordering on friable.

It is sectile.

The streak is shining.

It does not soil.

It is not particularly heavy.

Second Subspecies.

Radiated Red Cobalt-Ochre, or Cobalt-bloom.

Koboldblüthe, *Werner.*

Flos Cobalti, *Wall.* Syst. Min. t. ii. p. 181.—Koboldblüthe, *Wern.* Pabst. b. i. s. 206. *Id. Wid.* s. 939.—Cobaltic Germinations, Flowers of Cobalt, of some, *Kirw.* vol. ii. p. 278.—Koboldblüthe, *Emm* b. ii. s. 507.—Le Fleurs de Cobalt, ou Cobalt terreux rayonné rouge, *Broch.* t. ii. p. 403.—Cobalt arseniaté aciculaire, *Hauy,* t. iv. p. 217.—Strahlicher rother Erdkobold *Reuss,* b. iv s. 420.—Kobaltblüthe, *Lud.* b. i. s. 283. *Id. Mohs,* b. iii. s. 672.—Strahliger rother Erdkobold, *Leonhard,* Tabel. s. 77.—Cobalt arseniaté aciculaire, *Brong.* t. ii. p. 119.—Strahlige Kobaltblüthe, *Karsten,* Tabel. s. 72.—Cobalt arseniaté aciculaire, *Hauy,* Tabl. p. 108.—Strahlige Kobaltbluthe, *Haus.* Handb. b. iii. s. 1125.—Red Cobalt, *Aikin,* p. 60.

External Characters.

Its colour is peach-blossom red, often also cochineal and

and crimson red, pearl grey, greenish-grey, and by decomposition it passes into cherry-red.

It occurs massive, disseminated, often in membranes, small reniform, small botryoidal, and crystallised in the following figures:

1. Rectangular four-sided prism, bevelled on the terminal planes.
2. Acute double six-sided pyramid, in which the lateral planes of the one are set on the lateral planes of the other. Sometimes two opposite planes are much larger than the others, when the pyramid appears flat or compressed.

The crystals are generally acicular or capillary, and are scopiformly or stellularly aggregated.

Externally it is shining, passing into splendent.

Internally it is shining and glistening, and the lustre is pearly.

The fracture is narrow, and sometimes stellular and scopiform radiated: extremely seldom straight radiated, and sometimes it passes into fibrous.

The fragments are splintery and wedge shaped.

Sometimes it occurs in thin columnar concretions, which are collected into coarse and small granular distinct concretions.

It is more or less translucent; sometimes translucent on the edges.

Its colour is not changed in the streak.

It is soft

It is rather sectile.

It is easily frangible.

It is light; or not particularly heavy.

Chemical

Chemical Characters.

Before the blowpipe it becomes grey, and emits an arsenical odour, and tinges borax glass blue.

Constituent Parts.

Cobalt, - -	39
Arsenic Acid, -	38
Water, - -	23
	100

Bucholz, in J. d. Min. t. 25. p. 158.

Geognostic Situation.

It occurs in veins, in primitive, transition, and flœtz rocks, along with silver-white cobalt, tin white cobalt, grey cobalt, and other ores of cobalt; also with copper-nickel, nickel-ochre, copper pyrites, grey copper-ore, azure copper ore, ironshot copper green, native bismuth, brown ironstone, galena or lead-glance, and blende; the vein-stones are heavy-spar, calcareous-spar, brown-spar, hornstone and quartz.

Geographic Situation..

It occurs in veins in flœtz rocks at Alva in Stirlingshire; in limestone of the coal formation in Linlithgowshire; formerly in small veins in sandstone of the coal formation, along with galena and blende, at Broughton in Edinburgh; in the Clifton lead mines, near Tyndrum, already described; and at Dolcoath in Cornwall. On the Continent, it is met with at Modum in Norway, Riegelsdorf in Hessia; Schneeberg, Annaberg, and Saalfeld in Saxony; Kupferberg in Silesia; Wittichen in Furstemberg, and Alpersbach in Wurtemberg; Allemont in France; and in Salzburg and Hungary.

Observations.

A mixture of red cobalt-ochre, black cobalt-ochre, with ochre of nickel and native silver, occurs in the mines of Allemont and Schemnitz in Hungary, and is known to the miners by the name of *Goose-dung ore.* It is the *minera argenti mollior diversicolor*, Waller. t. ii. p. 346.; *Mine d'argent merde d'oie*, Delisle, t. iii. p. 150.; *Cobalt merde d'oie*, Brong.; *Cobalt arseniate terreux argentifere*, Lucas; and the *Gansekothigsilber & gansikothigerz* of the Germans. Some other mixtures of silver-ores have received the same name. This is the case with a mixture of native arsenic, red silver ore, earthy silver-glance: and slaggy yellow orpiment, is named *goose-dung ore*, in the Hartz.

Third Subspecies.

Slaggy Red Cobalt-Ochre.

Schlackige Kobaltblüthe, *Hausmann.*

Id. Haus. Syst. d. Unorgan. Natk. s. 140. *Id. Haus.* Handb. b. iii. s. 1126.

External Characters.

Its colours are muddy crimson-red, and dark hyacinth-red, which passes into chesnut-brown.

It

It occurs in thin crusts, and sometimes reniform.
Externally it is smooth.
The lustre is shining and resinous.
The fracture is conchoidal.
It is translucent.
It is soft.
It is brittle.

Geognostic and Geographic Situations.

It occurs in veins along with other ores of cobalt, in the mine of Sophia at Wittichen in Furstenburg.

Use.

All the three subspecies are valued as ores of cobalt, and are used in the preparation of blue colours.

ORDER XVII.—NICKEL.

This Order contains four species, viz. Native Nickel, Copper-Nickel, Black Ore of Nickel, and Nickel-Ochre.

1. Native Nickel.

Gediegen Nickel, *Klaproth.*

Gediegen Nickel, *Klaproth,* in d. Magaz. d. Berlin, Gesellschaft Natf. Freunde, b. i. s. 307.; also his Beit. b. v. s. 231. *Id. Karsten,* Tabel. s. 72.—Nickel natif, *Hauy,* Tabl. p. 84.—Gediegen Nickel, *Haus.* Handb. b. i. s. 117. *Id. Leonhard,* Taschenbuch für 1815, s. 442.

External Characters.

Its colour is bronze-yellow; but is frequently tarnished greenish-grey, steel-grey, ash-grey, and lead grey, and occasionally invested with a delicate crust of brown ironstone.

It occurs in capillary crystals, which are either promiscuously or scopiformly aggregated.

Externally it is shining and splendent; but the tarnished parts are only feebly glimmering or dull.

Internally it is splendent and metallic.

The cross fracture is even, passing into very flat conchoidal.

It is opaque.

Its lustre is increased in the streak.

It

It is semihard, passing into soft.
It is intermediate between brittle and sectile.
It is uncommonly easily frangible.
It is more or less elastic flexible.

Chemical Characters.

Before the blowpipe, on charcoal, it melts rather easily, and without any perceptible smell of sulphur or arsenic, into a metallic globule.

Constituent Parts.

It is pure nickel with a very minute quantity of cobalt and probably arsenic.

Geographic Situation.

It has been hitherto found only in the mine named Adolphus, at Johanngeorgenstadt in Saxony, and at Joachimsthal in Bohemia.

Observations.

It was formerly confounded with capillary pyrites; but that mineral, before the blowpipe, emits a strong sulphureous odour, and burns with a bluish flame, characters which sufficiently distinguish it from native nickel.

2. Copper-Nickel.

Kupfernickel, *Werner*.

Niccolum Ferro et Cobalto mineralisatum; Cuprum Niccoli, *Wall.* t. ii. p. 188.—Kupfernickel, *Romé de L.* t. iii. p. 135. *Id. Wern* Pabst. b. i. s. 206. *Id Wid.* s. 943 —Sulphurated Nickel, *Kirw.* vol. ii. p. 286.—Kupfernickel, *Emm.* b. ii. s. 513. *Id Lam.* t. i. p. 384.—Nickel arsenical, *Hauy*, t. iii p. 503. —Le Kupfernickel, *Broch.* t. ii. p 408.—Kupfernickel, *Reuss*, b. iv. s. 430. *Id. Lud* b. i. s. 289.—Nickelerz, *Suck.* 2ter th. s. 412.—Kupfernickel, *Bert.* s. 489. *Id. Mohs*, b. iii. s. 656. —Nickel arsenical, *Lucas*, p. 123.—Kupfernickel, *Leonhard*, Tabel. s. 77.—Nickel arsenical, *Brong.* t. ii. p. 209. *Id. Brard*, p. 463.—Kupfernickel, *Karsten*, Tabel. s. 72. *Id. Haus.* s. 74.—Nickel arsenical, *Hauy*, Tabl. p. 84.—Nickel alloyed with Arsenic, *Kid*, vol. ii. p. 213.—Copper-Nickel, *Aikin*, p. 60.

External Characters.

Its colour is copper red of different degrees of intensity; but tarnishes first grey, and then black.

It occurs most frequently massive and disseminated; seldom reticulated, dendritic, fruticose, small globular, botryoidal; and sometimes crystallised in six-sided tables *.

Internally it is glistening; sometimes shining, bordering on splendent, and the lustre is metallic.

The fracture is usually imperfect conchoidal, sometimes passing into coarse, small and fine grained uneven: The

* In Morgenbesser's collection at Vienna, there is a specimen of copper-nickel in six-sided tables.

The uneven has the least, the conchoidal the greatest degree of lustre.

The fragments are indeterminate angular and sharp-edged.

It is usually compact, sometimes also in coarse and small granular distinct concretions.

It is semihard in a high degree; it yields with difficulty to the knife.

It is rather brittle.

It is rather difficultly frangible.

It is uncommonly heavy.

Specific gravity, 7.560, *Gellert*; 6.6086—6.6481, *Brisson*.

Chemical Characters.

Before the blowpipe it gives out an arsenical vapour, and then fuses, though not very readily, into a dark scoria, mixed with metallic grains; is soluble in nitro-muriatic acid, forming a dark-green liquor, from which caustic alkali throws down a pale-green precipitate, whereas from a solution of copper the precipitate is dark-brown.

Constituent Parts.

It is a compound of Nickel and Arsenic.

Geognostic Situation.

It generally occurs in primitive rocks, such as gneiss, mica slate, syenite and clay-slate, along with tin-white cobalt-ore, and silver-white cobalt-ore; also in transition rocks and flœtz rocks, particularly the first flœtz limestone, and the limestone which occurs in one of the formations of black coal. The minerals with which it is most generally associated are nickel-ochre, tin-white and silver-white cobalt-ores, ores of copper, and of silver, along with calcareous-spar, brown-spar, heavy-spar, and quartz.

Geographic Situation.

Europe.—It occurs in small quantity in the lead-mines of Lead Hills and Wanlockhead; also in veins along with nickel-ochre, galena or lead-glance, brown-blende, and heavy-spar, in a bed of limestone in the coal-field of Linlithgowshire *. On the Continent, it occurs in veins in primitive rocks at Schneeberg and Johanngeorgenstadt in Saxony; at Joachimsthal in Bohemia; at Schladring in Upper Stiria; and Allemont in France. It is found in a bed along with native gold and ores of cobalt and copper, in porphyritic syenite, at Cravicza in the Bannat. It is met with in veins that traverse transition rocks at Andreasberg in the Hartz. In the county of Mansfeldt, it occurs in veins that traverse bituminous marl-slate. It is also found at Wittichen in Swabia; Salzburg, and Gistain in Arragon in Spain.

Asia.—Koliwan in Siberia.

Observations.

It very nearly resembles native copper, but its brittleness very readily distinguishes it from that mineral.

3. Black

* Vid. Dr Fleming, in Annals of Philosophy.

3. Black Ore of Nickel.

Nickelschwärze, *Hausmann.*

Id. Haus. Handb. b. i. s. 331.

External Characters.

Its colour is dark greyish-black, which inclines to brownish-black.

It occurs massive, disseminated, and in crusts.

It is dull.

The fracture is earthy.

It is soft.

It becomes shining and resinous in the streak.

It soils slightly.

Chemical Characters.

It forms an apple-green coloured solution with nitric acid, which lets fall a white precipitate of arsenic acid.

Constituent Parts.

It has not been analysed; but is conjectured to be a compound of oxide of nickel and oxide of arsenic.

Geognostic and Geographic Situations.

It occurs in veins that traverse bituminous marl-slate, along with copper-nickel, and nickel ochre, in the district of Riegelsdorf, particularly in the mine named Friedrich-Wilhelm *.

4. Nickel-

* Hausmann is of opinion, that this ore is formed by the decomposition of copper-nickel.

4. Nickel-Ochre.

Nickelocker, *Werner.*

Flos Niccoli, *Wall.* t. ii. p. 300.—Nickelocker, *Werner,* Pabst. b. i. s. 207. *Id. Wid.* s. 945.—Nickel Ochre, *Kirw.* vol. ii. p. 283.—Oxide de Nikel, *De Born,* t. ii. p. 210.—Nickelocker, *Emm.* b. ii. s. 516.—Oxide de Nikel, *Lam.* t. i. p. 383. —Nickel oxidé, *Hauy,* t. iii. p. 516.—L'Ocre de Nikel, *Broch.* t. ii. p. 411.—Nickelochre, *Reuss,* b. iv. s. 435. *Id. Lud.* b. i. s. 290. *Id. Suck.* 2ter th. s. 414. *Id. Bert.* s. 496. *Id. Mohs,* b. iii. s. 661.—Nickel oxidé, *Lucas,* p. 123. *Id. Brard,* p. 278.—Nickelocker, *Leonhard,* Tabel. s. 77.—Nickel oxidé, *Brong.* t. ii. p. 209.—Nickelocher, *Karsten,* Tabel. s. 72.—*Id. Haus.* s. 112.—Nickel oxydé, *Hauy,* Tabl. p. 84.—Nickelblüthe, *Haus.* Handb. b. iii s. 1129.—Native Oxyd of Nickel, *Kid,* vol. ii. p. 213.—Nickel Ochre, *Aikin,* p. 60.

External Characters.

Its colour is apple-green, seldom inclining to grass-green. On exposure to the air for some time, it becomes greenish white.

It occurs almost always as a thin coating or efflorescence; seldom massive or disseminated.

It is dull.

The fracture is sometimes splintery, passing on the one side into even, on the other into uneven: or it is coarse or fine earthy.

The splintery and conchoidal varieties are translucent on the edges; but those with earthy fracture are opaque.

It is very soft or friable.

It feels meagre.

The

The varieties with earthy fracture adhere to the tongue.

Chemical Characters.

It is infusible, without addition, before the blowpipe; with glass of borax it is reduced, and the glass acquires a hyacinth-red colour; and it is insoluble in cold nitric acid.

Constituent Parts.

It is conjectured to be an Arseniate of Nickel.

Geognostic Situation.

It occurs in veins in primitive and flœtz rocks, along with copper-nickel, and black nickel ore, and other ores.

Geographic Situation.

It occurs at Leadhills and Wanlockhead; at Alva in Stirlingshire, and in Linlithgowshire; at Andreasberg in the Hartz; Riegelsdorf; in Saxony; and at Allemont in France.

ORDER XVIII.

ORDER XVIII.—ARSENIC.

This Order contains five species, viz. Native Arsenic, Arsenical Pyrites, Orpiment, Oxide of Arsenic, and Pharmacolite.

1. Native Arsenic.

Gediegen Arsenik, *Werner.*

Arsenicum nativum, *Wall.* t. ii. p. 161.—Gediegen Arsenik, *Wern.* Pabst. b. i. s. 207. *Id. Wid.* s. 965.—Native Arsenic, *Kirw.* vol. ii. p. 255.—Arsenic testacé, *De Born,* t. ii. p. 194. —Gediegen Arsenik, *Emm.* b. ii. s. 548.—Arsenic natif, *Lam.* t. i. p. 353. *Id. Hauy,* t. iv. p. 220. *Id. Broch.* t. ii. p. 435. —Gediegen Arsenik, *Reuss,* b. iv. s. 494. *Id. Lud.* b. i. s. 297. *Id. Suck.* 2ter th. s. 442. *Id. Bert.* s. 500.—Arsenic natif, *Lucas,* p. 162. *Id. Brard,* p. 359.—Gediegen Arsenik, *Leonhard,* Tabel. s. 78.—Arsenic natif, *Brong.* t. ii. p. 80.—Gediegen Arsenic, *Karsten,* Tabel. s. 74. *Id. Haus.* s. 70.—Arsenic natif, *Hauy,* Tabl. p. 108.—Native Arsenic, *Kid,* vol. ii. p. 203. *Id. Aikin,* p. 63.

External Characters.

On the fresh fracture it is whitish lead-grey, inclining to tin-white; it however tarnishes very quickly, first yellowish and brownish, and then greyish black.

Besides massive and disseminated, it occurs also reniform, botryoidal, and reticulated, in plates, with pyramidal, cubical, and conical impressions.

Externally

Externally it is either rough or granulated, and very feebly glimmering.

Internally, on the fresh fracture, it is usually glistening, inclining to glimmering, sometimes shining, and the lustre is metallic.

The fracture is small and fine-grained uneven, sometimes imperfect, and curved foliated; rarely narrow, straight, and scopiform radiated.

The fragments are indeterminate angular, and rather sharp-edged.

It occurs in thin, curved lamellar distinct concretions, and seldom in small and fine granular concretions.

It is semihard in a high degree.

It is very difficultly frangible.

It is sectile.

The streak is shining and metallic.

When struck, it has a ringing sound, and emits an arsenical odour.

Specific gravity, 5.7249–5.7633, *Brisson ;* 5.670, *Kirwan.*

Chemical Characters.

Before the blowpipe it yields a white smoke, diffuses an arsenical odour, burns with a blue flame, is gradually and almost entirely volatilised, and deposites a white coating on the coal.

Constituent Parts.

It usually contains a small portion of iron, and when it occurs with gold or silver ores, a little gold or silver.

Geognostic

Geognostic Situation.

It occurs principally in veins in primitive rocks, as in gneiss, mica-slate and clay-slate. It is frequently associated with red silver-ore, silver-glance or sulphureted silver-ore, arsenical-pyrites, orpiment, and galena or lead-glance; sometimes also along with native silver, silver-white cobalt-ore, grey copper-ore, grey antimony-ore, copper-nickel, sparry ironstone, iron-pyrites, copper-pyrites, heavy-spar, calcareous spar, brown-spar, fluor-spar, and quartz.

Geographic Situation.

Europe.—It occurs at Kongsberg in Norway, along with ores of silver, cobalt, and antimony, at Andreasberg in the Hartz, and Allemont in France: in veins along with red silver-ore, in mica-slate, at Joachimsthal in Bohemia, and with the same mineral in gneiss, at Freyberg in Saxony, and at St e Marie aux Mines in France; and it is also found in Silesia, Swabia, Spain and Hungary.

Asia.—In large masses at the bottom of a silver-mine at Zmeof in Siberia.

America.—In Chili.

2. Arsenical Pyrites.

Arsenikkies, *Werner.*

This species is divided into two subspecies, viz. Common Arsenical Pyrites, and Argentiferous Arsenical Pyrites.

First

First Subspecies.

Common Arsenical Pyrites.

Gemeiner Arsenikkies, *Werner.*

Id. Wern. Pabst. b. i. s. 212. *Id. Wid.* s. 968.—Arsenical Pyrites, or Marcasite, *Kirw.* vol. ii. p. 256.—Gemeiner Arsenikkies, *Emm.* b. ii. s. 553.—La Pyrite arsenicale commune, *Broch.* t. ii. p. 438.—Fer arsenical, *Hauy,* t. iv. p. 57.—Gemeiner Arsenikkies, *Reuss,* b. iv. s. 505. *Id. Lud.* b. i. s. 298. *Id. Suck.* 2ter th. s. 446. *Id. Bert.* s. 501. *Id. Mohs,* b. iii. s. 314.—Fer arsenical, *Lucas,* p. 138. *Id. Brard,* p. 314.—Gemeiner Arsenikkies, *Leonhard,* Tabel. s. 78. *Id. Karsten,* Tabel. s. 74. *Id. Haus.* s. 73.—Fer arsenical, *Havy,* Tabl. p. 95.—Arsenic alloyed with Iron, *Kid,* vol. ii. p. 203.—Mispickel, *Aikin,* p. 63.

External Characters.

On the fresh fracture it is silver-white, but by exposure it acquires a yellowish tarnish; sometimes it has a pavonine, columbine, or iridescent tarnish, in its natural repository.

It occurs massive, disseminated, often also crystallised in the following figures:

1. Oblique four-sided prism, with lateral edges of 112° 37′ 11″. and 67° 22′ 49″, and in which the lateral faces are either perfect *, or bevelled either

* Fer arsenical primitif, Hauy. According to Hauy, the primitive form of this species is an oblique four-sided prism, with lateral edges of 111° 18′, and 68° 42′; whereas Bernardi considers it as a cube.—Vid. Annal. du Mus. t. xii. p. 306.; and Gehlen's Journal, t. iii. p. 8. The above measurement is that of Hausmann.

acutely *, or flatly †, on both extremities, and the bevelling planes set on the acute lateral edges.

2. The acutely bevelled prism bevelled on the bevelling edge ‡.
3. The flatly bevelled prism in which the angles on the acute edges are truncated ||.
4. Sometimes the two summits approach to each other, so as to form a figure which may be described either as a very flat or very acute octahedron, depending on the direction in which we view it.

The crystals are middle-sized, small, and sometimes very small; they have smooth lateral planes, but the bevelling planes are usually transversely ribbed, and are splendent externally.

Internally it is shining, seldom glistening, and the lustre is metallic.

The fracture is coarse and small grained, and sometimes inclines to scopiform radiated.

The fragments are indeterminate angular, and rather blunt edged.

It usually occurs compact; sometimes also in columnar distinct concretions, which are thin and straight, sometimes diverging or promiscuous. The surface of the concretions is smooth and shining. The concretions sometimes terminate in crystals.

It is hard.

It

* Fer arsenical primitif unitaire, Hauy.

† Fer arsenical ditetraedre, Hauy.

‡ Fer arsenical unibinaire, Hauy.

|| Fer arsenical quadrioctonal, Hauy.

It is brittle.

It is rather difficultly frangible.

It is heavy, approaching to uncommonly heavy.

When rubbed it emits an arsenical smell.

Specific gravity, 5.753, *Gellert;* 6.5223, *Brisson;* 5.600, *Lametherie.*

Chemical Characters.

Before the blowpipe it emits a copious arsenical vapour, which incrusts the charcoal white; and it leaves a reddish-brown oxide of iron behind. It colours borax blackish.

Constituent Parts.

Arsenic,	48.1	43.4	54.55
Iron,	36.5	34.9	45.46
Sulphur,	15.4	20.1	
	Thomson.	*Chevreul.*	*Berzelius.*

Geognostic Situation.

It occurs in beds and veins in primitive rocks, as gneiss, mica-slate, clay-slate, chlorite-slate, and serpentine. It is usually associated with tinstone, galena or lead-glance, black-blende, copper-pyrites and iron-pyrites, magnetic pyrites, and also quartz, brown-spar, fluor-spar, calcareous-spar, common hornblende, and garnet. It also occurs in transition-rocks, as grey-wacke, along with galena or lead-glance, grey copper-ore, sparry ironstone, and fluor-spar.

Geographic Situation.

It occurs at Alva in Stirlingshire *; and abundantly in Cornwall and Devonshire, accompanying ores of copper and tin. It occurs in beds in serpentine at Reichersdorf in Silesia; in beds at Kupferhügel, also in Silesia; at Gottesgab in Bohemia in beds in clay-slate, accompanied with tinstone, copper pyrites, magnetic pyrites, magnetic iron-stone, native silver-quartz, prase, garnet, and actynolite; at Joachimstahl in Bohemia, and Johanngeorgenstadt in Saxony in primitive mountains, along with tinstone, wolfram, galena or lead-glance, blende, sparry iron-ore, common iron-pyrites. It is also found at Kongsberg in Norway; Sahlberg in Sweden; in Salzburg, Stiria, Hungary and the Bannat.

Asia.—In Siberia, it is found along with beryl; it is also met with in China, and in the island of Sumatra †.

America.—In granite, in the vicinity of Boston in Massachussets.

Use.

It is from this ore that the White Oxide of Arsenic is principally obtained, and artifical Orpiment is also prepared from it.

Second

* Found at Alva by Dr Murray.

† Marsden's Sumatra, p. 137.

Second Subspecies.

Argentiferous Arsenical Pyrites.

Weiserz, *Werner.*

Id. Werner, Pabst. b. i. s. 216. *Id. Wid.* s. 970.—Argentiferous Arsenical Pyrites, *Kirw.* vol. ii. p. 257.—Weiserz, *Emm.* b. ii. s. 557.—La Pyrite arsenical argentifere, *Broch.* t. ii. p. 442. —Fer arsenical argentifere, *Hauy,* t. iv. p. 63.—Weiserz, *Reuss,* b. iv. s. 503. *Id. Lud.* b. i. s. 299. *Id. Suck* 2ter th. s. 449. *Id. Bert.* s. 503. *Id. Mohs,* b. iii. s. 321. *Id. Leonhard,* Tabel. s. 78.—Fer arsenical argentifere, *Brong.* t. ii. p. 150.—Edler Arsenikies, *Karsten,* Tabel. s. 74.—Weiserz, *Haus.* s. 73.—Fer arsenical argentifere, *Hauy,* Tabl. p. 96. —Argentiferous Mispickel, *Aikin,* p. 64.

External Characters.

Its colour is silver-white, inclining to tin-white, and is generally tarnished yellowish on the surface.

It seldom occurs massive, almost always disseminated, and in very small acicular oblique four-sided prisms.

Externally it is shining; internally it is glistening, sometimes glimmering, and the lustre is metallic.

The fracture is fine-grained uneven.

The fragments are indeterminate angular.

It sometimes shews a tendency to fine granular distinct concretions.

In the remaining characters, it agrees with the preceding subspecies.

Constituent Parts.

Besides arsenic and iron, it contains from .01 to 0.10 parts of silver.

Geognostic and Geographic Situations.

Its geognostic situation is the same as that of common arsenic pyrites, with which it is usually associated. It is also accompanied with dark red silver-ore, galena or lead-glance, and copper-pyrites; sometimes with white silver-ore, brown-blende, and generally with quartz and brown-spar.

It is a rare fossil, and has been hitherto found only at Braunsdorf and Freyberg in Saxony; Rathhausberg in Gastein in Salzburg; and in Chili.

Use.

It is used as an ore of silver.

Observations.

1. It is distinguished from the first subspecies by its inferior lustre, smallness of its crystals, fineness of the grain in the fracture, and its granular distinct concretions.

2. Hausmann describes as a distinct species, a sulphuret of iron, with 4 *per cent.* of arsenic. He names it *Arsenikal kies*, and considers it as synonymous with the *Minera arsenicalis flavescens* of Wallerius. It is found at Goslar in the Hartz.

3. Orpiment

3. Orpiment.

Rauschgelb, *Werner.*

This species is divided into two subspecies, viz. Red Orpiment, and Yellow Orpiment.

First Subspecies.

Red Orpiment, or Realgar.

Rothes Rauschgelb, *Werner.*

Sandaraca, *Pliny.*—Arsenicum risigallum, *Wall.* t. ii. p. 163.—Realgar, et Soufre rouge des Volcans, *Romé de L.* t. iii. p. 33. Rothes Rauschgelb, *Werner,* Pabst. b. i. s. 210. *Id. Wid.* s. 975.—Realgar, *Kirw.* vol. ii. p. 261.—Realgar, Sandarac, Rubine d'Arsenic, *De Born,* t. ii. p. 199.—Rothes Rauschgelb, *Emm.* b. ii. s. 562.—Arsenic sulphuré, *Lam.* t. i. p. 358.—Arsenic sulphuré rouge, *Hauy,* t. iv. p. 228.—Le Realgar rouge, *Broch.* t. ii. p. 447.—Rothes Rauschgelb, *Reuss,* b. iv. s. 516. *Id. Lud.* b. iv. s. 301.—Rother Schwefelarsenic, *Suck.* 2ter th. s. 425.—Roth Rauschgelb, *Bert.* s. 505. *Id. Mohs,* b. ii. s. 287.—Arsenic sulphuré rouge, *Lucas,* p. 163.—Rothes Rausehgelb, *Leonhard,* Tabel. s. 78.—Arsenic sulphuré realgar, *Brong.* t. ii. p. 88.—Arsenic sulphuré rouge, *Brard,* p. 362.—Dichtes Rauschgelb, *Karsten,* Tabel. s. 74.—Arsenic sulphuré rouge, *Hauy,* Tabl. p. 109.—Native Realgar, *Kid,* vol. ii. p. 205.—Realgar, *Aikin,* p. 64.

External Characters.

Its colour is aurora-red, which passes through scarlet-red and hyacinth-red into orange-yellow; sometimes with a pavonine tarnish on the surface.

 It

It occurs massive, disseminated, in membranes, reniform; and crystallised in the following figures;

1. Oblique four-sided prism, flatly acuminated on the extremities with four planes, which are set on the acute lateral planes *, Fig. 263.
2. The preceding figure, truncated on the acute lateral edges †.
3. No. 1. truncated on all the lateral edges ‡, Fig. 264.
4. No. 1. truncated on the obtuse lateral edges, and bevelled on the acute lateral edges §, Fig. 265.
5. The preceding figure, in which the edges formed by the meeting of the bevelling planes are truncated ||, Fig. 266.
6. Oblique four-sided prism, acuminated with four planes, which are set on the lateral planes, and all the angles formed by the meeting of the acuminating and lateral planes, truncated ¶.

The crystals are seldom middle-sized, usually small, very small and minute.

The crystals are smooth, and frequently longitudinally streaked and shining, passing into splendent.

Internally it is shining, and the lustre is intermediate between adamantine and resinous.

The fracture is coarse and small-grained uneven, sometimes passing into imperfect conchoidal.

The fragments are indeterminate angular, and blunt-edged.

It

* Arsenic sulphuré rouge emoussé, Hauy.

† Arsenic sulphuré rouge sexoctonal, Hauy.

‡ Arsenic sulphuré rouge dioctaedre, Hauy.

§ Arsenic sulphuré rouge octodecimal, Hauy.

|| Arsenic sulphuré rouge octoduodecimal, Hauy.

¶ Arsenic sulphuré rouge surcomposé, Hauy.

It is translucent, but the crystals are semi-transparent

It yields an orange-yellow coloured streak.

It is very soft.

It is brittle.

It is easily frangible.

Specific gravity, 3.223, *Muschenbröck;* 3.225, *Bergman;* 3.3384, *Brisson.*

Chemical Characters.

It melts immediately before the blowpipe, and burns with a blue flame, giving out arsenical and sulphureous vapours. It generally leaves a minute and earthy residue.

Physical Character.

It is idio-electric by friction, acquiring the resinous or negative electricity.

Constituent Parts.

	Bannat.
Arsenic, - -	69
Sulphur, - -	31
	100

Klaproth, Beit. b. v. s. 238.

Geognostic Situation.

It occurs in veins in primitive rocks, especially in gneiss and clay slate; also disseminated, with iron-pyrites or primitive dolomite, and in the craters and in the vicinity of volcanoes.

It is usually accompanied with native arsenic, light red silver ore, galena or lead-glance; sometimes also with silver-white cobalt-ore, iron-pyrites, grey copper-ore, brown-blende, grey and red antimony-ores, quartz, heavy-spar, and seldom cross-stone, zeolite, and mineral pitch.

Geographic Situation.

Europe.—It occurs in veins at Andreasberg in the Hartz; disseminated in dolomite on St Gothard; in beautiful crystals at Joachimsthal in Bohemia; Braundsdorf and Marienberg in Saxony; and at Kapnic in Transilvania; also along with volcanic substances at Vesuvius, Solfatara, and Puzzola.

Asia.—In the island of Japan *, and in the Burmah Dominions †.

West Indies.—Island of Guadaloupe.

America.—Neck, west territory of the United States ‡.

Uses.

It is used as a pigment. The Chinese cut it into vessels and figures of different shapes.

Observations.

It is distinguished from *red silver-ore* by its inferior specific gravity, and its orange-coloured streak; from *red lead-ore*, by its inferior specific gravity; from *cinnabar*, by the colour of its streak, that of cinnabar being scarlet-red. The strong smell of garlic, and the white fumes which it emits before the blowpipe, are characters which readily distinguish it from those minerals with which it might be confounded.

Second

* Thunberg's Travels, vol. iii. p. 203.

† Ainslie's Materia Medica, p. 53.

‡ I observed fine specimens of the American red orpiment in Dr Murray's Mineralogical Cabinet.

Second Subspecies.

Yellow Orpiment.

Gelbes Rauschgelb, *Werner.*

Αῤῥενικον of *Theophrastus;* by the later Greeks written Αρσενικόν. —Arsenicum, *Plin.* Hist. Nat. xxxiv. 18. s. 56. (ed. Bib. v. 269.)—Arsenicum auripigmentum, *Wall.* t. ii. p. 163.—Orpiment, Orpin, *Romé de L.* t. iii. p. 39.—Gelbes Rauschgelb, *Wern.* Pabst. b. i. s. 210. *Id. Wid.* s. 972.—Orpiment, *Kirw.* vol. ii. p. 260.—Oxide d'Arsénic sulphuré jaune, *De Born,* t. ii. p. 202.—Gelbes Rauschgelb, *Emm.* b. ii. s. 559.—Arsenic sulphuré jaune, *Hauy,* t. iv. p. 234.—Le Realgar jaune, *Broch.* t. ii. p. 444.—Gelbes Rauschgelb, *Reuss,* b. iv. s 512. *Id. Lud.* b. i. s. 300.—Gelber Schwefelarsenic, *Suck.* 2ter th. s. 450.—Gelb-rauschgelb, *Bert.* s. 504. *Id. Mohs,* b. ii. s 233. —Arsenic sulphuré jaune, *Lucas,* p. 164.—Gelbes Rauschgelb, *Leonhard,* Tabel. s. 78.—Arsenic sulphuré, Orpiment, *Brong.* t. ii. p. 89.—Arsenic sulphuré jaune, *Brard,* p. 363. —Blättriges Rauschgelb, *Karsten,* Tabel. s. 74.—Arsenic sulphuré jaune, *Hauy,* Tabl. p. 109.—Native Orpiment, *Kid,* vol. ii. p. 206.—Orpiment, *Aikin,* p. 64.

External Characters.

Its colour is perfect lemon-yellow, which sometimes runs into red, green, and brown colours.

It occurs massive, disseminated, stalactitic, reniform, botryoidal, in crusts, and minute crystals.

On the fresh fracture it is splendent, and the lustre is intermediate between adamantine and semimetallic

The fracture is curved foliated, with a single cleavage.

The

The fragments are indeterminate angular, and blunt-edged in the great, but slaty in the small.

It usually occurs in large, coarse and small longish angulo granular distinct concretions; also in concentric lamellar concretions.

It is translucent; but in small leaves transparent.

Its colour is not altered in the streak.

It is soft.

It is sectile.

It is flexible; but not elastic.

It splits easily.

Specific gravity, 3.313, *Muschenbröck;* 3.315, *Bergman;* 3.048 to 3.435, *Kirwan.*

Constituent Parts.

		Turkey.
Arsenic,	-	62
Sulphur,	-	38
		100

Klaproth, Beit. b. v. s. 238.

Geognostic Situation.

It occurs very rarely in primitive mountains, principally in flœtz rocks, where it is in veins along with copper-pyrites, iron-pyrites, quartz, and calcareous-spar.

Geographic Situation.

Europe.—It occurs, along with red silver-ore, in granite, at Wittichen in Swabia: In the Hartz; at Moldawa and Saska in the Bannat; Nagyag, and Felsobanya in Transilvania; Neusohl in Hungary; Wallachia; and Servia.

Asia.

Asia.—In Natolia and China.

America.—Zimapan in Mexico; and the north-west territory of the United States *.

Observations.

1. Its foliated fracture distinguishes it from *sulphur.*

2. Hausman, at page 209. of his *Mineralogy*, describes a third subspecies of orpiment, under the name *Slaggy Orpiment*, and which he says has a conchoidal racture, and glistening resinous lustre, and is found at Andreasberg in the Hartz.

3. This substance appears to differ from the arsenic of the ancients. It differs from the substance commonly called Arsenic at the present day, in containing a portion of sulphur; and in being consequently of a yellow colour; whereas our arsenic is perfectly white.

Pliny and Theophrastus describe arsenic as having a yellow colour. Thus Pliny says, that the best arsenic is "coloris in auro excellentis." Theophrastus says, on account of its resemblance in colour, *ochra* (ὤχρα) is used instead of arsenic; but the term ὤχρα itself is apparently derived from its yellow colour; and that it was of this colour, appears further probable, from its being changed to a red by calcination, which is mentioned by Theophrastus; and being thus converted into the substance called μιλτος, which answers exactly to our red ochre. Of *Sandaraca*, which is used as a synonym for realgar or red orpiment, Pliny says, "melior quo magnis rufescit." The term Αρσενικὸν, from which our word Arsenic is derived, was an epithet applied by the ancients to those natural substances, the properties of which were found to be of a strong, and, as it were, *masculine* character; and as the poisonous quality of arsenic was soon found to be remarkably

* Also in Dr Murray's collection.

ably powerful, the term was especially applied to that form of it which was most commonly met with. The arsenic of commerce of the present day is in some instances of a yellow colour, owing to its containing a portion of sulphur.—*Kid*, Min. v. ii. p. 206, 207.

4. Oxide of Arsenic.

Arsenikblüthe, *Karsten*.

Arsenic oxydé, *Hauy*, t. iv. p. 225.—Arsenikblüthe, *Reuss*, b. iv. s. 522. *Id. Karsten*, Tabel. s. 74.—Arsenic oxydé, *Hauy*, Tabl. p. 108. *Id. Lucas*, t. ii. p. 447.—Arsenikblüthe, *Haus*. Handb. b. iii. s. 805.

It is divided into three subspecies, viz. Common, Capillary, and Earthy.

First Subspecies.

Common Oxide of Arsenic.

Gemeine Arsenikblüthe, *Hausmann*.

Id. Haus. Handb. b. iii. s. 805.

External Characters.

Its colours are snow-white or milk-white, and sometimes tinged accidentally reddish, yellowish, or greenish.

It occurs in crystalline or stalactitic crusts ; sometimes in small, adhering, tabular or prismatic crystals.

Internally

Internally it is shining or glistening, and the lustre is intermediate between vitreous and adamantine; sometimes pearly.

The fracture is uneven, more or less inclining to radiated, and foliated.

It alternates from opaque to semitransparent.

It is soft.

Second Subspecies.

Capillary Oxide of Arsenic.

Haarförmige Arsenikblüthe, *Hausmann.*

Id. Haus. Handb. b. iii. s. 806.

External Characters.

Its colour is snow-white.

It occurs in very delicate capillary crystals, which are sometimes scopiform, sometimes globularly aggregated; and are often so delicate, that the whole appears often like the finest mould.

The lustre is silky and shining.

Third Subspecies.

Earthy Oxide of Arsenic.

Erdige Arsenikblüthe, *Hausmann.*

Id. Haus. Handb. b. iii. s. 806.

External Characters.

Its colour is yellowish and greyish white.

It

It seldom occurs massive; more frequently in crusts, and stalactitic.

It is dull.

The fracture is fine earthy.

It is opaque.

It sometimes occurs in curved lamellar concretions.

It is friable.

Geognostic and Geographic Situation of the Species.

It occurs at Andreasberg in the Hartz, along with native arsenic, red silver-ore, antimonial silver-ore, galena or lead-glance, yellow orpiment, and corroded quartz; at Biber, along with sulphat of cobalt; at Joachimsthal with orpiment. It is also found at Gistain in the Pyrenees, and at Saint Marie aux Mines in France; and in the Island of Guadaloupe.

Observations.

It very much resembles Pharmacolith, with which, indeed, it has been often confounded. An obvious chemical character may be used for distinguishing them;—the oxide of arsenic is soluble in water, which is not the case with the pharmacolith.

5. Pharmacolite,

5. Pharmacolite, or Arsenic-bloom.

Arsenikblüthe, *Werner*.

Pharmakolith, *Karsten*, Tabel. (1. Ausg.) 36. 75.—Chaux arseniatée, *Hauy*, t. ii. p. 293.—Pharmakolith, Nordeutsche Beit. z. Berg. und Huttenk. iii. s. 116. *Id. Karsten*, Tabel. (2. Ausg.) s. 74. *Id. Haus.* Handb. b. iii. s. 860. *Id. Aikin*, p. 65.

External Characters.

Its colours are reddish-white, snow-white, yellowish-white, and milk-white.

It occurs as a coating, in small balls, small reniform and botryoidal, with a drusy surface; frequently in very delicate capillary shining crystals, which are scopiformly or stellularly aggregated.

Externally it is glimmering, and the lustre is silky.

Internally it is shining or glistening, and silky on the radiated, but dull on the earthy fracture.

The fracture is very delicate, straight, scopiform and stellular radiated, and sometimes passes into fibrous, also earthy.

The fragments are indeterminate angular, and also wedge-shaped.

It occurs in coarse and small granular distinct concretions.

It alternates from semi-transparent to opaque, which latter occurs in the varieties with earthy fracture.

It is very soft, passing into friable.

It is easily frangible.

It soils.

Specific gravity, 2.536, *Selb.*; 2.640, *Klaproth.*

Chemical

Chemical Characters.

Before the blowpipe it is almost entirely dissipated, with a dense white arsenical vapour.

Constituent Parts.

	Wittichen.	Andreasberg.
Lime, -	25.00	27.28
Arsenic acid,	50.54	46.58
Water, -	24.46	23.86
	100	96.82
	Klaproth, Beit. b. iii. s. 281.	*John* in Gehlen's Journ. f. Chem. & Phys. b. iii. s. 539.

Geognostic Situation.

It occurs in veins along with tin-white cobalt-ore, native arsenic, and frequently earthy-red cobalt-ochre.

Geographic Situation.

It is found in veins in granite in the mine named Sophia near Wittichen in Furstenburg ; at Andreasberg in the Hartz ; Riegelsdorf ; and Glucksbrunn in the Forest of Thuringia.

ORDER XIX.

ORDER XIX.—TUNGSTEN, OR SCHEELIUM *.

This Order contains two species, viz. Tungsten, and Wolfram.

1. Tungsten.

Schwerstein, *Werner.*

Minera Ferri lapidea gravissima, *Wall.* t. ii. p. 254.—Wolfram de couleur blanche, *Romé de L.* t. iii. p. 264.—Schwerstein, *Werner,* Pabst. b. i. s. 222.—Weisser Tungsten, *Wid.* s. 980. —Tungsten, *Kirw.* vol. ii. p. 314.—Tungstate calcaire, [illegible] ine d'Etaine blanche, *De Born,* t. ii. p 230.—Schwerstein, *Emm.* b. ii. s. 570.—Tungstene, *Lam.* t. i. p. 402.—Scheelen calcaire, *Hauy,* t. iv. p. 320.—La Pierre pesant, ou le Tungstene, *Broch.* t. ii. p. 453.—Scheelerz, *Reuss,* b. iv. s. 534. *Id. Lud.* b. i. s. 303.—Kalk-Scheel, *Suck.* 2ter th. s. 459.—Schwerstein, *Bert.* s. 509. *Id. Mohs,* b. iii. s. 623.—Scheelin calcaire, *Lucas,* p. 183.—Scheelerz, *Leonhard,* Tabel. s. 81.—Scheelin calcaire, *Brong.* t. ii. p. 93. *Id. Brard,* p. 389.—Scheelerz, *Karsten,* Tabel. s. 74.—Scheelin calcaire, *Hauy,* Tabl. p. 118. —Tungsten, *Kid,* vol. ii. p. 225. *Id. Aikin,* p. 65.

External Characters.

Its colours are usually yellowish and greyish-white, which sometimes verge on snow white; from these it passes

* Werner gave the name *Scheele* to this genus, in honour of the illustrious chemist Scheele, who discovered the peculiar metal which characterises it.

passes into yellowish-grey, and light yellowish brown, which approaches to orange-yellow.

The white varieties are sometimes tarnished on the surface, either dark pearl-grey or plum-blue.

It occurs massive, disseminated, and crystallised in the following figures:

1. Acute octahedron *, fig. 267. in which the edges of the common base are 130° 20. This figure is sometimes truncated on the lateral edges, and on the summits. When the truncations on the lateral edges become so large as to cause the original faces to disappear, then there is formed
2. A less acute octahedron †, in which the edges on the common base are 113° 36′, and those formed by the meeting of the lateral planes 107° 26′; which is sometimes truncated on the summits, or flatly acuminated on the summits with four planes, which are set on the lateral planes, as in fig. 268. This octahedron is sometimes so deeply truncated on the summits, that there is formed
3. A four-sided table, in which the terminal planes are bevelled, as in fig. 269. In this figure the lateral planes correspond to the truncating planes of the octahedron, and the bevelling planes to the lateral planes of the octahedron.
4. Sometimes the angles formed by the meeting of the acuminating planes, fig. 269. are truncated, and the truncating planes set on the lateral edges.

The

* Scheelin calcaire primitif, Hauy.

† Scheelin calcaire unitaire, Hauy.

The crystals are middle-sized, small and very small, and are sometimes heaped on one another.

The lateral planes of the crystals are smooth, the bevelling planes are slightly transversely streaked ; they are shining and splendent, and their lustre inclines to adamantine.

Internally it is shining, and the lustre is intermediate between vitreous and resinous, and sometimes inclines to adamantine.

The fracture is imperfect foliated, with a fourfold cleavage parallel with the sides of an octahedron.

The fragments are indeterminate angular, and rather blunt-edged.

It sometimes occurs in large, coarse, and small granular distinct concretions, with streaked and shining surfaces.

It is more or less translucent, seldom semitransparent.

It yields pretty easily to the knife.

It is rather brittle.

It is easily frangible.

Specific gravity 5.800 to 6.028, *Kirwan.* 6.0665, *Brisson.* 6.000, *Gellert.* 6.015 and 5.570, *Klaproth.*

Chemical Characters.

It crackles before the blowpipe and becomes opaque, but does not melt; with borax it forms a transparent or opaque white glass, according to the proportions of each.

Constituent Parts.

		Schlakenwald.	Cornwall.
Acid of Tungsten,	65	77.75	75.25
Lime, -	31	17.60	18.70
Silica, -	4	3.00	1.56
Oxide of Iron, -	—	—	1.25
Oxide of Manganese,	—	—	0.75
	100	98.35	97.45
	Scheele, in n. Abhand. d. Schwd. Akad. 1781, 2, 89.	*Klaproth*. Beit. b. iii. s. 47. & 51.	

Geognostic Situation.

It occurs along with tinstone, magnetic ironstone, and brown ironstone. In the tinstone repositories it is associated with wolfram, quartz, mica, fluor-spar, steatite, &c.

Geographic Situation.

It occurs along with wolfram and tinstone, &c. in Cornwall: in Sweden, in a bed of magnetic ironstone; in Bohemia and Saxony.

Observation.

It is distinguished from the white varieties of *tinstone*, by its shape, intensity, and kind of lustre, hardness, greater weight, and its becoming yellow when thrown into nitric acid; from *white lead-ore*, by not effervescing with acids, and by not being blackened by an alkaline sulphuret, from *heavy-spar*, by its greater weight, and by the yellow colour it assumes when thrown into nitric acid.

2. It is named Tungsten, which signifies *heavy stone* from its great specific gravity.

2. Wolfram.

2. Wolfram.

Wolfram, *Werner*.

Magnesia cristallina; Spuma Lupi, *Wall.* t. ii. p. 344.—Wolfram, *Wern.* Pabst. b. i. s. 223. *Id Wid.* s. 983. *Id. Kirw.* vol. ii. p. 316. *Id. De Born,* t. ii. p. 227. *Id. Emm.* b. ii. s. 574. *Id. Lam.* t. i. p. 404. *Id. Broch.* t. ii. p. 456.—Scheelen ferruginé, *Hauy,* t. iv. p. 314.—Wolfram, *Reuss,* b. iv. s. 541. *Id. Lud.* b. i. s. 303.—Eisen-Scheel, *Suck.* 2ter th. s. 461.—Wolfram, *Bert.* s. 509. *Id. Mohs,* b. iii. s. 618.—Scheelin ferruginé, *Lucas,* p. 182.—Wolfram, *Leonhard,* Tabel. s. 81.—Scheelin ferruginé, *Brong.* t. ii. p. 94. *Id. Brard,* p. 388.—Wolfram, *Karsten,* Tabel. s. 74.—Scheelin ferruginé, *Hauy,* Tabl. p. 118.—Wolfram, *Kid,* vol. ii. p. 226. *Id. Aikin,* p. 66.

External Characters.

Its colour is intermediate between dark greyish-black and brownish-black, which sometimes inclines to velvet-black.

It occurs massive, disseminated, in plates, and crystallised in the following figures:

1. Flat rectangular four-sided prism, acuminated on both extremities with four planes, which are set on the lateral edges; sometimes the summits of the acuminations are truncated *, Fig 270.
2. The lateral edges of the preceding figure are truncated, as in Fig. 271 †.

Mm 3 3. When

* Scheelin ferruginé epointé, Hauy.

† Scheelin ferruginé unibinaire, Hauy.

3. When the truncating planes in Fig. 271. become so large as to obliterate the original planes, there is formed an oblique four-sided prism, which is acuminated with four planes, set on the lateral planes, and the summits bevelled, as in Fig. 272 *.

4. Rectangular four-sided table, in which the two opposite lateral planes are bevelled, and the angles truncated.

The crystals are middle-sized and large, and occur imbedded, or intersecting one another, but are seldom distinct.

The lateral planes are usually longitudinally streaked and glistening.

The principal fracture is shining or splendent; the cross fracture is glistening; the lustre is metallic, inclining to adamantine.

The principal fracture is foliated; with a distinct cleavage in the direction of the smaller lateral planes of the prism, and a less distinct one at right angles to the former, and parallel with the larger planes. The cross-fracture is coarse and small-grained uneven.

The fragments are indeterminate angular, and blunt-edged.

It seldom occurs in angulo-granular or prismatic concretions; more frequently in thick or thin lamellar concretions; which are either fortifications-wise bent, or concentrically curved, and in which the surfaces are transversely streaked.

It is opaque.

It yields a dark reddish-brown coloured streak.

It

* Scheelin ferruginé progressif, Hauy.

It yields readily to the knife.

It is brittle.

It is uncommonly heavy.

Specific gravity, 6.835, *Elhuyar*; 7.130, *Gellert*; 7.1195, *Brisson*; 5.705, *Gmelin*; 7.000, *Leonardi*; 7.3333, *Hauy*; 7.006, *Kirwan*; 6.955, *Hatchett*; 6.857, *Ullmann*.

Chemical Characters.

It decrepitates before the blowpipe, but is infusible without addition. It colours glass of borax reddish, when exposed to the exterior flame of the blowpipe.

Constituent Parts.

Tungstic Acid,	64.0	67.00
Oxide of Manganese,	22.0	6.25
Oxide of Iron, -	13.5	18.10
Silica, -	—	1.50
	99.5	92.75
	D'Elhuyar, Mem. d. l'Acad. d. Toulouse. ii.	*Vauquelin*, in Journ. d. Min. N. 19. 18.

Geognostic Situation.

It occurs in primitive rocks, and generally along with tinstone, and wolfram; less frequently in veins in greywacke, along with galena or lead-glance, grey copper-ore, sparry ironstone, and quartz.

Geographic Situation.

It occurs in gneiss in the island of Rona, one of the Hebrides;

Hebrides*; at Herland, Pednandre, Huel Fanny, Cligga and Kit-hill, in Cornwall; in the Hartz it is met with in veins that traverse grey wacke; in primitive rocks at Ehrenfriedersdorf, Altenberg and Geyer, in Saxony; Zinnwald and Schlackenwald in Bohemia; and Puy les Mines in France.

Observations.

1. It is distinguished from *tinstone*, among other characters, by its streak, which is reddish brown; whereas that of tinstone is grey.

2 This mineral was originally mistaken for antimony, which by the alchemists was called the *wolf;* probably because it acted violently upon, and, as it were, devoured the base metals, in the process of refining gold; hence arose the term *spuma lupi*; the word *ram*, which signifies spuma, being commonly applied by the Germans to substances of a laminated texture.—*Kid.* vol. ii. p. 227.

ORDER XX.

* Macculloch.

ORDER XX.—URANIUM.

This Order contains three species, viz. Pitch-ore, Uran-mica, and Uran-ochre.

1. Pitch-Ore.

Pecherz, *Werner.*

Id. Wern. Pabst. b. i. s. 170.—Pech-Blende, *Wid.* s. 987.—Sulphurated Uranite, *Kirw.* vol. ii. p. 305.—Pech-Blende, ou Blende de Poix, *De Born,* t. ii. p. 159.—Schwarz Uranerz, *Emm.* b. ii. s. 580.—Mine d'Uranit sulphuré, *Lam.* t. i. p. 408. Urane oxydulé, *Hauy,* t. iv. p. 280.—Le Pecherz, ou L'Urane noir, *Broch.* t. ii. p. 460.—Pecherz, *Reuss,* b. iv. s. 551. *Id. Lud.* b. i. s. 307.—Uranpecherz, *Suck.* 2ter th. s. 466.—Pecherz, *Bert.* s. 511. *Id. Mohs,* b. iii. s. 716.—Uran oxydulé, *Lucas,* p. 176.—Pecherz, *Leonhard,* Tabel. s 80.—Uran oxydulé, *Brong.* t. ii. p. 102. *Id. Brard,* p. 378.—Pecherz, *Karsten,* Tabel. s. 74.—Uran oxydulé, *Hauy,* Tabl. p. 113. —Pech-Blende, *Kid,* vol. ii. p. 220.—Pech-Uran, *Haus.* Handb. b. i. s. 325.—Pitch-Blende, *Aikin,* p. 69.

External Characters.

Its colour is velvet-black, or dark greyish-black, which inclines to iron-black.

It occurs massive, disseminated, small reniform, and small botryoidal.

Internally it is shining, inclining to glistening, and the lustre is resinous.

The

The fracture is imperfect and flat conchoidal, which passes into uneven.

The fragments are indeterminate angular, and sharp-edged.

It sometimes occurs in curved lamellar distinct concretions.

It yields readily to the knife.

In the streak, neither colour nor lustre is changed.

It is brittle.

It is uncommonly heavy.

Specific gravity, 6.3785, *Guyton*; 6.5304, *Hauy*; 7.500, *Klaproth*.

Chemical Characters.

It is completely infusible, without addition, before the blowpipe. With soda or borax it forms a grey, muddy, slaggy-like globule; with phosphoric salts a transparent green bead. It dissolves imperfectly in sulphuric and muriatic acids; but it is nearly completely dissolved in nitric and nitro-muriatic acids; and from this solution, which has a pale orange yellow colour, the uranium is precipitated brownish-red by prussiate of potash, and yellow by the alkalies.

Constituent Parts.

	Joachimsthal.
Oxide of Uranium, -	86.5
Black Oxide of Iron, -	2.5
Galena or Lead-glance, -	6.0
Silica, - - -	5.0

Klaproth, Beit. b. ii. s. 221.

Geognostic

Geognostic Situation.

It occurs in veins in primitive rocks, along with lead and silver ores.

It is usually accompanied with galena or lead-glance, copper pyrites, iron-ochre, calcareous-spar, quartz, lamellar heavy-spar; seldomer with silver-white cobalt-ore, red cobalt-ochre, silver glance or sulphureted silver-ore, uran-mica and uran ochre; also associated with native-silver, red silver-ore, corneous silver ore, and native arsenic.

Geographic Situations.

It is found at Tol Carn and Tincroft in Cornwall; in mica-slate at Johanngeorgenstadt, Schneeberg and Wiesenthal in Saxony; in granite at Joachimsthal in Bohemia; and Kongsberg in Norway.

Observations.

It is distinguished from *brown blende* by colour, specific gravity, fracture and streak; from *wolfram* by its streak and fracture.

2. Uran-

Uran-Mica.

Uran-Glimmer, *Werner.*

Chalkolith, *Wern.* Pabst. b. i. s. 290.—Grün Uranerz, *Wid.* s. 990.—Micaceous uranitic Ore, *Kirw.* vol. ii. p. 304.—Grün Uranerz, *Emm.* b. ii. s. 584.—Oxide d'Uranit avec Cuivre, *Lam.* t. i. p. 410.—Urane oxidé, *Hauy,* t. iv. p. 283.—L'Urane micacé, *Broch.* t. ii. p. 463.—Uranglimmer, *Reuss,* b. iv. s. 556. *Id. Lud.* b. i. s. 308. *Id. Suck.* 2ter th. s. 469. *Id. Bert.* s. 511. *Id. Mohs,* b. iii. s. 721.—Uran oxydé, *Lucas,* p. 177.—Uranglimmer, *Leonhard,* Tabel. s. 81.—Uran oxidé micacé, *Brong.* t. ii. p. 103. *Id. Brard,* p. 379.—Uranglimmer, *Karsten,* Tabel. s. 74.—Uran oxydé, *Hauy,* Tabl. p. 115. —Micaceous Uranite, *Kid,* vol. ii. p. 221.—Uran-oxyd, *Haus.* Handb. b. i. s. 327.—Uranite, *Aikin,* p. 69.

External Characters.

Its chief colour is grass-green, which passes on the one side into apple-green, emerald green and leek-green, and on the other into siskin green, sulphur-yellow and wax-yellow.

It sometimes occurs massive, and also in membranes; but more frequently crystallised, and in the following figures:

1. Rectangular four-sided prism, which is often truncated on the terminal edges.
2. Octahedron, generally truncated on the summits, which is formed when the truncating planes of the preceding figure becomes so large as to meet in a point, and the prism disappears.
3. Rectangular four sided table, bevelled on the terminal planes; and sometimes the angles formed by the meeting of the bevelling planes are truncated.

4. Rectangular

4. Rectangular four-sided table, bevelled on the terminal planes, and the edges of the bevelment truncated.
5. Rectangular four sided table, truncated on the terminal edges, thus forming an eight-sided table
6. Rectangular four-sided table, bevelled on the terminal edges, and the edges formed by the meeting of the bevelling planes truncated.
7. Long six sided table.
8. Long eight-sided table.

The crystals are small and very small, and superimposed, and form druses.

The terminal planes of the table are streaked, but the lateral planes are smooth.

Externally it is usually shining and sometimes splendent.

Internally it is shining, approaching to glistening, and the lustre is pearly.

The fracture is foliated, with a fourfold cleavage; but of the cleavages only one is distinct, which is that parallel to the base of the prism.

The massive varieties occur in granular distinct concretions.

It is transparent and translucent.

It is soft.

It is sectile.

It is easily frangible.

Specific gravity, 3.1212, *Champeaux*; 2.190, 3.3, *Gregor*.

Chemical Characters.

It decrepitates violently before the blowpipe on charcoal; loses about 33 *per cent* by ignition, and acquires a brass-

brass-yellow colour; with borax it yields a yellowish green glass; it dissolves in nitric acid without effervescence, and communicates to it a lemon-yellow colour.

Constituent Parts.

	Cornwall.
Oxide of Uranium, with a trace of Oxide of Lead, - -	74.4
Oxide of Copper, - -	8.2
Water, - - -	15.4
Loss, - - -	2.

Gregor, in Annals of Phil. vol. v. p. 284.

Geognostic Situation.

It generally occurs in ironstone veins in granite, and is very frequently accompanied with ochry and compact brown ironstone, compact red ironstone, pitch-ore, uran-ochre, iron-flint, jasper, quartz, hornstone, indurated clay, rarely with olivine-ore, and black and yellow cobalt-ochre.

Geographic Situation.

It occurs at Carharrak, Tincroft, Tol-carn, near Redruth, Huel Jewel Stenna-gwyn near St Austle, at Gunnislake, near Callington, in Cornwall; at Johanngeorgenstadt, Eibenstock, and Schneeberg in Saxony; in veins in granite at St Symphorien near Autun; and in the same species of rock at St Yrieux, near Limoges, in France.

3. Uran-Ochre.

Uranocker, *Werner.*

This species contains two subspecies, viz. Friable Uran-ochre, and Indurated Uran-ochre.

First

First Subspecies.

Friable Uran-Ochre.

Zerreibliche Uranocker, *Werner.*

Uran oxydé pulverulent, *Hauy,* t. iv. p. 285.—Uranocher, *Reuss,* b. iv. s. 561. *Id. Leonhard,* Tabel. s. 81. *Id. Karsten,* Tabel. s. 74.

External Characters.

It colour is straw-yellow, which passes sometimes into sulphur-yellow and lemon-yellow; sometimes into yellowish brown and orange-yellow, and occasionally borders on aurora-red.

It occurs usually as a coating or efflorescence on pitch-ore.

It is friable, and composed of dull, dusty, weakly soiling, and weakly cohering particles.

It feels meagre, and is

Not particularly heavy.

Geognostic Situation.

It occurs always on pitch-ore.

Second Subspecies.

Indurated Uran-Ochre.

Feste Uranocker, *Werner.*

Verhärtete Uranocher, *Karsten,* Tabel. s. 74.

External Characters.

It has the same colours as the preceding subspecies.

It

It occurs massive and disseminated.

Internally it is dull, but in some varieties it passes into glimmering and glistening.

The fracture is small grained uneven, which sometimes passes into earthy, sometimes into small conchoidal.

It is opaque.

It is soft and very soft, sometimes passing into semi-hard.

It is brittle.

It soils very little.

Specific gravity, 3.1500, *La Metherie* ; 3.2438, *Hauy*.

Chemical Characters.

According to Klaproth, the yellow varieties are pure oxide of uran, but the brownish and reddish contain also a little iron.

Geognostic and Geographic Situations.

It occurs along with the other ores of uranium.

It is found at Joachimsthal, and Gottesgab in Bohemia; Johanngeorgenstadt in Saxony; and also in France.

Observations.

La Metherie, in his Leçons de Mineralogie, t. i. p. 316. describes a mineral under the name *Mine d'Urane Silicieuse*, which has a brownish-black colour, conchoidal shining fracture, and is easily frangible. It contains, according to Lampadius, Uranium, 32.; Silica, 56.; Iron, 7.50, and Alumina, 3.50. It was found at Siebenlehn near Freyberg, imbedded in greenstone.

ORDER XXI.

ORDER XXI.—TANTALUM.

This Order contains three species, viz. Tantalite, Yttrotantalite, and Gadolinite.

1. Tantalite.

Tantalit, *Karsten.*

Columbite, *Hatchett.*

Tantalit, *Eckeberg,* Kongl. Vetensk. Acad. Handl. 1802, Q. 1. p. 68.—83.—Tantalite, *Reuss,* b. ii. 4. s. 685.—Columbeisen, *Reuss,* b. ii. 4. s. 632.—Tantalit, *Leonhard,* Tabel. s. 83.—Tantal oxydé ferro-manganesifere, *Hauy,* Tabl. p. 120.—Tantalit, *Haus.* Handb. b. i. s. 310. *Id. Aikin,* p. 72.

External Characters.

Its colour is iron-black, sometimes with a shade of blue.

It occurs in imbedded angular pieces, from the size of a pea to that of a hazel-nut: also crystallised, in the form of an acute octahedron, with a square base.

The surface of the angular pieces is uneven; that of the crystals sometimes smooth, sometimes streaked.

Externally it is glistening, internally shining, and the lustre is metallic, inclining to resinous.

The fracture is coarse-grained uneven, inclining to conchoidal or concealed foliated.

 The

The fragments are indeterminate angular, and sharp-edged.

It scratches glass, and gives few sparks with the steel.

It is opaque.

The streak is dull; the powder brownish-black.

It is brittle.

It is difficultly frangible.

Specific gravity, 5.918, *Hatchett*. 7.15 to 7.65, *Wollaston*. 7.953, *Eckeberg*. 7.3, *Klaproth*.

Chemical Characters.

Before the blowpipe, without addition, it suffers no other change than a diminution of lustre. It is insoluble in glass of borax.

Constituent Parts.

	Finland.	Finland.	Finland.	N. American or Columbite.
Oxide of Tantalum,	83	85	88	80
Oxide of Iron,	12	10	10	15
Oxide of manganese,	8	4	2	5
	103	99	100	100
	Vauquelin in Hauy, Tabl. p. 308.	*Wollaston*, in Ph. Tr. 1809.	*Klaproth*, Beit. b. v. s. 5.	*Wollaston*, in Ph. Tr. 1809.

Geognostic and Geographic Situations.

It occurs disseminated in a coarse red granite, at Brokärns Zinnsgute in the parish of Kemito in Finland; and the specimen examined by Mr Hatchett is said to be from Massachusets Bay in North America.

Observations.

Observations.

This species bears a considerable resemblance to several other minerals, particularly to magnetic ironstone, tinstone, wolfram, yttrotantalite, and gadolinite. It is distinguished from *magnetic ironstone*, by its not affecting the magnetic needle, and its greater specific gravity ; from compact *black tinstone*, by its metallic lustre, brownish-black powder, and also by the action of the blowpipe, for tinstone, when exposed, on charcoal, to the reducing flame of the blowpipe, is reduced ; and from *wolfram*, by the absence of the foliated fracture, and the action of the blowpipe, as wolfram, along with glass of borax, when exposed to the exterior flame of the blowpipe, becomes of a reddish colour. The characters that distinguish it from *yttrotantalite* will be given in the account of the other species. It is distinguished from *gadolinite* by its higher specific gravity, uneven fracture, and infusibility.

2. Yttrotantalite.

Ytter-Tantal, *Karsten.*

Yttertantalit, *Eckeberg,* Kongl. Vetensk. Acad. Handl. 1802, Q. 1.—Yttertantal, *Reuss,* Min. ii. 4. s. 637. *Id. Leonhard,* Tabel. s. 83. *Id. Karsten,* Tabel. s. 74.—Tantal oxydé yttrifère, *Hauy,* Tabl. p. 120.—Yttertantalit, *Haus.* Handb. b. i. s. 312.—Yttrotantalit, *Aikin,* p. 72.

External Characters.

Its colour is dark iron black.

It occurs imbedded in angular pieces, sometimes as large as a hazel-nut ; and also crystallised :

 1. Oblique

1. Oblique four-sided prism, with lateral edges of 95° and 85° ?
2. Six-sided prism, with two lateral edges of 95°, and four others of 132° 30′; or with two lateral edges of 95°, and two others of 144° 50′, and two others of 120° 10′.

Internally it is shining or glistening, and the lustre is resinous, sometimes inclining to metallic.

It occurs in granular distinct concretions.

It is opaque.

It scratches glass, but does not afford sparks with steel.

It yields a grey-coloured powder.

It is brittle.

Specific gravity, 5.130, *Ekeberg*.

Chemical Characters.

On the first application of the flame of the blowpipe it decrepitates; but if the heat is increased, it melts into a greenish-yellow slag.

Constituent Parts.

Oxide of Tantalum,	-	45
Oxide of Iron and Yttria,		55
		100

Vauquelin, Hauy, Tabl. p. 309.

Geognostic and Geographic Situations.

It occurs along with gadolinite, in a bed of flesh-red felspar, in gneiss at Ytterby near Roslagen in Sweden.

Observations.

Observations.

It is very nearly allied both to gadolinite and tantalite. It is distinguished from *gadolinite* by its concretions, higher specific gravity, and the effect produced on it by the action of the blowpipe; and it is distinguished from *tantalite* by its granular concretions, specific gravity, and its imperfect fusibility.

3. Gadolinite.

Gadolinit, *Karsten.*

Gadolinit, *Geyer,* in V. Crell's Chem. Annal. 1788, b. i. s. 229. —Gadolin, in K. Sv. Acad. n. Handl. 1794, 11.—Gadolinite, *Hauy,* t. iii. p. 141. *Id. Reuss,* ii. 2. 7. *Id. Karsten,* Tabel. s. 22. *Id. Hauy,* Tabl. p. 47. *Id Haus.* Handb. b. ii. s. 608. *Id. Aikin,* p. 128.

External Characters.

Its colour is velvet-black, sometimes greenish-black, less frequently brownish-black, and very rarely hyacinth-red.

It occurs massive, disseminated; and rarely crystallised in oblique four sided prisms; sometimes also in six-sided prisms; and it is said also in rhomboidal dodecahedrons.

Internally it is shining, and the lustre is resinous, inclining to vitreous.

The fracture is generally conchoidal; seldom uneven.

It appears sometimes to occur in distinct concretions, which are granular or prismatic; and the surfaces of the concretions have frequently a whitish or bluish aspect, and vary from glistening to dull.

It

It is faintly translucent on the thinnest edges, and then it appears blackish green.

It scratches glass, and affords very few sparks with steel.

It is brittle.

It is difficultly frangible.

When pure, it does not appear to affect the magnet.

Specific gravity, 4.2230, *Geyer*. 4.0280, *Gadolin*. 4.2370, *Klaproth*. 4.0497, *Hauy*.

Chemical Characters.

It intumesces very much before the blowpipe, and at length melts into an imperfect slag, which is magnetical. It loses its colour in nitric acid, and gelatinises.

Constituent Parts.

Yttria, -	59.75	35.0	55.5	60.0
Silica, -	21.25	25.5	23.0	22 0
Oxide of Iron,	17.50	25.0	16.5	16 5
Glucina, -			4.5	
Alumina, -	0.50			
Lime, -		2.0		
Oxide of manganese, -		2.0		a trace.
Water or other volatile matters, -	0.50	10.5	0.5	0.5
	99.50	100	100	99
	Klaproth, Beit. b. iii. s. 65.	*Vauquelin*, Ann. d. Ch. xxxvi. p. 143.	*Ekeberg*, K. Vet. Acad. n. H. 1802. s. 76.	*Klaproth*, Beit. b. v. s. 175.

Geognostic

Geognostic and Geographic Situations.

It occurs along with yttrotantalite, at Ytterby near Waxholm in Roslagen, in beds of a coarse granular red felspar, which alternate with layers of mica; at Finbo near Fahlun, also in Sweden, in a coarse granular granite, along with pyrophysalite and tinstone? In both places the gadolinite is invested with an ochre-yellow earthy crust, which appears to be hydrate of iron. This mineral is said to have been also found in the Island of Bornholm, and in flesh-red felspar at Afvestad in Sweden.

Observations.

1. The following characters given by Hausmann, as distinguishing marks of the Gadolinite, Yttrotantalite, and Tantalite, will be found useful.

Gadolinite: conchoidal fracture; specific gravity = 4.2; intumesces before the blowpipe, and melts into a magnetic slag.

Yttrotantalite: fine granular distinct concretions; specific gravity rather more than 5.1; melts into a greenish-yellow slag, without intumescing.

Tantalite: uneven fracture; specific gravity 7.963; infusible, without addition, before the blowpipe.

2. As the Gadolinite is very nearly allied to the species of the tantalum order, not only in external characters, but also in geognostic relations, I have for the present arranged it in this part of the system.

3. The metal which characterises this order, was first discovered by Hatchett, in that variety of the tantalite named by him *Columbite.*

ORDER XXII.

ORDER XXII.—CERIUM.

This Order contains two species, viz. Cerite and Allanite.

1. Cerite.

Bastnäs Tungsten, *Cronstedt*, (first description), in the Abh. d. Schwed. Akad. 1751, s. 235.—Cerit, *Hisinger & Berzelius*, S. Cerium en ny metall, funnen i Bastnäs Tungsten frau Riddarhyttan i Westmannland, Af. Hisinger & Berzelius, Stockholm, 1804, 8.—Cerit, *Hisinger & Berzelius*, in Afh. i Fys. Kem. och Min. 1. 58. *Id. Leonhard*, Tabel. s. 83. *Id. Karsten*, Tabel. s. 74.—Cerium oxydé silicifere, *Hauy*, Tabl. p. 120.—Cerite, *Aikin*, p. 72.

External Characters.

Its colour is intermediate between crimson-red, clove-brown, and reddish brown; also dark or pale flesh-red, and very rarely inclines to yellow.

It occurs massive, and disseminated.

Internally it is glimmering and resinous.

The fracture is fine splintery.

The fragments are indeterminate angular, and not particularly sharp-edged.

It is opaque.

It scratches glass with difficulty, and affords few sparks with steel.

Its streak is greyish-white.

It

It is brittle.

Specific gravity 4.988, *Cronstedt.* 4.660, *Klaproth.* 4.619 and 4.489, *Hisinger* and *Berzelius.*

Chemical Characters.

It is infusible before the blowpipe; but when pulverised and heated, its colour changes from grey to yellow.

Constituent Parts.

Oxide of Cerium,	54.50		67
Silica, -	34.50		17
Oxide of Iron,	3.50		2
Lime, -	1.25		2
Water, -	5.00	and Carbonic Acid,	12
	98.75		100
	Klaproth, Beit. b. iv. s. 147.		*Vauquelin* in Annal. d. Mus. t. v. p. 412.

Geognostic and Geographic Situations.

It occurs in a bed of copper-pyrites, along with bismuth-glance, or sulphuretted bismuth, molybdena, wolfram? hornblende, actynolite, and mica. The bed is situated in gneiss near Riddarhytta in Westmannland in Sweden.

Observations.

The peculiar metal which characterises this order, was first detected by Hisinger and Berzelius, who bestowed on it the name Cerium, from the planet Ceres, discovered by Piazzi.

2. Allanite.

2. Allanite.

Allanite, *Thomson.*

Allanite, Edinburgh Phil. Trans. vol. vi. p. 371.—Cerium oxydé silicifere noire, *Lucas,* t. ii. p. 498.—Cerium allanite, *Delam.*

External Characters.

Its colour is brownish-black.

It occurs massive, disseminated, and crystallised in the following figures:

1. Oblique four-sided prism.
2. Six-sided prism, acuminated with four planes, set on the lateral planes.

Externally it is dull.

Internally it is shining, and resino-metallic.

The fracture is small conchoidal.

The fragments are indeterminate angular, and sharp-edged.

It is opaque.

It affords a greenish-grey coloured streak.

It scratches glass and hornblende.

It is brittle.

It is easily frangible.

Specific gravity, 3.523 to 4.001.

Chemical Characters.

Before the blowpipe it froths, and melts imperfectly, into a black scoria. Gelatinates in nitric acid *. In a strong heat it loses 3.98 *per cent.* of its weight.

Constituent

* This is doubted by Hauy.

Constituent Parts.

Oxide of Cerium,	-	33.9
Oxide of Iron,	-	25.4
Silica,	- -	35.4
Lime,	- -	9.2
Alumina,	- -	4.1
Moisture,	- -	4.0
		112.0

Thomson, in Edin. Phil. Trans. vol. vi. p. 385.

Geognostic and Geographic Situations.

It occurs in a granite rock in West Greenland, where it was first discovered by Professor Giesecké of Dublin, —an intelligent naturalist,—and a gentleman of great worth,—who, with a rare zeal and intrepidity, and in defiance of the horrors and miseries of that forlorn region, courageously devoted many years of his life to the investigation of its natural history *.

Observations.

1. It was first described and analysed by Dr Thomson, who named it Allanite, in honour of Thomas Allan, Esq.

* One of Mr Giesecké's collections of minerals was captured during the late war, and brought to Leith, where it was purchased by Colonel Imrie and Mr Allan. A particular account of it has been published in the first volume of Annals of Philosophy by Mr Allan. The public anxiously expect Mr Giesecké's promised account of West Greenland, which, with that of both Greenlands, announced by a promising navigator and naturalist Mr Scoresby, will make a valuable addition to our knowledge of these remote and hitherto but imperfectly known regions.

Esq. who was first aware of its being a particular and undescribed species.

2. In external aspect it very nearly resembles Lievrite, and it also bears some resemblance to Gadolinite.

3. In the year 1811, Hisinger discovered a new combination of oxide of cerium with silica, lime, and oxide of iron, which he named *Cerin.* It is associated with the cerite in the mine of Bastnaes. It is considered to be the same species as that named Allanite by Dr Thomson.

FINIS.

INDICES.

ENGLISH INDEX.

Ammoniac

Bismuthic

Cinnabar,

Felspar,

Limestone,

Prehnite,

GERMAN INDEX.

A

B

Chlorit,

Galmei,

K

K

L

M

P

Salzkupfererz,

* In the text, at p. 333. the Eisenpecherz of WERNER is, by mistake, given as a synonym of Phosphor Mangan.

Tafelspath,

FRENCH

FRENCH INDEX.

B

C

Magnesie

FINIS.

DIRECTIONS TO THE BINDER.

Plates I. II. III. IV. and V. to be placed at the end of Vol. I.

Plates VI. VII. and VIII. at the end of Vol. II.

Plates IX. X. XI. XII. and XIII. at the end of Vol. III.

EDINBURGH, MARCH 1816.

On Monday the 4th of December was Published,

Handsomely printed in Quarto, with Fifteen Plates, engraved (from Original Drawings) by the First Artists of London and Edinburgh,

PRICE L. 1, 5s. BOARDS,

VOLUME FIRST, PART FIRST,

OF

SUPPLEMENT

TO THE

FOURTH AND FIFTH EDITIONS

OF THE

ENCYCLOPÆDIA BRITANNICA.

EDINBURGH:

PRINTED FOR ARCHIBALD CONSTABLE AND CO. EDINBURGH; SOLD BY GALE AND FENNER, LONDON; THOMAS WILSON AND SONS, YORK; ROBINSON, SON, AND HOLDSWORTH, LEEDS: JOHN RODFORD, HULL; AND JOHN CUMMING, DUBLIN.

THE ENCYCLOPÆDIA BRITANNICA forms a General Dictionary, not only of Arts and Sciences, but of every branch of Human knowledge. The present SUPPLEMENT was undertaken with the view of supplying the omissions in the two last Editions of that Work, and of exhibiting the Arts and Sciences in their latest state of improvement. Though more immediately connected with these Two Editions, it must also prove a valuable sequel to the *Third;* and, indeed, it is arranged upon a Plan calculated to render it a valuable acquisition to the public at large; as it will of itself, within the limits of *Five Volumes*, afford a comprehensive view of the progress and present state of all the more interesting branches of Human Knowledge.

In the composition of this Work, which is Edited by MACVEY NAPIER, Esq. F.R.S.E. the following Gentlemen have engaged to contribute their assistance.

Reverend ARCHIBALD ALISON, LL.B. F.R.S. L. & E.
JOHN AIKEN, M.D. F.L.S.
THOMAS ALLAN, Esq. F.R.S. L. and E.
JOHN BARROW, Esq. F.R.S. one of the Secretaries of the Admiralty.
WILL. THO. BRANDE, F.R.S. L. & E. Professor of Chemistry to the Royal Institution of London.
H. D. BLAINVILLE, M.D. Professor of Zoology and Comparative Anatomy, Paris.
Mr DAVID BUCHANAN.
WILLIAM ARCHIBALD CADELL, Esq. F.R.S. L. & E.
Mr ALEXANDER CHALMERS, F.S.A.
JOHN COLQUHOUN, Esq. Advocate.
Reverend GEORGE COOK, D.D.
ANDREW DUNCAN, Junior, M.D. F.R.S.E. Professor of Medical Jurisprudence in the University of Edinburgh.
JOHN GRAHAM DALYELL, Esq. Advocate.
JOHN DUNLOP, Esq. Advocate.
DANIEL ELLIS, Esq. F.R.S.E.
WILLIAM FERGUSON, M.D. Inspector of Hospitals to the Forces.
Mr FAREY, Jun. Draughtsman, London.
Reverend JOHN FLEMING, D.D. F.R.S.E.
JAMES GLASSFORD, Esq. Advocate.
JOHN GORDON, M.D. F.R.S.E. Lecturer on Anatomy and Physiology.
JAMES HORSBURGH, Esq. F.R.S. Hydrographer to the East India Company.
ALEXANDER HENDERSON, M.D. London.
ALEXANDER IRVINE, Esq. F.R.S.E. Advocate, Professor of Civil Law in the University of Edinburgh.

James Ivory, A.M. F.R.S. Member of the Royal Society of Gottingen.

Francis Jeffrey, Esq.

Robert Jameson, F.R.S.E. Professor of Natural History in the University of Edinburgh.

Charles Köenig, Esq. F.R.S.—L.S. Mineralogist to the British Museum.

John Leslie, F.R.S.E. Professor of Mathematics in the University of Edinburgh.

Stephen Lee, Esq. F.R.S. London.

William Elford Leach, M.D. F.R.S.—L.S. Zoologist to the British Museum.

William Lawrence, Esq. F.R.S. Surgeon, London.

Joseph Lowe, Esq.

W. Lowry, F.R.S. London.

Charles Mackenzie, Esq. F.R.S.—L.S.

Robert Muschet, Esq. Royal Mint, London.

James Mill, Esq. London.

Joshua Milne, Esq. Actuary to the Sun Life Assurance Society.

Reverend Robert Morehead, A.M. late of Bal. Col. Oxford.

Hugh Murray, Esq. F.R.S.E.

John Playfair, F.R.S.L. and E. Professor of Natural Philosophy in the University of Edinburgh.

James Pillans, F.R.S.E. Rector of the High School, Edinburgh.

James Cowles Pritchard, M.D. F.L.S.

Peter M. Roget, M.D. F.R.S. London.

Walter Scott, Esq.

Sir James Edward Smith, M.D. F.R.S. President of the Linnæan Society.

Henry Salt, Esq. F.R.S.—L.S.

Mr Sylvester, Derby.

Mr Stoddart, London.

William Stevenson, Esq. London.

Thomas Thomson, M.D. F.R.S. L. and E.

John Taylor, Esq. Civil Engineer, London.

Rev. William Turner, Newcastle.

William Wallace, F.R.S.E. one of the Professors of Mathematics, Royal Military College, Sandhurst.

To the first Half Volume, which appeared in December last, there is prefixed a Discourse, exhibiting a view of the progress of *Metaphysical, Ethical,* and *Political Philosophy*, since the revival of Letters in Europe, by Mr Dugald Stewart; and to the first Half of the *Second* Volume, there will be prefixed a similar Discourse on the progress of *Mathematical* and *Physical Science*, by PROFESSOR PLAYFAIR. Besides the *Preliminary Discourse* by Mr STEWART, the first Half Volume contains, among a variety of others, the following Articles: viz. ABYSSINIA, ACHROMATIC GLASSES, ACOUSTICS, AERONAUTICS, AFRICA, AGRICULTURE, AMERICA, ALUM, AMMONIAC (SAL), and ANATOMY.

The *second* half of the *First* Volume, illustrated with *twenty-two* Engravings, executed by the first Artists, will appear in April; and will contain, besides a variety of articles in *Biography*, *Topography*, *Statistics*, and *Miscellaneous Literature*, the following extensive treatises: viz. ANATOMY (VEGETABLE), ANNUITIES, ANNULOSA, (including, under that head, a view of several of the most interesting classes of Animals), ARITHMETIC (its history and theory), ARTS, ASIA, ASSAYING, ASSURANCES, ASTRONOMY, ATOMIC-THEORY, ATTRACTION, and AUSTRALASIA.

In order to give the Public all due confidence in the care and ability with which the Work will be composed, the names of those who have contributed to each Volume will be given on its conclusion, with references to their particular contributions.

FOLIATED RED C

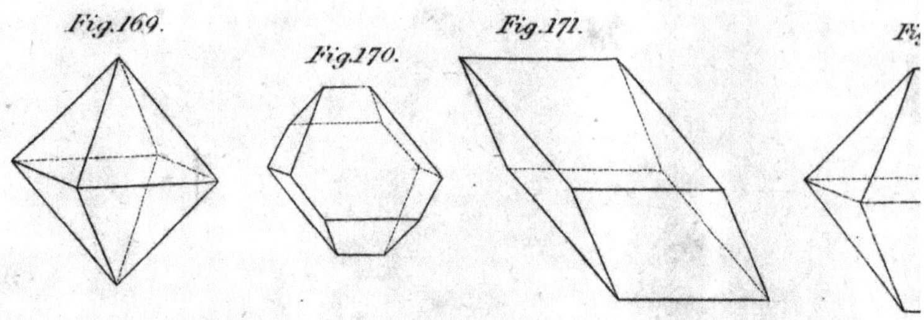

FOLIATED RED C

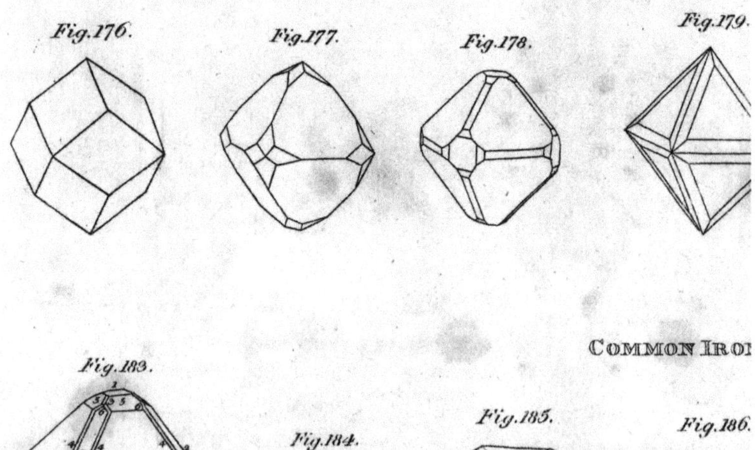

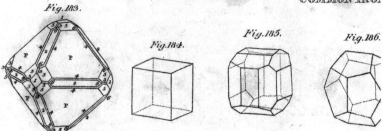

COPPER ORE.

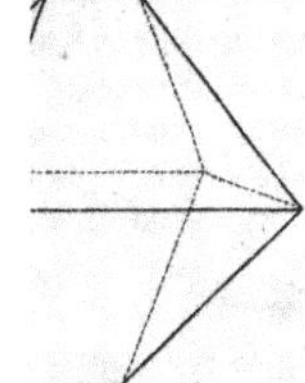

'ig.172.

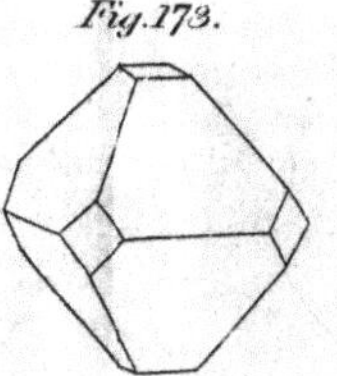

Fig.173.

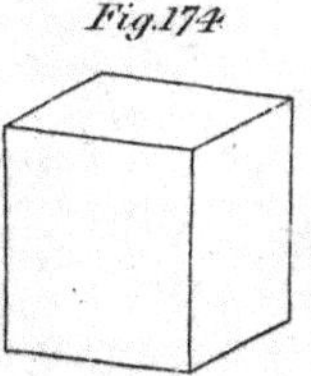

Fig.174.

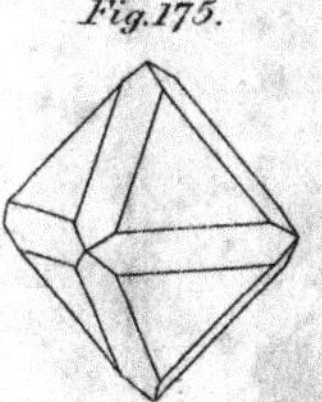

Fig.175.

COPPER ORE.

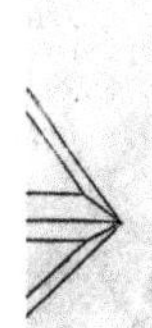

).

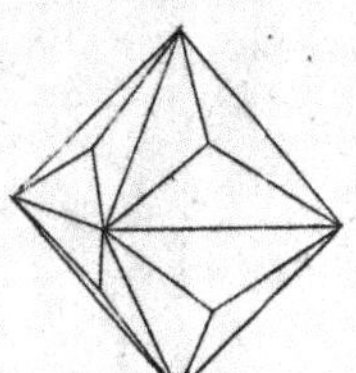

Fig.180.

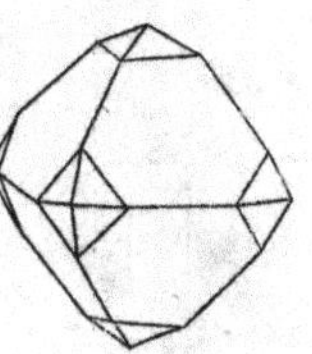

Fig.181.

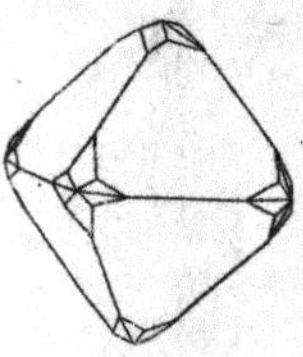

Fig.182.

ON PYRITES.

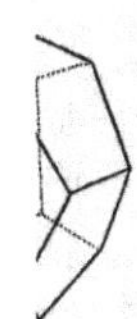

6.

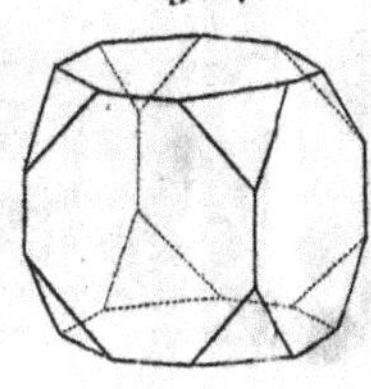

Fig.187.

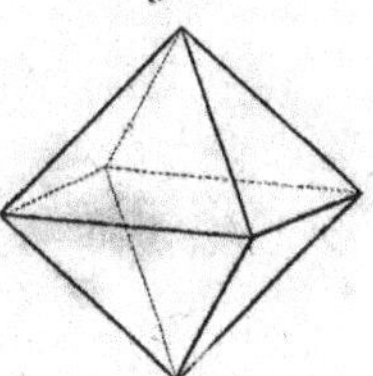

Fig.188.

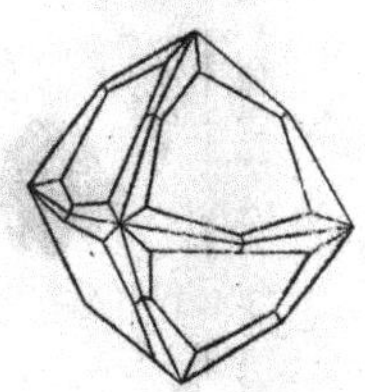

Fig.189.

E. Mitchell sculpt.

COMMON IRO[illegible]

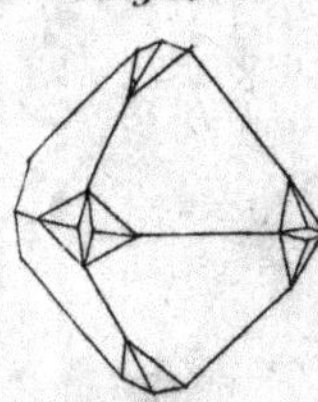
Fig. 190.

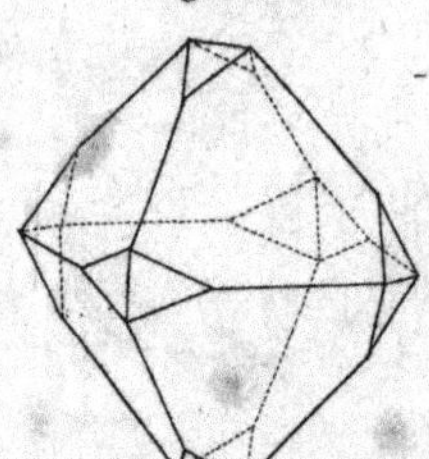
Fig. 191.

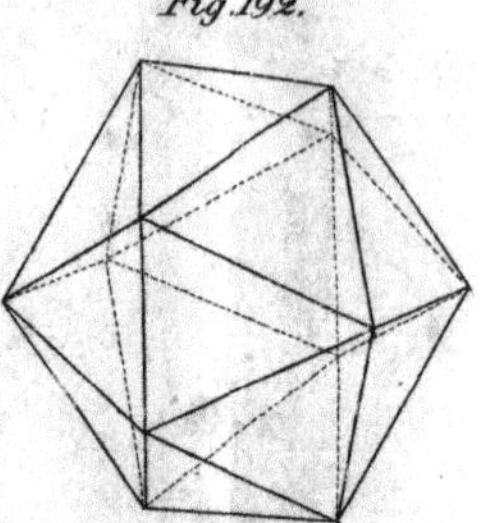
Fig. 192.

RADIATED PYRITES.

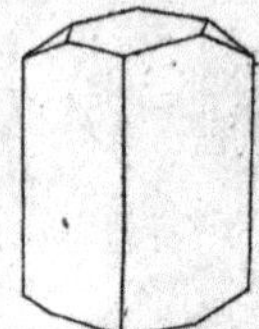
Fig. 196.

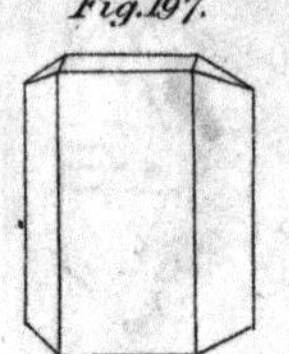
Fig. 197.

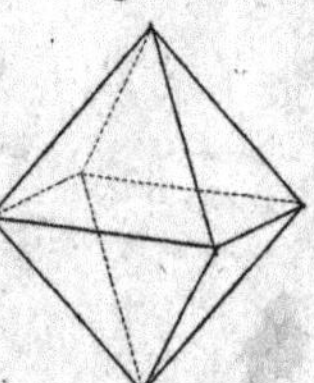
Fig. 198.

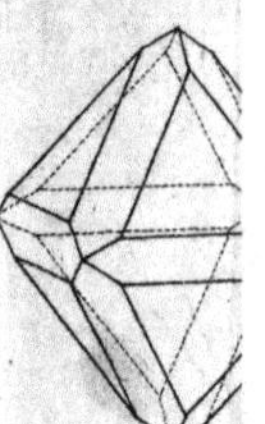
Fig. 199.

SPECULAR IRON ORE, OR CO[illegible]

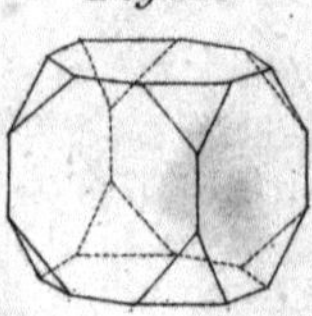
Fig. 203.

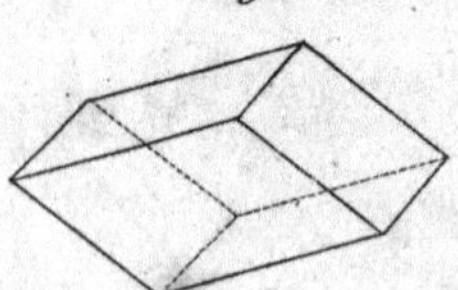
Fig. 204.

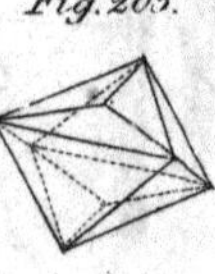
Fig. 205.

N PYRITES.

Fig. 193.

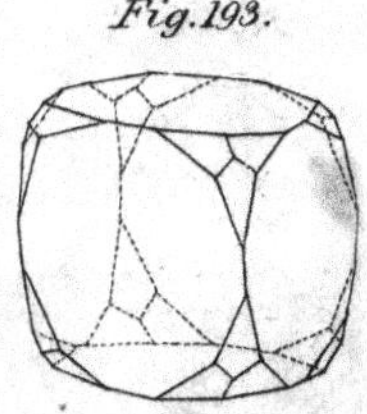

Fig. 194.

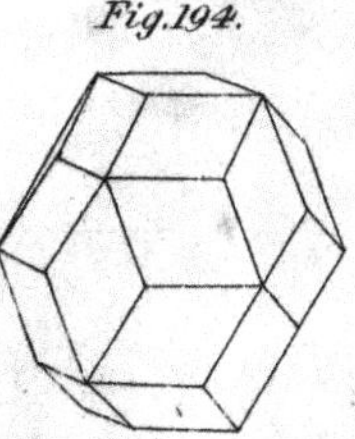

Fig. 195.

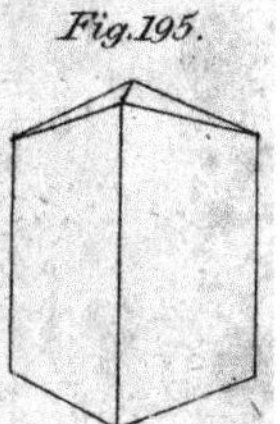

COMMON MAGNETIC IRON STONE.

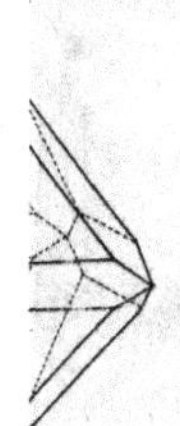

Fig. 200.

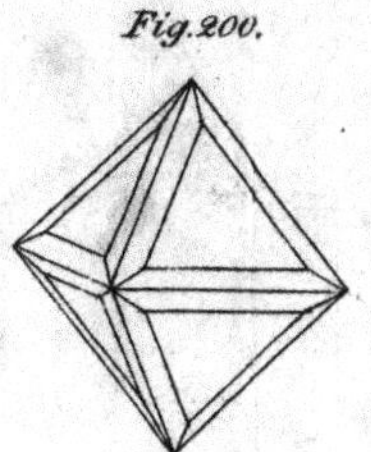

Fig. 201.

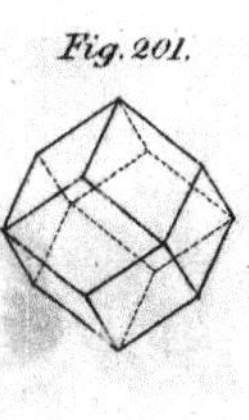

Fig. 202.

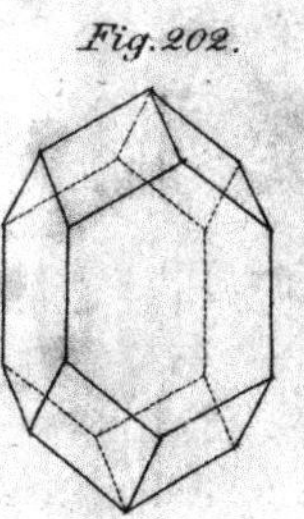

MMON IRON GLANCE.

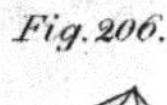

Fig. 206.

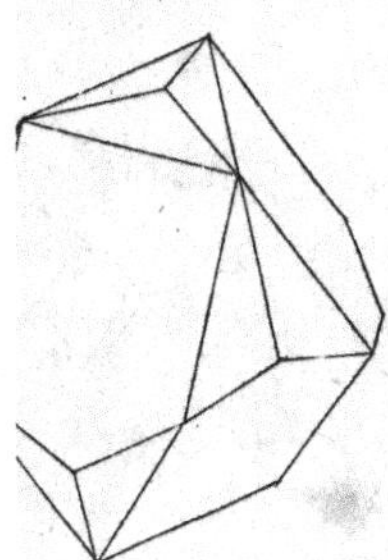

Fig. 207.

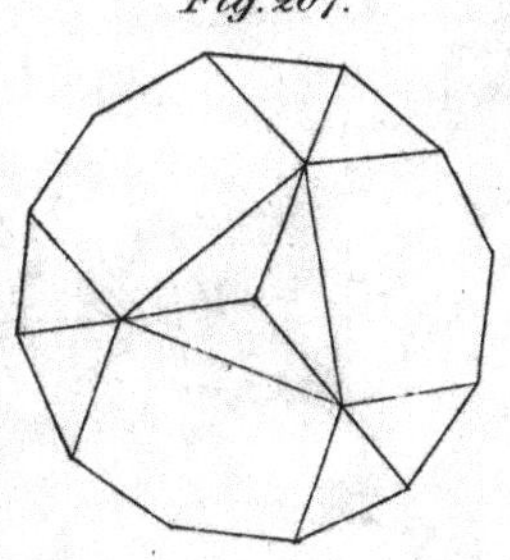

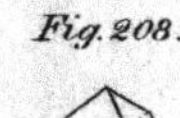

Fig. 208.

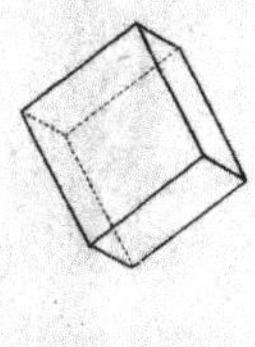

E. Mitchell sculp.t

SPECULAR IRON ORE OR IRON

Fig. 209.

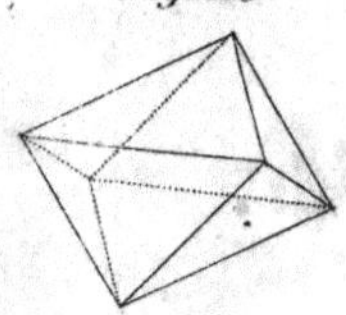

Fig. 210.

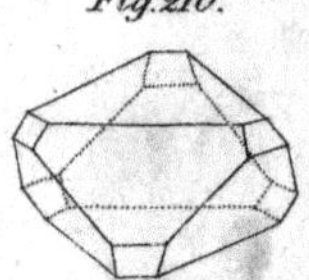

Fig. 211.

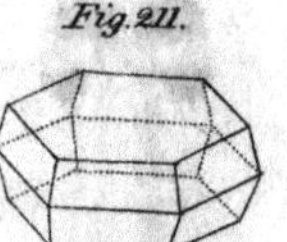

RUTILE.

COMMON SPHENE.

Fig. 216.

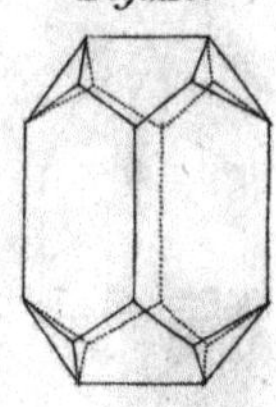

Fig. 217.

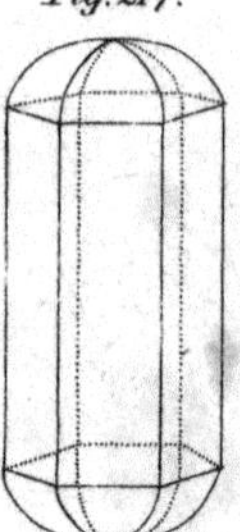

Fig. 218.

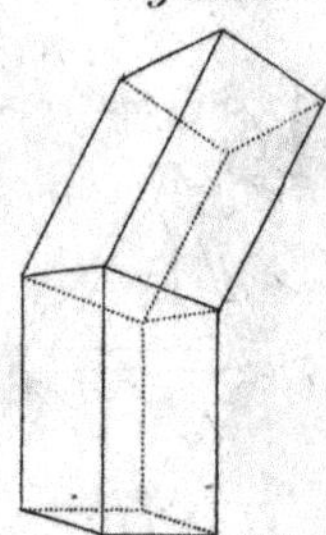

Fig. 219.

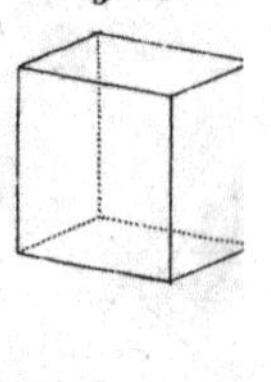

GALENA OR LE

Fig. 223.

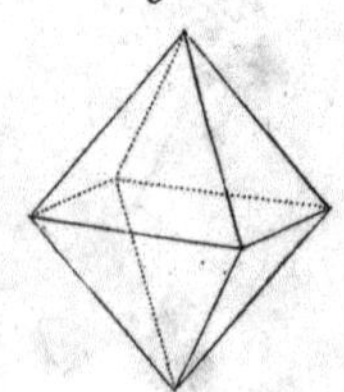

Fig. 224.

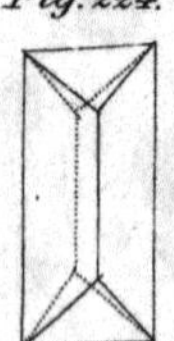

Fig. 225.

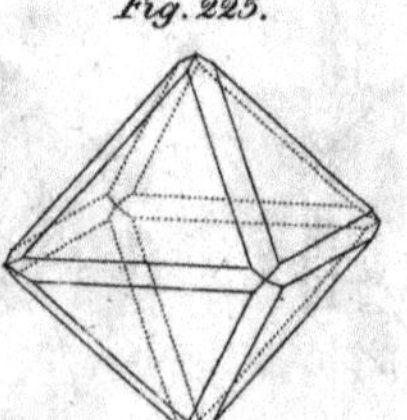

Fig. 22

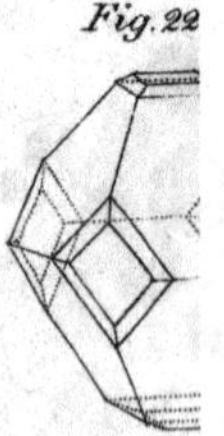

n Glance.

Fig. 212.

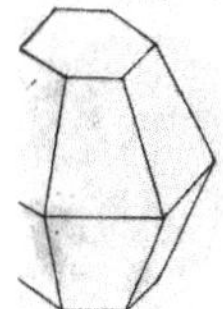

Fig. 213.

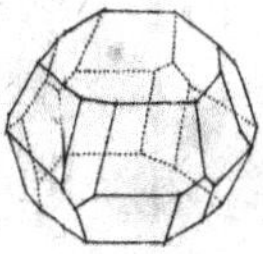

Fig. 214.

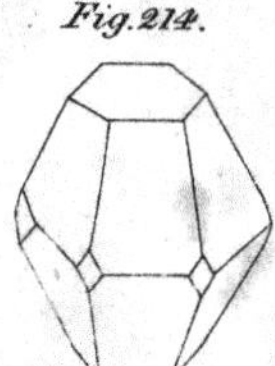

Common Sphene.

Fig. 215.

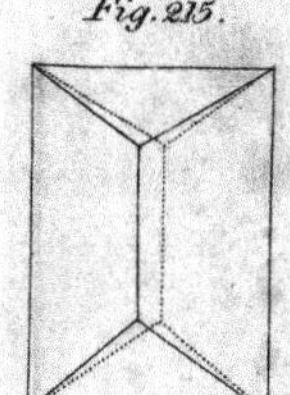

Galena or Lead Glance.

Fig. 220.

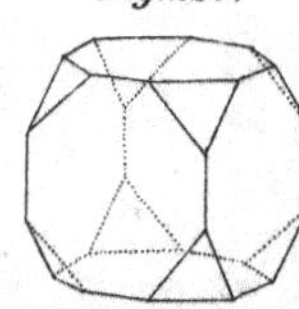

Fig. 221.

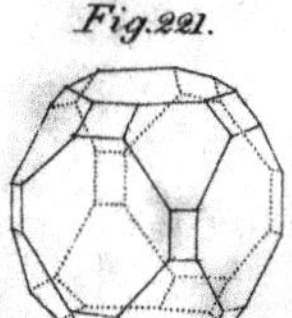

Fig. 222.

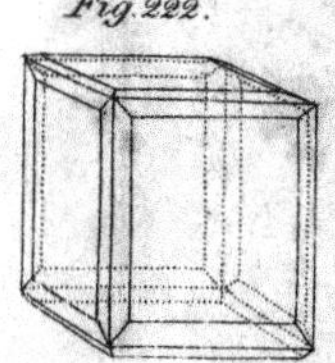

ead Glance.

26.

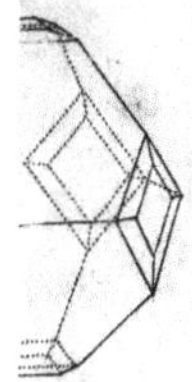

Fig. 227.

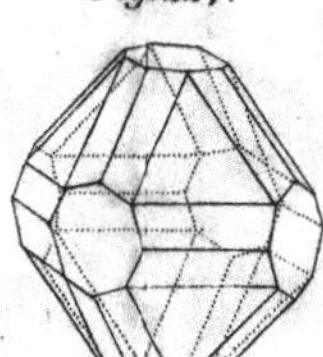

Fig. 228.

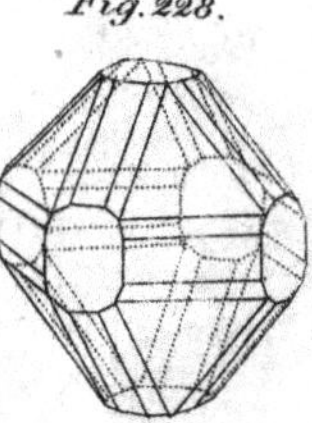

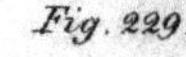

Fig. 229.

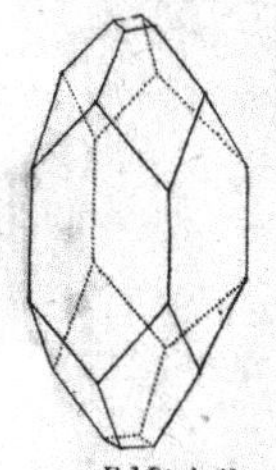

E. Mitchell sculp!

COMMON WHITE LEAD OR

Fig. 230.

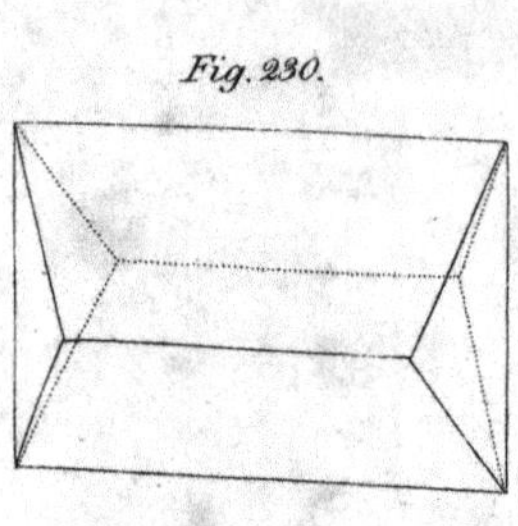

Fig. 231.

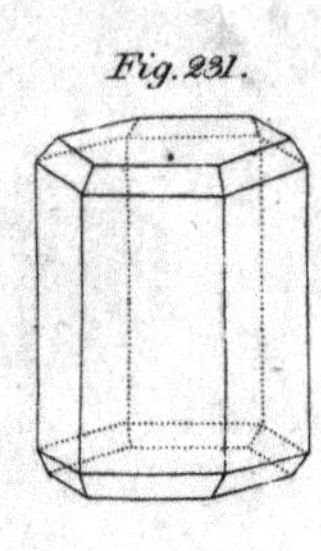

Fig. 232.

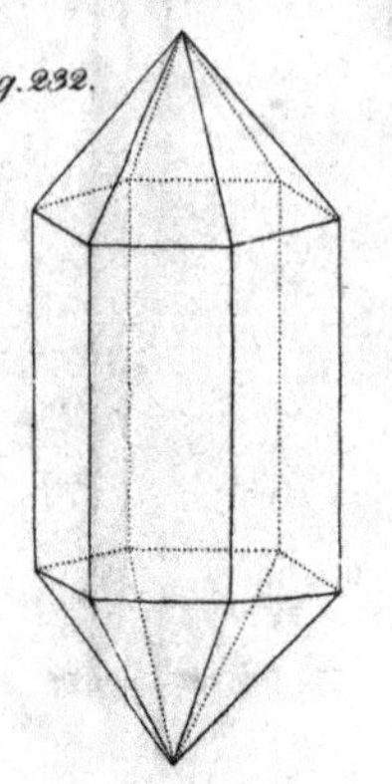

GREEN PHOSPHAT OF LEAD.

Fig. 237.

Fig. 238.

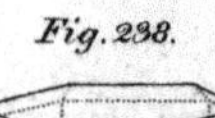

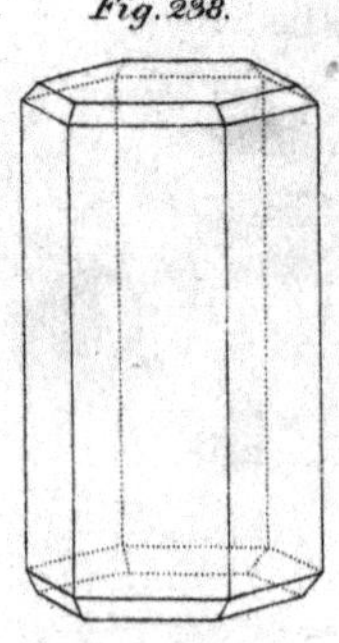

SULPHAT OF L

Fig. 239.

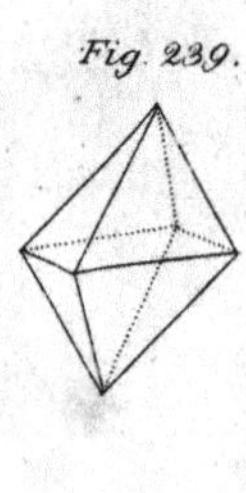

Fig. 240.

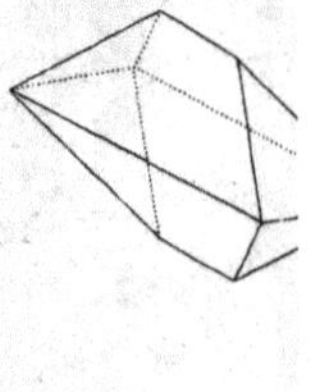

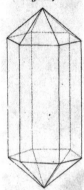

YELLOW LEAD ORE.

Fig. 244.

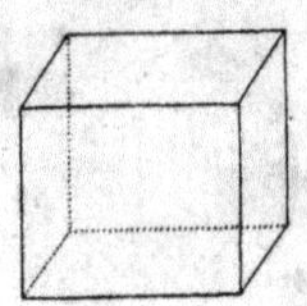

Fig. 245.

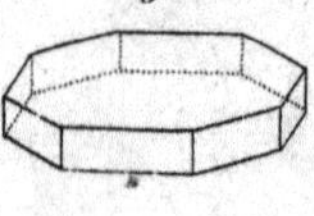

Fig. 246.

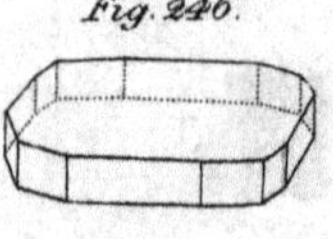

Fig. 24

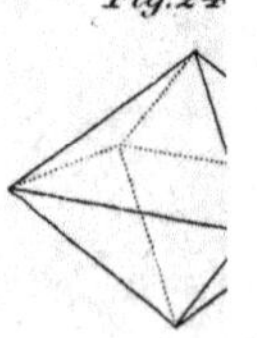

RE.

PLATE XII.

GREEN PHOSPHAT OF LEAD.

Fig. 233.

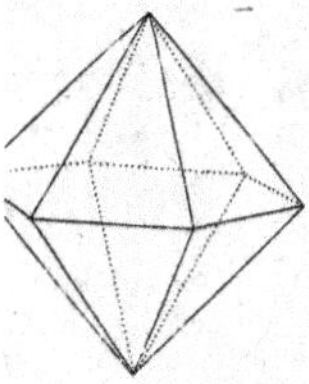

Fig. 234.

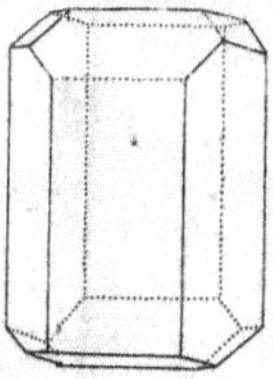

Fig. 235.

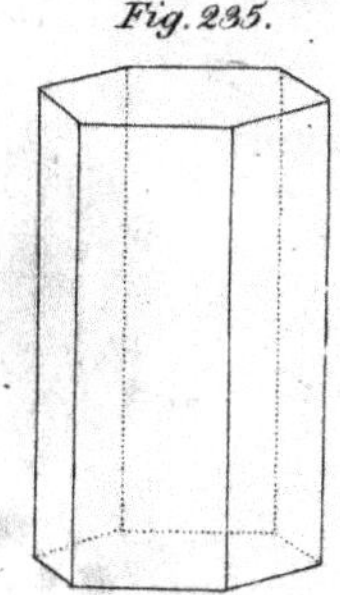

Fig. 236.

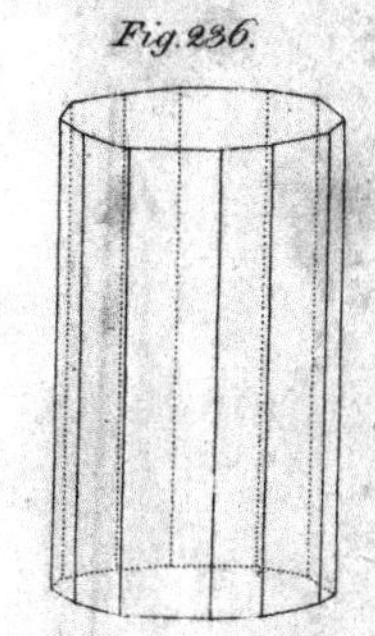

LEAD.

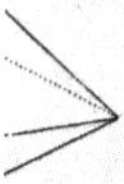

YELLOW LEAD ORE.

Fig. 241.

Fig. 242.

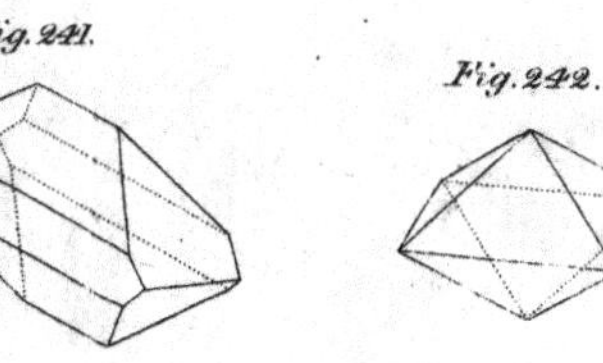

Fig. 243.

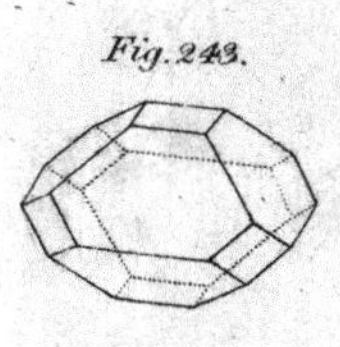

TIN STONE.

47.

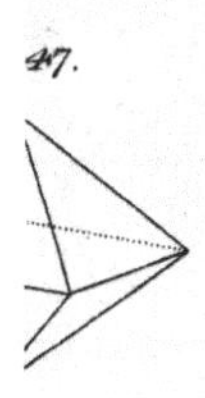

Fig. 248.

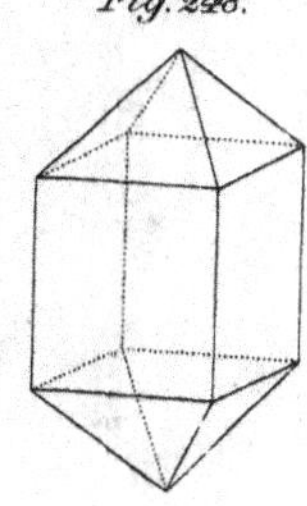

Fig. 249.

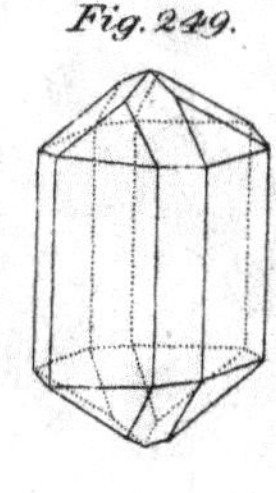

Fig. 250.

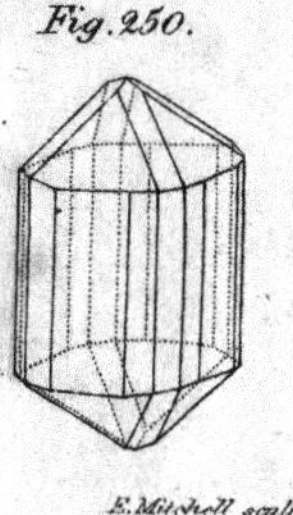

E. Mitchell sculp.t

Tin Stone.

Fig. 251.

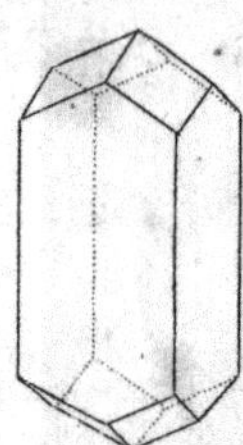

Fig. 252.

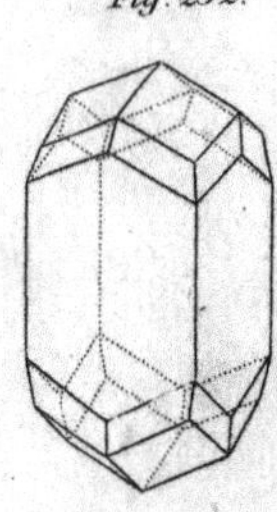

Fig. 253.

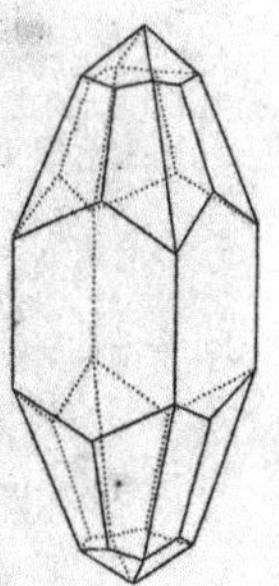

Fig. 254.

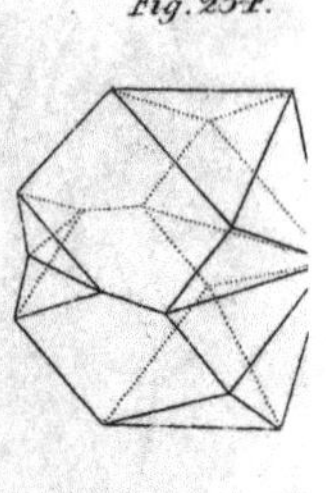

Radiated Grey Antimony.

Fig. 259.

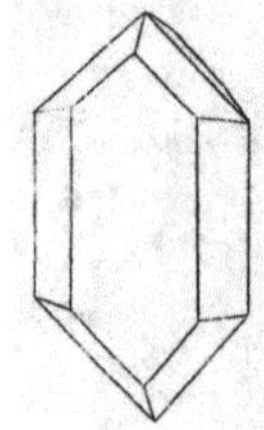

Fig. 260.

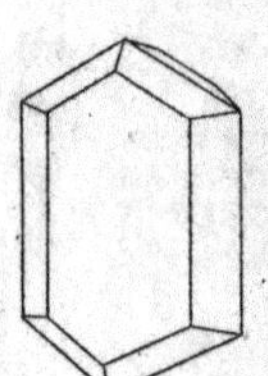

Molybdena.

Fig. 261.

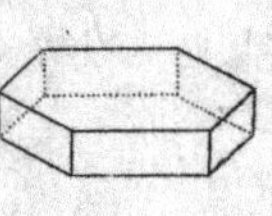

Fig. 262.

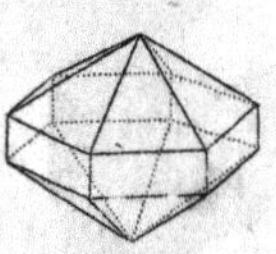

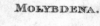

Red Orpiment.

Fig. 266.

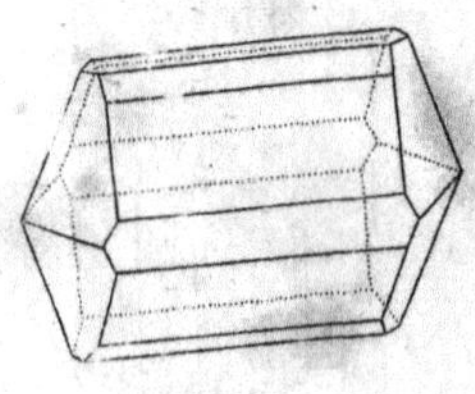

Fig. 267.

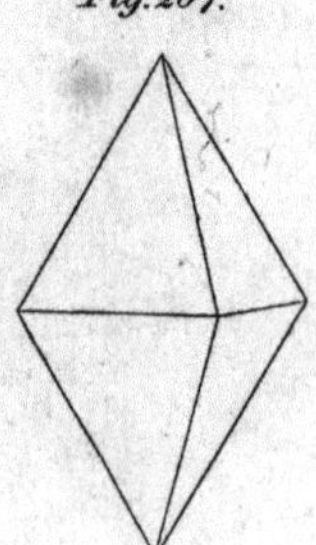

Tungsten.

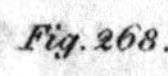

Fig. 268.

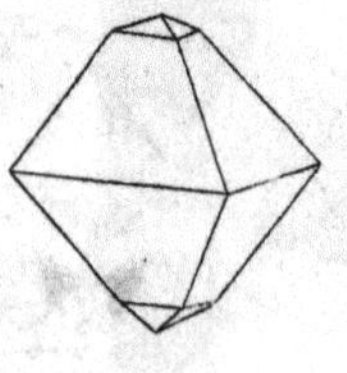

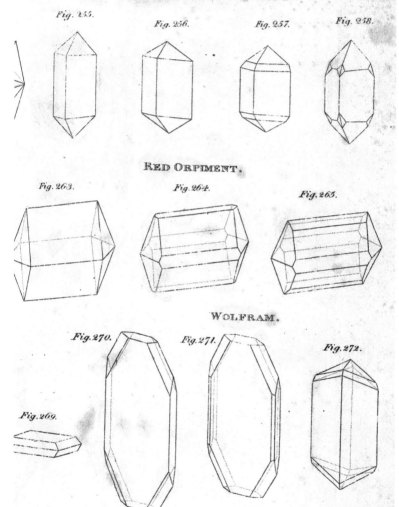

RADIATED GREY ANTIMONY.

Fig. 255.

Fig. 256.

Fig. 257.

Fig. 258.

RED ORPIMENT.

Fig. 263.

Fig. 264.

Fig. 265.

WOLFRAM.

Fig. 270.

Fig. 271.

Fig. 272.

Fig. 269.

E. Mitchell sculp.

Printed in the United States
By Bookmasters